[Out of print 9/11/2000]

AHDB LIBRARY
☎ 02476 478839
ACC.NO. 741 (1) PO NO -
PRICE: £ - DATE: 7/8/1970
CLASS.NO. 636.3 SPE

D1355871

SHEEP PRODUCTION AND GRAZING MANAGEMENT

TO

MY WIFE

Sheep Production and Grazing Management

C. R. W. SPEDDING
M.Sc., Ph.D., D.Sc. (Lond.), F.I.Biol.
Head of Ecology Division
The Grassland Research Institute
Hurley, Berkshire

Second Edition

LONDON
BAILLIÈRE, TINDALL AND CASSELL

First Edition 1965

Second Edition 1970

© 1970 Baillière, Tindall & Cassell

7–8 Henrietta Street, London WC 2

ISBN 0–7020–0326–3

M.L.C. LIBRARY
ACC. NO. ML 205
CLASS. NO. 636·3
SOURCE. B·H. BLACKWELL LTD
PRICE. £4 - 10 - 0
DATE. 7. 8. 70

Published in the United States of America by the Williams & Wilkins Company, Baltimore
Printed in Great Britain by Morrison and Gibb Ltd., London and Edinburgh

CONTENTS

LIST OF PLATES

PREFACE TO THE SECOND EDITION

In the preface to the first edition, which will be found on a succeeding page, I endeavoured to set out and explain my purpose in writing a book dealing with the basic principles of the biology of sheep production and how this fundamental knowledge can be applied within the restrictions which economics impose at any one time. Biological principles are in the main permanent and are thus of great importance from the long-term economic point of view. Since the first edition was written, a great deal of material has been published that is highly relevant to a discussion of sheep production. It is encouraging to see the way in which the more important gaps are being filled by a vigorous research effort in many different countries.

This rapid accumulation of information has, however, presented some problems in bringing the first edition up to date. Attempting to maintain the same level of citation has increased the risk of obscuring the main issues in a wealth of detail. An additional chapter (XVI) entitled 'Systems of Sheep Production' has therefore been added, in which the practical implications to current systems of sheep production are discussed. This also has the effect of balancing the possible future projections contained in Chapter XV dealing with 'The Potential for Sheep Production' (which has been completely rewritten).

Since a number of different disciplines are involved, each with its own technical language, a glossary has been added to define those terms which are not adequately explained in the text.

Advances in the control of reproduction and in the artificial rearing of lambs have been so great that these subjects have been removed from the discussion of the future to the main chapters entitled 'The Ewe' (Chapter IV) and 'The Lamb: its growth and development' (Chapter V).

Although the application of model-building techniques to agriculture (and, indeed, to applied biology generally) proceeds rapidly, it is not yet possible to express the sheep production process (see the largely rewritten Chapter VI entitled 'The Production Process') in an adequate model form. It will not be long before this is possible, however, and will then allow a more elegant expression of the way in which the most important factors in the process interact.

The most difficult area remains the grazing situation and there is a great need for research into whole grazing systems, not inhibited by concern with transient costs and prices but giving due weight to those economic factors which play a large part in the business of farming. The urgent need in agricultural research is always for biological information and understanding within the areas of greatest economic importance.

C. R. W. SPEDDING

The Grassland Research Institute
Hurley, Berkshire
January 1970

PREFACE TO THE FIRST EDITION

This book attempts to assemble a coherent picture of the agricultural ecology of the grazing sheep. It is primarily addressed to those of the veterinary profession and those in the many branches of agriculture who recognize the increasing convergence of their interests. An attempt is made to build a model of the production process which will illustrate the way in which the major factors influence the conversion of herbage to sheep products of agricultural value. It is hoped, however, that the attitude of mind behind this study will find a more general echo. It is (surely) the business of agricultural research to elaborate and test models of this kind, and from the knowledge so gained to make an increasing contribution to agricultural and veterinary teaching. The book is also written in the hope that it will be helpful to the farmer and his advisers. This synthesis of health, growth and productivity of both animal and pasture has always had to be made on the farm, with little guidance except experience.

It is a difficult task to put together the parts of agriculture, or even of one agricultural system, and to integrate the knowledge contributed by many disciplines. It requires familiarity with the components and, above all, knowledge of the principles governing their interactions. Such agricultural ecology has been neglected as a field for research and, in consequence, ill-represented in teaching. This is understandable but also more serious in relation to the grazing situation where the animal interacts so intimately with its food supply. Where an animal lives on its food as literally as does the ruminant, the need to regard health, disease, nutrition and husbandry as parts of one subject is obvious.

This book is not primarily about sheep or about sheep farming; it comes somewhere between the two. It is primarily about sheep production as a biological process. Knowledge of sheep is certainly required to understand this process but so also is knowledge of many other kinds. Sheep farming is based on the biological process, but economic criteria must be adopted when assessing the results. Clearly, economic criteria may at times make no biological sense, and in any event are liable to change quite rapidly. Biological principles, on the other hand, are relatively permanent. Little attempt has been made in this book to attach economic significance to any part of the production process except in the sense that economic efficiency may be linked to biological efficiency. This seems to be a more realistic approach than to introduce economic values which may be transient, but their exclusion does not imply the total omission of an economic outlook. Sheep production is carried out by man for his own purposes, and it is likely that processes directed to such ends must satisfy both biological and economic criteria simultaneously, as long as one product is being considered. It is true that the economic value of fat, for example, may bear no relation to its biological cost. Wool and meat differ in these respects, and the ultimate economic value of the sheep's digestive system certainly does not reflect its essential role in the production of the edible parts of the carcase. All this is quite understandable but should not obscure the probability that, for a specified product, biological and economic costs and efficiencies are likely to move in parallel.

This book, then, is about the biology of sheep production, partly because this fundamental knowledge can be applied in the directions that economics dictate at any one time. It is these biological principles that matter from an economic point of view; the advocacy of particular methods of sheep husbandry, although superficially closer to the farmer, is on less sure economic ground.

C. R. W. SPEDDING

ACKNOWLEDGEMENTS

I wish to thank Professor E. K. Woodford for facilities made available to me within the Grassland Research Institute and to acknowledge the help I have received from many colleagues during the preparation of the second edition.

In particular, I am most grateful to R. V. Large, M.Sc., J. E. Newton, M.A., M.Sc., P. D. Penning, Miss J. M. Walsingham, B.Sc., T. T. Treacher, B.Sc., Ph.D., I. A. N. Wilson, Miss S. J. M. Burton and Miss R. Harbour, all of the Ecology Division.

Thanks are due to Ripper Robots Ltd. and to the editors of the following journals for permission to reproduce tables or figures: *Agricultural Progress; Journal of Agricultural Science; Journal of the British Grassland Society; Empire Journal of Experimental Agriculture; Farm and Country; Farming Facts; Outlook on Agriculture; Proceedings of the 1st World Conference on Animal Production; Proceedings of the 8th International Grassland Congress; Proceedings of the Nutrition Society; Journal of the Science of Food and Agriculture; The Veterinary Record.*

The new figures were again drawn by Mr. B. D. Hudson and I am grateful to Mrs. H. L. B. Stone for her accurate typing of the manuscript. I am indebted to my wife for considerable assistance with the index and to Miss J. M. Walsingham for help in preparation of the glossary.

C.R.W.S.

I

AGRICULTURAL ECOLOGY

Many of the most important agricultural subjects cut across the accepted boundaries that separate one discipline from another. Yet there are good reasons for many of these boundaries and any alternative organization of knowledge would suffer from disadvantages similar to those that exist at present. This is because each major agricultural topic must draw upon different disciplines, each to a different extent, according to the purposes for which such knowledge is to be used. There are thus as many subjects as there are different agricultural purposes. Each subject is, in fact, characterized by its purpose, for example egg-production or the production of rice, and the need to combine information from different disciplines is entirely determined by this. Production within a grazing environment probably has to draw on a much wider range of knowledge than, for example, production processes which involve no animals or those in which the environment can be closely controlled, as with production from battery hens. This combination of factors for a particular agricultural purpose must be based on principles related to their interactions (36). I have used the term agricultural ecology (27) to describe the study of these principles. There is no reason why agricultural purposes should not be defined in very broad terms, such as animal production per unit of land, but the systematic study of such broad topics would be more easily tackled if the component processes had already received the same kind of attention.

In succeeding chapters we shall be concerned with only one such component, sheep production. Even so, there are many possible ways of approaching the subject, particularly in the extent to which the process is regarded as involving the environment. Sheep farming is the business of the economic application of the sheep production process within some given environment and cannot be entirely divorced from this context.

The Need for Synthesis of Information

Experimentally, it is often necessary to study parts of the process in relative isolation but the information so obtained is rarely of immediate value to the farmer. Equally, an analysis of current practices and the profitability associated with them cannot result in marked practical progress unless it takes into account recent technical advances. It is not sufficient, however, simply to know how to increase the technical efficiency of a particular activity. Candler & Sargent (4) have pointed to the difficulties of making recommendations on the basis of technical efficiency only: 'When changed management is being recommended a decrease in efficiency is as likely to be profitable as an increase'. This is because the aspect considered is only one facet of a complex system. This is easily illustrated by changes in the lambing percentage of a flock. There is good reason to suppose that, in certain circumstances, a ewe with twins may be more productive and more profitable than a ewe with a single lamb (see Chapters IV and VI). These circumstances may include a situation where the ewe considered is the only one with twins in a flock of ewes with single lambs (see Chapter III). An increase in the number of ewes with twins, which receive less milk, require more grass and may be subject to a greater degree of worm-infestation (see Chapter V), may so change the situation that the ewes with singles would become the more profitable, simply because the twins failed to grow satisfactorily. Thus the implication, from economic analysis of the first situation, that profitability would be increased by increasing the number of ewes with twins, would be quite unsound if, in doing so, important aspects of the environment would also be changed. Only a knowledge of these associated changes can enable full use to be made of the initial analysis.

As Moule, Braden & Lamond (21) have recently stressed, 'field situations are ecological' and involve a dynamic balance between grazing animals and the pasture they graze. Thomas (32) has pointed to the need for 'more careful and systematic documentation of research and practical farm results, in order to provide an accumulating fund of knowledge'. This clearly requires synthesis and organization of data rather than mere collection and Thomas (32) stated that 'the necessity for co-operation between economists and scientists on such work is so obvious that one is appalled at the serious lack of it in practice'. 'The economist', he wrote, 'is perhaps asking too many questions of the data which he collects, the scientists too few questions at the same time.' A seminar which included both economists and scientists from many countries (17), confirmed, in its conclusions, the lack of co-operation and drew attention to the difficulty of putting together in a coherent form the often widely-scattered results of agricultural research.

It is arguable that it is incumbent upon the agricultural scientist to attempt to set the results of his own work in perspective, in relation to the pattern which previous work has established. Unfortunately, it is often this very pattern which is unclear.

Economic interpretation

Economic interpretation needs to take account of the consequences which may follow a change in such things as the proportion of single and twin lambs. The combinations and permutations of ewes with 0, 1, 2 and 3 lambs and all the variables of pasture species, fertilizer practice and management, however, are formidable. It is particularly important, therefore, to avoid unnecessary complexity and, at some point, to suggest how situations might with advantage be simplified. It will be argued, for example (see Chapter V), that ewes with single, twin and triplet lambs differ so greatly from each other in so many of the important features in sheep production, that they would be better kept separately. The separation of ewes with one and two lambs is not, of course, unknown in practice, but it can hardly be said to be common. Yet it would limit the possible situations to 4, ewes with 0, 1, 2 or 3 lambs respectively, instead of the hundreds of possibilities which exist between the lambing percentages of 0 and 300. Such a change in practical farming would not merely test the suggested hypothesis: it could greatly add to practical experience by presenting the observer, farmer, adviser, economist, etc., with better defined situations for study.

Significance of the environment

Changes of the kind described must be based on a wide range of scientific activities, at all levels of complexity. The production process, for example, is intimately concerned with the sheep but not with it alone. At the very least it is concerned with the relationship between the sheep and its food; this, at its simplest, is a nutritional relationship. Considered on a flock basis, however, it becomes a great deal more than this. If the animals' food is expanded to mean the food supply, its distribution and the factors affecting it, the production process takes on something of the complexity to be found in sheep farming. It would be quite legitimate to restrict the definition of this process so as to exclude the environment in which it is carried out, but this would also be less helpful, I believe, in the present context. There is no reason why sheep should not be kept, like some other forms of livestock, under controlled environmental conditions and fed on controlled diets. Indeed, there are already some trends in this direction. The vast majority of sheep, however, are kept in a wide variety of *grazing* situations under a wide variety of very variable environments. The contrast between the environment of the Welsh mountain sheep, grazing on the sheep-walks of Western Snowdon at 1·27 per acre (16) and that of the Romney in New Zealand at five or six times this stocking rate (34), is very marked, but innumerable illustrations of this sort could be cited. In these circumstances, discussion of the production process without reference to the total environment is somewhat unrealistic. It is true that the basic principles underlying it are unaltered, but their significance depends, to some extent, on the problems attendant on their application. Nor is the question of practical application one that can be dealt with simply. In the kinds of environment in which sheep are commonly grazed, it is as essential to understand the principles underlying the application of the process, as it is to understand the process itself. This wider view of the subject, including an understanding of both the process and the context in which it is carried out, may be regarded as characteristic of agricultural ecology.

Sheep Husbandry

Fraser (9), in the 1951 edition of his book *Sheep Husbandry*, emphasized that the term 'husbandry' included more than any mere system. 'It implies', he wrote, 'all the traditional store of knowledge of sheep and of the ways of sheep.' This is true and does not conflict with the view that research should delve into topics such as animal behaviour in order to improve on the traditional store. Indeed, the work on sheep behaviour in relation to environment (2, 12, 22), is a recognition of the value of this kind of knowledge (see Chapter VII). Fraser went further and suggested that sheep husbandry should embrace, not only sheep breeds, sexual physiology, production and nutrition, but also disease and its control. This he acted on in a later edition (10), and there has been a widespread increase in the recognition of the need for veterinary science and husbandry to move closer together (10, 13, 14, 15, 24, 26, 29).

Fraser (9) stated that 'there is as yet no real science of sheep husbandry'. He went on to describe the 'synthesis of special scientific subjects into one practical subject' that ought to form such a science.

Such a subject embraces the interaction of the whole sheep population with its total environment. Because it is concerned with the synthesis of knowledge derived from a considerable number of disciplines it must be highly selective and employ unifying concepts in order to handle the mass of relevant information. In the past, this kind of synthesis has, to a very large extent, been left to the farmer and his advisers. Progressive farming has been expected to include a considerable amount of what has been appropriately called 'trial and error'. The well-recognized difficulties of interpretation in relatively controlled field experiments make this procedure on the farm appear inefficient at best and, at its worst, gravely misleading. This is not to decry innovation on the farm and, in any event, this is quite clearly the farmer's own affair. Nor is it intended to suggest that no experiments can be efficiently conducted on farms. It is suggested, however, that the experimental synthesis of data from different disciplines is a major field of research which, *in the main*, cannot be undertaken on commercial farms. The reasons for this are numerous: the two most important are the lack of information and lack of control that are characteristic of most grazing situations on the farm. What is required is experimentation, of a very high order, within husbandry contexts. In relation to the need, very little has been done. Two recent reports (5, 25) concerned with the sheep industry in Britain, have indicated the major gaps in research which remain to be filled. They have both been greatly concerned with the efficiency of production in relation to breed, size, prolificacy, management and environment. At all levels of the synthesis involved in sheep production at pasture the required information is frequently lacking. Little knowledge is available on the interaction of ewes, lambs, pasture plants, parasites and the weather: yet the farmer lives with the problem of putting them together for profit. Few data are available as to the food input/product output relationships of either ewes or lambs (28), and the seriousness of this situation is not altered by the fact that food intake is difficult to measure, and therefore rarely known, under grazing conditions. Very little can be said about the significance of the weight or size of the ewe in relation to the efficiency with which it milks or produces lamb at one level of nutrition or another. Yet farmers are constantly faced with a wide choice of breed and type, with little more than tradition to guide them.

The choice of breed

The choice of breed ought to be of major consequence since wide differences exist in the performance of different breeds. The sheer number of breeds in the world may have discouraged critical evaluation. The large number in Britain alone, about 40, presents a formidable task to the scientist or flockmaster who wishes to compare them experimentally. Simple comparisons of all these breeds with each other, over a limited range of pasture, management and fertilizer conditions, would require an impracticable number of experiments. Excellent descriptions of these breeds, in terms of what is known about them, are available in the books of Fraser (9, 10) and Thomas (31). Breeds that have died out, or almost so, are surprisingly numerous (11) and many breeds common in one country are virtually unknown in another. Ecologically, the need is to try to describe sheep in terms of their important attributes and to see whether the different breeds available really fulfil similar roles or not (see Chapter XII). The different purposes for which different breeds are required (5) may be numerous and vary greatly with geographical location, management practices and market requirements. For detailed descriptions of these aspects of sheep management, both textbooks and practical guides are available (1, 7, 8, 18, 23, 30, 37, 39) in addition to those already mentioned.

In trying to achieve a synthesis of information from all the relevant disciplines, it is necessary to refer to many subjects without being able to deal with them in great detail.* This is especially so since, in writing from one kind of experience, some sections may tend to loom too large. The least that can be done is to try to ensure that nothing of importance is ignored, albeit without overwhelming the reader with detail.

* The numerous references in this book are partly to enable such subjects to be pursued, if desired, in greater detail. A bibliography on Sheep Husbandry and Research has been compiled by Cresswell (6).

Ecology of sheep parasites

A study of grazing sheep generally involves populations of certain parasites, for example, although these may not be inevitable in sheep in other circumstances or on pasture in the absence of ruminants. They are primarily important as a consequence of combining the sheep and the sward and their importance will vary with the precise way in which these two are put together. It is necessary, therefore, to have some acquaintance with such parasites but only the briefest description of them can be given here. Excellent textbooks are available for reference, including those of Cameron (3), Lapage (19), Mönnig (20) and Soulsby (38).

More important in many ways than the knowledge itself is the attitude to its discovery, its interpretation and its application. Peters (24), in an inaugural lecture, outlined the concept of a 'pure' parasitology (a subject in its own right, rather than a mere handmaid of human or veterinary medicine) that embraces problems of economic importance, without being solely concerned with the control of parasites. He supported an ecological view of parasitology, where the emphasis is placed on understanding the relationships between the parasite, its host and the rest of the environment. This view enables the total environment to be taken into account. When applied to agriculture, it suggests that a close study of the relevant agricultural context should be the basis of conclusions as to which parasites are important and how their populations behave. An example may be taken from the interactions of parasites and grazing management (see Chapter IX). In systems of grazing management, based upon the systematic movement of sheep from one area of pasture to another, the number of parasitic larvae that survive on the herbage, from one grazing period to another, is of considerable importance. Yet we are not simply concerned with their survival rate. For one thing, the pasture is also a living and changing population. Before asking how long a parasite survives on the grass leaves, it must be asked how long the grass leaves themselves live: in agricultural contexts this may be only a few weeks. Chiefly, however, it is important to know how many parasitic larvae will be present so many weeks later. This is not simply a matter of how many of those that were present at the beginning of the period are still alive at the end of it, though this would be a valid assessment of the survival rate. Much greater numbers of larvae may appear during the period, which were embedded as eggs in faeces or soil at the beginning, having been deposited by sheep at any time in the preceding months. Thus a desirable period to rest a pasture from grazing in order to control parasite numbers, may have little to do with larval survival rate. In fact, what we need to know, for a given set of conditions, is the number that will be present at the end of any particular rest period. This is not to say that we should be content with an empirical answer but that the whole context must be considered when trying to synthesize a system from its component parts.

BIOLOGICAL MODELS

The problem of synthesis is not, of course, primarily a lack of detailed information but a lack of information on the interaction of different parts of the whole system under consideration. It is not profitable, however, to wait until all the information required is available before trying to construct a pattern.

In attempting to achieve some practical purpose on the farm it is obviously better to have a rough idea of the biological processes involved than to have no idea at all. The number of facets to be taken into account and the variety of disciplines concerned make these processes difficult to discern without some organization of the available knowledge.

Furthermore, only by attempting a description of the process is it possible to indicate with any certainty what further information is required. As usual, specialization and fragmentation will be necessary in the pursuit of this new knowledge but further synthesis must precede its application in practice. The principal difficulty is that it is not just one picture of a production process that has to be constructed. Rather, it is a basic model with a great many parts, for each of which a number of alternatives exists. These are the major variables of animal size, prolificacy, climatic conditions, growth rate of the sheep and the pasture, disease incidence, stocking rate and so on. Different combinations of these variables, fitted together in a basically similar fashion, may be appropriate or not, according to the purpose, the time and the place. It is the possibilities for variation in the basic model that concern the economist, the farmer and the agricultural scientist. No one system of sheep farming is likely to be equally appropriate to all circumstances. Each farmer may have to select a particular form of the production process for his special circumstances. The economist needs to know the alternatives, in biological terms, to which he can attach the economic significance appropriate to the time.

There is thus a need for choice and flexibility and for a basic model within which predictable changes can occur. As Whitlock (33) has pointed out, the structure of a biological problem can be depicted by either words or drawings which may serve as simple models. When problems are very complex it is extremely helpful to employ mathematical symbols to facilitate the handling of the numerous variables and their possible values (35). Full use of mathematical techniques, however, requires a certain amount of information and, equally important, an interested mathematician. Whitlock stated the problem in this way: 'For the effective application of statistics, it is necessary to match an adequate statistical model to an adequate biological model.' Such terminology may not seem familiar to many engaged in sheep production: an illustration may clarify the point, however.

It is obvious that when discussing 'sheep', we have no particular breed in mind; even less have we a particular individual sheep in mind. Nevertheless, we do visualize a picture or model of a generalized sheep, which could not be anything else but which could accommodate all the possible variations in size, shape, wool cover, colour and so on, that we may wish to consider. It is a kind of skeleton, which can be clothed in a variety of ways, governed by certain rules. It would be extremely difficult to discuss sheep problems without such a basic model. In a similar way, it is helpful to construct a model of the production process, to list its component parts and to describe their interactions with each other. The attempt would be expected to expose gaps in our knowledge which require attention: it should also suggest some facets of the sheep and its environment which may be relatively unimportant. Particular versions of the model, which may be new and appear promising, should be regarded as hypotheses to be tested under husbandry conditions. If they prove to have merit but are not without some disadvantages, long-term husbandry experiments could be used to state the problems associated with them, in terms which make them susceptible to systematic solution.

The number of possible combinations of the more important components would be very large and a firm sense of direction will be required to guide this kind of exploration. It can hardly begin, however, without the basic model and it is one of the purposes of the following chapters to consider to what extent this can be constructed with the knowledge currently available.

REFERENCES

(1) BELSCHNER, H. G. (1959). *Sheep Management and Diseases.* 6th ed. Sydney, Angus and Robertson.
(2) BURNS, M. (1962). Observations on the behaviour of fell sheep and their crosses. *Anim. Prod.* *4* (2), 298–9.
(3) CAMERON, T. W. M. (1951). *The Parasites of Domestic Animals.* 2nd ed. London, Adam and Charles Black.
(4) CANDLER, W. & SARGENT, D. (1962). Farm standards and the theory of production economics. *J. agric. Econ.* *15* (2), 282–90.
(5) COLBURN, O. (1963). *Report of the Sheep Panel.* Summary of Proc. Brit. Livestock Breeding Conf. London, H.M.S.O.
(6) CRESSWELL, E. (1963). *Sheep Husbandry and Research.* A bibliography. London, N.S.B.A.
(7) DIGGINS, R. V. & BUNDY, C. E. (1958). *Sheep Production.* N.J., U.S.A., Prentice-Hall Inc.
(8) FRASER, A. (1950). *Sheep Farming.* 5th ed. London, Faber and Faber.
(9) FRASER, A. (1951). *Sheep Husbandry.* 2nd ed. London, Crosby Lockwood.
(10) FRASER, A. & STAMP, J. T. (1961). *Sheep Husbandry and Diseases.* London, Crosby Lockwood.
(11) GOLDSWORTHY, M. T. (1961). *Breeds that have almost gone.* Yearbk., 40–42. London, N.S.B.A.
(12) GRIFFITHS, J. G. (1962). Observations on the use of natural shelter by sheep according to weather conditions. *Anim. Prod.* *4* (2), 298.
(13) GRUNSELL, C. S. & WRIGHT, A. I. (1961). *Preface to Animal Health and Production.* Proc. 13th Symposium, Colston Res. Soc. London, Butterworth.
(14) HARROW, W. T. (1956). The changing pattern of animal health. *Outlook on Agric.* *1*, 32–3.
(15) HERRICK, J. B. (1962). Current developments in veterinary nutrition. *Vet. Med.* *57* (10), 864.
(16) HUGHES, R. E. (1955). Ecology of Snowdonia. *Nature (Lond.)* *176*, 595–6.
(17) Interdisciplinary co-operation in technical and economic agricultural research (1961). O.E.C.D. Pub. No. 50.
(18) KAMMLADE, W. G. (Sr.) & KAMMLADE, W. G. (Jr.). (1955). *Sheep Science.* New York, J. B. Lippincott Co.
(19) LAPAGE, G. (1956). *Veterinary Parasitology.* London, Oliver and Boyd.
(20) MÖNNIG, H. O. (1962). *Veterinary Helminthology and Entomology.* 5th ed. revised by G. Lapage. London, Baillière, Tindall and Cox.
(21) MOULE, G. R., BRADEN, A. W. H. & LAMOND, D. R. (1963). The significance of œstrogens in pasture plants in relation to animal production. *Anim. Breed. Abstr.* *31* (2), 139–57.
(22) MUNRO, J. (1962). The use of natural shelter by hill sheep. *Anim. Prod.* *4* (3), 343–50.
(23) PEARSE, E. H. (1950). *Sheep, Farm and Station Management.* 6th ed. Sydney, The Pastoral Review Pty. Ltd.
(24) PETERS, B. G. (1956). Inaugural Lecture, Univ. of Lond.
(25) REPORT OF THE N.S.B.A. (1961). *Sheep Research in Britain.* Yearbk., 50-57. London, N.S.B.A.

(26) RITCHIE, Sir J. (1962). The future of the veterinary profession. Preventive medicine. *Brit. vet. J.* *118* (12), 499.

(27) SPEDDING, C. R. W. (1962). The agricultural ecology of sheep grazing. *Brit. vet. J.* *118*, 461.

(28) SPEDDING, C. R. W. (1963). The efficiency of meat production in sheep. *World Conf. Anim. Prod.*, Rome, 1963.

(29) STEERE, J. H. (1961). The ecological approach to veterinary medicine in the U.S.A. *Vet. Ann.*, 17–31. Ed. by W. A. Pool. Bristol, Wright & Sons Ltd.

(30) STEVENS, P. G. (1958). Sheep, Pt. 1. *Sheep Husbandry.* New Zealand, Whitcombe & Tombs Ltd.

(31) THOMAS J. F. H. & others (1955). *Sheep.* London, Faber and Faber.

(32) THOMAS W. J. (1960). Some economic aspects of grassland production and utilization. *Proc. VIII int. Grassl Congr.* (Reading), p. 27.

(33) WHITLOCK, J. H. (1961). Parasitology, biometry and ecology. *Brit. vet. J.* *117*, 337.

(34) WANNOP, A. R. (1958). Sheep farming in New Zealand. *Proc. Brit. Soc. Anim. Prod.*, 101–103.

(35) ARCUS, P. L. 1963). An introduction to the use of simulation in the study of grazing management problems. *Proc. N.Z. Soc. Anim. Prod.* *23*, 159–68.

(36) MCCLYMONT, G. L. (1963). Improvement of animal production education. *Proc. World Conf. Anim. Prod.*, Rome, *2*, 293–306.

(37) COOPER, M. M. & THOMAS, R. J. (1965). *Profitable Sheep Farming.* Ipswich, Farming Press (Books) Ltd.

(38) SOULSBY, E. J. L. (1965). Textbook of veterinary clinical parasitology. *Vol. 1, Helminths.* Oxford, Blackwells.

(39) THOMAS, J. F. H. (1966). *Sheep Farming Today.* London, Faber and Faber.

II

THE PASTURE

If the sheep is regarded primarily as a grazing animal, then it is reasonable to regard the pasture as an integral part of the production process or, at least, the background against which it takes place. An outline of sheep production and the principles on which it is based may reasonably begin, therefore, with a description of the pasture. To the specialist this cannot, of course, be done properly in one chapter. This will be true of every part of this book, dealing as it does with the *ecology* of the sheep; that is, with the relationships between one sheep and another, and between the sheep and the rest of its environment. What cannot be ignored is the fact that anyone, whether farmer or adviser, who is concerned to apply scientific knowledge to agricultural practice, has to create a synthesis of fractions from many separate disciplines. This requires a condensation of each contributory subject, not to its own bare bones, so to speak, but down to the essential minimum necessary to the understanding of the whole. The selection of information must be towards this end and it will differ radically according to the precise nature of the end. The pasture knowledge that can be regarded as essential in order to understand soil conservation, for example, or botanical succession, will be quite different from that needed for an understanding of sheep production. The latter is the criterion adopted in this chapter.

NATURAL AND SOWN GRASS

It is sometimes said that grass is only grown to feed animals and it is clearly felt that in so underlining its essential purpose the whole question is put in some kind of perspective. If it were completely true, it would be even more remarkable that such a relatively large proportion, up to 50 per cent, it has been estimated (42, 45), of the grass grown in Britain, for example, is never actually consumed by animals, but is wasted in one way or another. In many parts of the world, grassland may be regarded as a natural 'climax' vegetation. British grasslands can be described as 'man-made', however, in the sense that they are either sown, created or maintained by the activities of man, notably through his grazing animals. Wild grazing animals, such as the rabbit, can have the same effect but, in the main, the grasslands of Britain depend on man for their existence: without him they would proceed to a forest vegetation. The idea that grass just grows is therefore doubly erroneous. It grows by itself, i.e. without controlled activity by farmer and stock, only in the most limited sense and, left strictly to itself, it would probably cease to be grassland at all. There is here a point of fundamental importance, however. In many parts of the world grassland can be kept in being by a slight activity, such as maintaining a very low grazing pressure, and this kind of grassland, on the hills and uplands of Britain, for example, or the natural pastures of Southern Brazil, may represent an enormous national asset. The simplest way to exploit such a natural resource is to convert the herbage into meat, wool and hides. In this the sheep has a large part to play. But clearly the sheep are present primarily to convert the grass; the grass is not grown in order to feed them. By contrast, in the lowlands, pastures may be specially sown to grow sheep food: in this case the grass is only there to feed the sheep. This is not simply a distinction between upland and lowland, sown and unsown pastures. In the drier, eastern part of England, grass is often sown as a 'break' between arable crops. The reasons for this are often associated with the control of pests and disease in the arable crops or with the maintenance of soil 'fertility'—a term which may include plant nutrient status, water-holding capacity and soil structure. As with the hills, the pasture is there and the problem is how best to use it. The situation is not always as clear as in these examples. The sheep production process must be viewed from a somewhat different standpoint according to whether the grass is grown for the sheep or the sheep are there to exploit the grass. One difference in the attitude to the pasture in these two cases is that, in the former, the aim is to achieve the most profitable relationship between the input of resources and the output of products. In the latter, on the other hand, it is frequently the object to exploit the natural resource as efficiently as possible while keeping the inputs to a minimum. Even if these problems do not differ very greatly in biological terms, they may certainly do so in economic and social terms.

These two quite different ways of looking at the pasture should be borne in mind when considering pasture types, their growth and the ways in which this may be influenced.

Pasture Types

The 'natural' grasslands of the world have been classified as savannah, steppe, desert scrub and alpine. Within Britain, grasslands may be conveniently considered as either (*a*) uncultivated, or (*b*) cultivated (44).

(*a*) The uncultivated grasslands include much of the hill and rough grazings. Large areas of these are dominated by *Nardus* and *Molinia* and they include, for example, the heather moors (dominated by *Calluna* or *Erica*) and the boggy areas of cotton-grass (*Eriophorum* spp.), rush (*Juncus* spp.), sedge (*Carex* spp.) and moss (*Sphagnum* spp.). Within this category are the large tracts of bracken (*Pteridium* sp.) or bilberry (*Vaccinium* spp.) and the downlands with brome (*Zerna erecta*), fine-leaved fescue (*Festuca* spp.) and tor grass (*Brachypodium pinnatum*). Such grasslands can be altered and their agricultural value greatly improved often by relatively small inputs which initiate long-term changes. Although soil, altitude and climate may set limits to the species which occur, the management or use to which the pasture is put has an enormous influence also. Thus changes in the botanical composition of these areas have no permanence unless accompanied by activity, generally by man and his grazing stock, conducive to the vigour of the new species.

These uncultivated grasslands supply animal food of very variable quality and in a very seasonal fashion. The seasonality of herbage supply can occasionally be exploited but, in general, it is perhaps as much the non-uniformity of the supply as the relatively small total quantity that restricts its stock-carrying capacity. Exploitation by grazing in these circumstances may require an animal population whose needs fit this seasonal supply fairly closely (125).

(*b*) The cultivated grassland includes 'permanent' pasture and 'leys', and is for the most part found in relatively small enclosures. Permanent pasture* may have been sown initially but its composition

* A distinction is sometimes made between 'pasture', which is grazed, and 'meadow', which is cut for hay. 'Pasture' is used here for all grassland.

really reflects its management. With few exceptions it is characterized by the large number and variety of species composing it. This is illustrated in Table I by a botanical analysis of a very good permanent pasture on the lias clay of the Midlands of England. Altogether, no

TABLE I

BOTANICAL ANALYSIS OF A PERMANENT PASTURE ON LIAS CLAY—SPRING 1949

SPECIES	COMMON NAME	% *
Lolium perenne	Ryegrass	31·5
Festuca rubra	Red Fescue	13·2
Dactylis glomerata	Cocksfoot	10·4
Phleum pratense	Timothy	8·8
Poa trivialis	Rough-stalked Meadow Grass	6·6
Festuca pratensis	Meadow Fescue	5·2
Agrostis stolonifera	Common Bent	1·2
Alopecuris pratensis	Meadow Foxtail	1·2
Helictotrichon pubescens	Hairy Oat	1·2
Holcus lanatus	Yorkshire Fog	1·1
Anthoxanthum odoratum	Sweet Vernal Grass	< 1·0
Arrhenatherum elatius	False Oat	1·0
Bromus mollis	Soft Brome Grass	1·0
Cynosurus cristatus	Crested Dog's-tail	1·0
Festuca arundinacea	Tall Fescue	1·0
Hordeum secalinum	Meadow Barley	1·0
Poa pratensis	Meadow Grass	1·0
Trisetum flavescens	Golden Oat	1·0
Zerna erecta	Upright Brome Grass	1·0
Trifolium repens	White Clover	7·8
Trifolium pratense	Red Clover	< 1·0
Ranunculus repens	Creeping Buttercup	2·5
Galium verum	Lady's Bedstraw	< 1·0
Heracleum sphondylium	Hogweed	1·0
Lathyrus pratensis	Meadow Vetchling	1·0
Lotus corniculatus	Birdsfoot-trefoil	1·0
Luzula campestris	Field Woodrush	1·0
Potentilla reptans	Creeping Cinquefoil	1·0
Poterium sanguisorba	Salad Burnet	1·0
Primula veris	Cowslip	1·0
Rumex acetosa	Sorrel	1·0
Taraxacum officinale	Dandelion	1·0
Vicia sativa	Common Vetch	1·0

* percentage volume by point-quadrat

(Source : B. N. George, unpublished data)

less than sixty-five species of grasses, clovers and herbs were recorded. Only the more important of the herbs are listed in the table. Permanent grassland occupied 31 per cent of the agricultural area of England in 1959 (2) and is classified into six grades. The best contain 30 per cent or more of *Lolium perenne*; poorer grades contain less ryegrass and increasing proportions of *Agrostis* or *Festuca*.

A ley has been defined by Dr William Davies as 'a grass-legume sward explicitly sown as part of a pre-designed rotation of crops, the intention being to plough up again after a pre-determined number of years. The ley remains a ley so long as the species of grasses and clovers which it contains are directly attributable to the seeds that were initially sown.' The value of leys in crop rotation has been discussed by Williams (130): the benefits of ley farming to the arable crops grown after grass may include disease control and effects on the physical condition of the soil (35) and on its 'fertility'. Much of the 'fertility' effect may be attributable to soil nitrogen status (27, 28, 131). In recent years particularly, leys have been characterized by the small number of species sown. Since the creation of a ley provides the opportunity to establish a particular flora, leys are usually sown for specified purposes and to last specified periods, from, say, one to seven years. It is natural to sow varieties, such as those bred at Aberystwyth, which are specially designed to be leafy, persistent, productive and with early or late growth characteristics. It is desirable to sow species that will provide animal food at the right times and some seed mixtures are designed to include species or varieties which grow in a complementary manner. Others are specifically designed to provide grazing at some critical time, such as lucerne (46) for summer drought, or tall fescue for very early growth in the spring. A great deal of work has been done on the provision of out-of-season forage by the use of different species (66), by the application of nitrogenous fertilizers to encourage early growth of herbage in the spring (3, 5) and by management to conserve grass *in situ* as 'foggage' (69). Hughes (65) summarized a whole series of experiments largely concerned with the production and utilization of winter grass by cattle: he underlined the importance of integrating a system of winter utilization with summer production. Baker, Chard & Hughes (4) reported on the effects of management on winter grass production at forty-seven sites in England and Wales from 1954 to 1958. They found that local factors, such as aspect, altitude and sward vigour, were

TABLE II

THE MORE COMMONLY SOWN GRASSES AND LEGUMES

With brief notes on their normal usage

Species		Normal Usage
Perennial ryegrass	*Lolium perenne*	Permanent pasture; long term leys, either as single species or in mixtures often accompanied by white clover.
Italian ryegrass	*Lolium multiflorum*	Short-term leys, typically one year. Often accompanied by red clover.
Timothy	*Phleum pratense*	Long-term leys and permanent pasture. In mixtures, often with meadow fescue or white clover. As single species where winter soil moisture is high.
Cocksfoot	*Dactylis glomerata*	Long-term leys; in mixtures; grown as single species to give late autumn keep or winter foggage.
Meadow fescue	*Festuca pratensis*	In general purpose leys; compatible with timothy or white clover.
Tall fescue	*Festuca arundinacea*	Winter green, some strains giving good winter growth; used in special purpose leys.
Red fescue	*Festuca rubra*	Under less fertile conditions, where ability to grow in acid or alkaline soils which dry out in the summer is an advantage. Also sown at high altitudes and in exposed pastures.
Crested dog's-tail	*Cynosurus cristatus*	Sown in mixtures under less fertile conditions.
Smooth-stalked meadow grass	*Poa pratensis*	Not sown as single species but frequently included in mixtures and found in widely different situations.
Rough-stalked meadow grass	*Poa trivialis*	Not sown as single species but frequently included in mixtures and found in widely different situations.
White clover	*Trifolium repens*	The legume most frequently sown in combination with grass for a long-term ley or for permanent pasture.
Red clover	*Trifolium pratense*	Less persistent in the sward than *T. repens*. A rapid grower and highly compatible with *Lolium multiflorum*.
Alsike clover	*Trifolium hybridum*	More persistent than *T. pratense*; sown under soil conditions of dampness and acidity where neither *T. repens* nor *T. pratense* is entirely satisfactory.
Lucerne	*Medicago sativa*	Sown as a pure stand or with a companion grass, often in alternate rows. More often found in the east of England where it is particularly useful in mid-summer under dry conditions. Restricted to calcareous soils.
Sainfoin	*Onobrychis viciifolia*	Grown in one-year leys for hay; persistent strains may be grown for a longer time and grazed. Often sown in soils which become dry in summer. Restricted to calcareous soils.

(compiled by D. D. Kydd)

generally more important than latitude in determining the level of yield in December, but that there were greater losses of dry matter from December to January in the north than in the south. The use of simple seed mixtures (64, 67) has been associated particularly with attempts to even out the supply of herbage and to lengthen the grazing season. Remarkably few species can be regarded as of major consequence in British sown pastures: they are listed in Table II. The importance of strain or variety within a species was stressed by Sir George Stapledon, whose work represented 'an important milestone in the history of grassland thought and technique' (37). Whatever the species involved, however, the management imposed on the sown pasture is of major significance (38, 41, 42), as is the recognition that grass should be regarded as a crop (44).

The concept of grass as a crop has been taken further by Woodford (168), who has suggested that the word 'ley' is both imprecise and unhelpful in this context. There is little doubt that a more specific description of the species sown becomes more appropriate as the purposes and functions of the crop are better defined, especially where only one species is involved.

The Pasture Population

Perhaps nothing is more deceptive visually than pasture, often looking much the same to the casual eye on successive occasions. It is not immediately obvious that it represents a dynamic population of individual plants in which the 'turnover' of leaves and plant units is comparatively rapid. Populations may vary enormously in the number and variety of species present. In general, they consist of grasses, clovers and broad-leaved weeds. The last will be considered separately later in this chapter. The grasses may be annual, biennial or perennial but they may all be regarded as annual in terms of their basic plant units. These are the 'tillers', which, arising as branches from axillary buds, develop roots and become independent plant units. The new tiller is dependent on the parent plant until it has about two leaves of its own. The rate at which tillers are produced and their survival have an important influence on the composition, density and productivity of the sward. Langer carried out an intensive study of 'tillering' in grass and showed that frequency of defoliation could affect the number of tillers, especially in timothy (74). Individual tillers may live for over a year and produce 30 leaves in that time, at the rate of, perhaps, one a week during the summer period. Others may die before they have achieved a separate existence. Each leaf grows and expands, continues to photosynthesize and then senesces, unless harvested in one way or another. The length of time for which leaves survive varies with environmental conditions but few leaves live longer than about eight weeks. The legumes vary greatly in growth habit but, in general, have relatively large, flat leaves; some species spread by stoloniferous growth. Their outstanding value is due to their ability to fix atmospheric nitrogen: this they do by means of bacteria which are found in nodules on their roots. Legumes are therefore independent of the soil nitrogen supply and in a mixed grass-legume association can supply nitrogen to the grass as well. Frequently, a first step in the improvement of poor natural grassland is to encourage the vigorous development of legumes.

For any population, there is an optimum size in relation to the available source of nutrients and energy. The pasture population grows and produces in response to the same factors that influence individual plant growth and the density and structure of the pasture have an influence on the individuals that compose it. The density of the population can be greater or less than optimal in relation to nutrient supply, including water, or to light. Light is probably the most important factor in this context. A dense sward can have the effect of sharing out the incident light in such a way that no plant receives enough: in these circumstances plants photosynthesize at a low rate relative to their rate of respiration and lose more energy than they gain. They are then referred to as being 'below their compensation point'. Too low a plant density, on the other hand, results in incident light being wasted on bare ground.

Seasonal Growth of Pasture

The factors affecting growth will be considered in the next section. Clearly the rate of plant growth may vary enormously according to the availability of nutrients but, because it is so greatly affected by light, temperature and water, it is possible to generalize to some extent for a particular region. Thus, the broad pattern of herbage growth in Britain can be described as very little in the winter, an enormous increase in the late spring and early summer, followed by

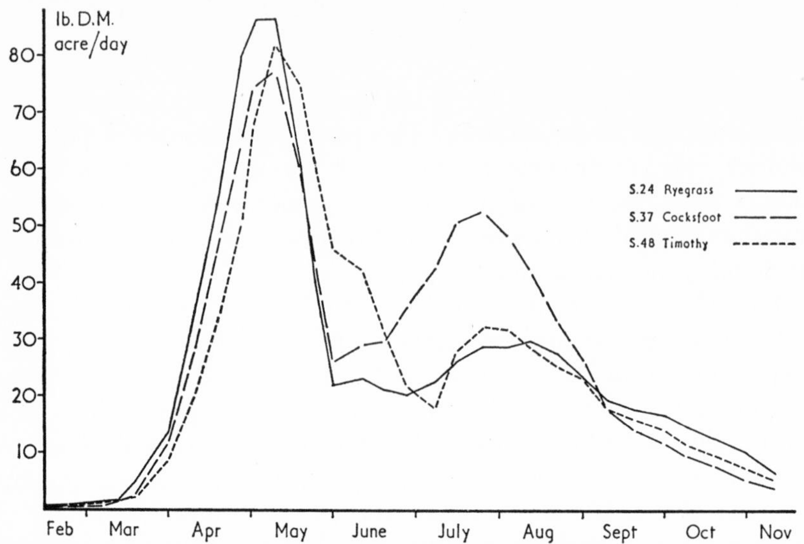

FIG. 2.1—*Production curves of three grasses. Since rate of growth is plotted against time, production is represented by the area below the curve*

(after Anslow & Green (137))

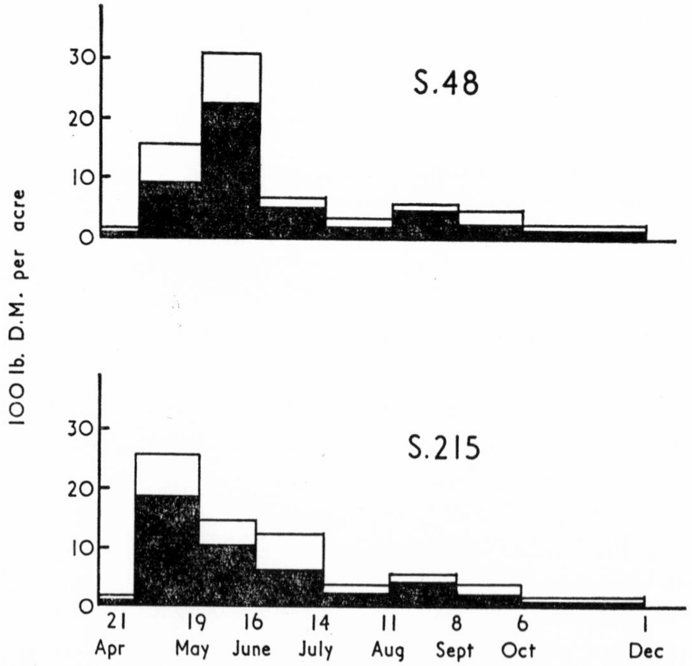

FIG. 2.2—*Dry Matter (D.M.) yields from S.48 timothy and S.215 meadow escue, treated with either 52 lb N/acre (shaded areas) or 156 lb N/acre (unshaded areas) and harvested at monthly intervals to approximately 2 in. above soil surface. The harvest on Dec. 1 represents 2 months' growth*

(Source : O. R. Jewiss, unpublished data, Grassland Research Institute, Hurley)

some reduction of growth rate in late summer and, very often, another, smaller increase in the autumn. This pattern is illustrated in Figs. 2.1 and 2.2. Fig. 2.1 shows an example of the change in *growth rate* which occurs from one part of the year to another. Fig. 2.2 shows the variation in yield of dry matter (D.M.) by two grass species, harvested at monthly intervals during the growing season.

Looked on as a supply of food, the most striking feature of herbage growth is the extreme seasonality of its distribution (39, 55, 120). Superimposed on the major seasonal fluctuations there may be great variations over short periods, which are relatively unpredictable. The great value of herbage conservation is to transfer major surpluses to times of major shortage. One effect of increasing the total amount of grass grown by, for example, the use of fertilizers, may be to increase the disparity between summer and winter herbage production. Thus conservation is even more essential at high levels of production and, if it is not possible, e.g. in some upland areas, a high level of food requirement must be met by supplementary feeding of material grown elsewhere.

HARVESTING

Conservation requires a means of harvesting and a means of preservation of the cut material. Preservation can take several forms, the most common being as hay, silage and dried grass.

Haymaking

Hay has suffered greatly in the past from bad weather and through being made from overmature herbage. The practical reasons for doing this are, first, that the maximum amount of dry matter is grown when the herbage is allowed to grow to a stage of maturity at which quality is reduced and, secondly, that long, stemmy material is easier to handle and dry. There have, however, been many recent changes and very high quality hay can now be made more easily: with modern techniques, such as barn hay drying, even the weather can be circumvented. Two features of haymaking deserve special mention. The quality of the hay closely reflects that of the grass from which it is made, provided the better parts, such as leaves, are not lost in the process. However, some changes do occur, notably a considerable loss of carotene and the formation of vitamin D in the presence of sunlight, though some vitamin D may occur in green leaves (61). There are obvious practical advantages of hay, since it is easy and pleasant to handle.

The production of dried grass is not simply an extension of the haymaking process. It is essentially rapid, whereas haymaking is a slow-drying procedure. When herbage is cut, the cells are not immediately killed; the herbage continues to respire and lose weight, not merely water, until the cells die. The object of rapid drying of the herbage by high temperatures is to reduce losses due to respiration.

In both hay and dried grass the high dry-matter content may be an advantage at times. When an animal's intake of bulk is limited, as it may be, for example, in late pregnancy, it is sometimes felt that a food with a high dry-matter content may allow a higher daily intake. In very cold weather, it may be a disadvantage for a sheep to have to heat up the large quantities of water taken in with a low dry-matter food. The heat required has to be provided by the animal and food may have to be used to produce it.

Ensilage

Ensilage involves a fermentation process, with or without additives (87, 123, 126); the resulting silage is preserved by the products of

TABLE III

THE DIGESTIBILITY OF RYEGRASS AND SILAGE

The relationship between the digestibility of H.1. ryegrass at different stages of maturity and that of silage made from similar grass

(from 'The effect of ensiling on crop digestibility'. C. E. Harris & W. F. Raymond, (1963) *J. Brit. Grassl. Soc. 18* (3), 204)

Date Cut	% Digestibility of Dry-Matter	
	Grass	Silage
8th April, 1960	77·4	74·0
28th April, 1960	78·1	78·6
13th May, 1960	72·0	71·6
24th May, 1960	67·6	67·6
30th May, 1960	65·4	65·2

TABLE IV
FERMENTATION

Examples of satisfactory and unsatisfactory fermentation resulting from the ensilage of two grasses by similar methods (i.e. in towers, without additives)

GRASS CHARACTERISTICS				SILAGE CHARACTERISTICS					
Variety	*%Crude Protein**	*%Soluble Carbo-hydrates*	*Moisture Content %†*	*pH*	*% Lactic acid*	*% Butyric acid*	*% Acetic acid*	*% NH_3*	*Quality of Product*
H.1. ryegrass	12·9	19·0	81	3·8	11·4	0·6	3·8	0·4	Satis-factory
Germinal cocksfoot	19·6	11·0	79	5·4	0·2	5·2	5·9	1·7	Unsatis-factory

(Source : C. E. Harris, unpublished data Grassl. Res. Inst.)

* Crude protein = Nitrogen × 6·25.
† All other percentages are expressed on a dry-matter basis.

TABLE V
LOSSES IN SILAGE MAKING AND FEEDING

Two harvests (A and B) taken on the same occasion (Sept. 1962) from separate paddocks of ryegrass with a little white clover. Each cut was stacked without consolidation or additives and covered completely with a plastic sheet. The two lots of silage were fed to separate groups of sheep between December 1962 and February 1963.

	A	B
Total weight of fresh herbage (lb.)	38,300	28,700
% D.M. in the fresh herbage	19·8	15·6
*% loss of original D.M.**		
(a) in ensilage	21·6	26·6
(b) not offered to sheep because it was unsuitable	12·5	9·4
(c) rejected by sheep	7·4	5·7
(d) total uneaten for both the above reasons	19·9	15·1
(e) total losses as a % of total ensiled	41·5	41·7
% consumed by sheep, of the D.M. ensiled	58·6	58·2

(from Expt. H.202, Grassland Res. Inst., 1962)

* D.M. determinations were based on oven-drying at 100° C: for silage this involves some losses of volatile components.

fermentation. Silage also varies in quality with the quality of the herbage from which it is made (see Table III), but the form in which the nutrients occur tends to be similar even when the nutrients in the original herbages were present in different forms. Table IV gives a typical chemical analysis for silages of both good and poor quality and that of the grass from which they were made. In considering the food value of silage and hay, it should be remembered that the losses during the making and feeding of these products can be very high (see Table V). Silage can be difficult and unpleasant to handle and, for these reasons, self-feeding techniques have been developed which allow animals to help themselves, without excessive wastage.

The process of ensilage depends on a controlled bacterial fermentation, mainly anaerobic, reaching a stable situation in which no further decomposition occurs. Stabilization is obtained when the sugars in the fresh grass have been converted to produce about 2 per cent lactic acid in the silage. The amount of sugar varies with the crop, its stage of growth, the amount of nitrogen used and the weather; the amount needed depends also on the degree of wilting after cutting.

Additives are often used to assist in silage making, either by adding sugar (e.g. as molasses) or acid (as in the A.I.V. method).

Alderman (134) has recently summarized the losses associated with various forms of conservation. The estimates vary from total losses (in field and in store) of 5 per cent for dried grass, to 15 per cent for barn-dried hay and tower silage, to 30 per cent of the dry matter for clamp silage and field-cured hay.

The modern development of vacuum silage (149) clearly aims to keep air out and this proper sealing with plastic sheets will do, whether a vacuum is applied or not (though trying to achieve a vacuum may be the best method of testing whether proper sealing exists).

All aspects of conservation have been discussed recently by Wilkins (167), including future trends.

Grazing

This harvesting by the animal is, at present, of paramount importance in the establishment, growth, production and utilization of the grass crop.

Herbage plants change their growth-form, size and shape in response to grazing management, and pasture populations change their botanical composition. It cannot be too strongly emphasized that any particular.

desirable mixture of plants, however it is achieved, will only remain if the total management favours that particular combination. Starting from the same point, very different swards can be produced in a remarkably short time by varying the grazing pattern alone (89). The act of grazing has many far-reaching consequences for the pasture. Some of these will be considered in later chapters (VII, IX and XI): the possibilities of zero-grazing (104) should be borne in mind, however. One of the distinctive attributes of grasses is their tolerance to

FIG. 2.3—*Stoloniferous behaviour of a prostrate tiller of perennial ryegrass.* (*Drawn by D. D. Kydd from plants developed under heavy sheep grazing; July, 1959.*)

frequent defoliation. In extreme cases, such as very heavy sheep stocking on a perennial ryegrass/white clover sward, up to twenty-three separate defoliations have been possible within a nine-month grazing season. Some herbage plants, e.g. lucerne, would probably not survive such treatment. The pattern of grazing may thus affect survival and production, both of individual plants and of the whole pasture population. This it does by altering both the amount of herbage present and also the rate of removal of plant material.

The individual plant derives its nutrients, i.e. minerals and water, from the soil, via its root system (128, 129). This may take the form of a long tap-root, as in lucerne, or a mass of adventitious, fibrous roots, as in the grasses. A deep-ranging tap-root has clear advantages in times of drought but it is not always appreciated that, although some 90 per cent of grass roots (by weight) generally occur within the top 3 inches of soil, some penetrate to remarkable depths (sometimes even to 6 ft). It is of interest that Kernick (72) found that the roots of cocksfoot recovered substantial amounts of nitrogen from depths of up to 2 ft. The requirement for water is considerable. It acts as a transport system for nutrients and metabolites within the plant, quite apart from its function as a constituent of the living cells, and is also required for transpiration. In both plants and animals a remarkably high proportion of the tissues is water. Above ground the plant has a stem (the true stem is normally very short in the grasses) bearing leaves. The leaves are the main site for the synthesis of simple elements into more complex substances: the energy for this is derived from incident light and the harnessing of light for this purpose (photosynthesis) requires chlorophyll. The leaves are also the main areas from which water is lost. This process of transpiration may be thought of as purely physical, closely akin to evaporation and governed chiefly by the prevailing temperature and relative humidity.

Since photosynthesis can occur only in the green parts of the plant exposed to light, after any complete defoliation re-growth of leaves must depend upon reserves. Whether these are in the roots or in stem bases (7) they are obviously not inexhaustible. Consequently, very frequent removal of all the green parts must weaken the plant and ultimately kill it. Frequent defoliation can be tolerated, therefore, if (*a*) the plant reacts by adopting a growth-form which renders some green parts relatively unavailable to the grazing animal (for example, rosette plants and prostrate tillers in grasses do this (see Fig. 2.3)) or (*b*) the rate of growth of the plant is such that it has a sufficient period of photosynthesis even in the short interval available, to replace its reserves. The latter clearly demands a high level of plant nutrition and, indeed, of all the factors influencing growth.

FACTORS AFFECTING PRODUCTION

Production refers to the amount grown by the pasture and is thus the result of growth by all members of the plant population. In fact, production has to be measured over some period and the difference between initial and final amounts present must represent nett production, that is the amount grown less the amount which has disappeared in the interval. Disappearance can be the result of death and decay, removal by earthworms, or defoliation by insects, slugs, rabbits, etc. In addition, it is worth remembering that most methods of measuring the amount present leave some part of the sward unharvested and changes can occur in this fraction. The pattern of uninterrupted growth can be built up by measuring the quantity present at intervals, a separate area being used for each sample. Such growth has a rather restricted meaning, particularly when expressed in terms of dry matter, but this measurement avoids the complication of the influence of method and frequency of harvesting on the amount grown. It is perhaps better, then, to consider the factors that affect crop growth rate*: the effect on total production must take the harvesting method into account (99, 100). It takes much careful work to measure the growth rate of a grass species and Fig. 2.1 is the result of such effort.

First, it is necessary to mention the soil. It might have seemed logical to start with the soil, but for the present purpose, it may be regarded as acting chiefly through the herbage. It may be an important limiting factor on herbage production but it can be greatly modified. Nutrients can be added, excess water can be drained. These activities can have considerable consequences to the sheep, however. Thus, changes in acidity and water level can greatly affect the numbers of snails harbouring liver fluke, for example. Different soils take longer or shorter periods to warm up in the spring, they may 'poach' more or less in wet conditions and become a differential influence on the oxygen supply to plant roots.

Temperature

During vegetative growth grasses have their growing points close to the soil surface and soil temperature is therefore important. Grasses such as ryegrass and cocksfoot appear to grow most rapidly at a temperature between 13° C and 18° C, but it would be a gross oversimplification to consider only mean temperatures and to ignore the large variations that occur within a day and also between day and night. Temperature frequently has a very great influence on herbage growth rate: Table VI gives examples of this and illustrates the differences that exist in the optimum temperatures for different species.

* Crop growth rate here refers to the rate of D.M. increase of a crop per unit area of land per unit of time.

Water supply

This can be a serious limiting factor but its importance varies enormously with geographical location (24). Our knowledge of irrigation

TABLE VI

THE EFFECT OF TEMPERATURE ON PLANT GROWTH

Growth is expressed as the weight of tissue formed on one tiller per day (mg)

SPECIES	45° F	55° F	65° F	75° F	85° F	95° F
Perennial ryegrass	2·8	4·6	4·3	3·8	3·2	0·8
Short-rotation ryegrass	2·5	5·2	5·0	4·7	3·9	1·1
Cocksfoot	5·0	8·2	8·8	8·0	7·1	1·2
Yorkshire Fog	3·8	8·2	9·1	8·3	6·7	0·2
Paspalum	0·7	4·0	8·5	17·3	26·8	22·0
Brown top *	1·1	1·7	1·6	1·3	1·2	0·0
Subterranean clover	2·3	6·3	6·9	6·3	5·0	2·5
White clover	1·2	2·5	4·1	5·0	4·3	2·1
Lotus major (now *L. uliginosus*)	0·5	1·5	3·5	3·8	2·9	1·3

* Agrostis tenuis

After Mitchell, K. J. (84)

needs and effects on crop growth owes much to the work of Penman (93). Broadly speaking, the growth of grass is significantly reduced when the 'water deficit' reaches about 2 inches. Water shortage is expressed in this way because it represents the number of inches of rain or irrigation water required to restore the pasture to 'field capacity'. The latter is the situation when the soil is holding all the water it can, against the force of gravity; if more is applied it drains away. The influence of water shortage is illustrated in Fig. 2.4. One use of irrigation is to even out the supply of pasture over the season (78).

Light

This is obviously of major importance; its incidence varies within and between days and seasons but is beyond the control of man, at least at the present time in relation to pasture. Before discussing it

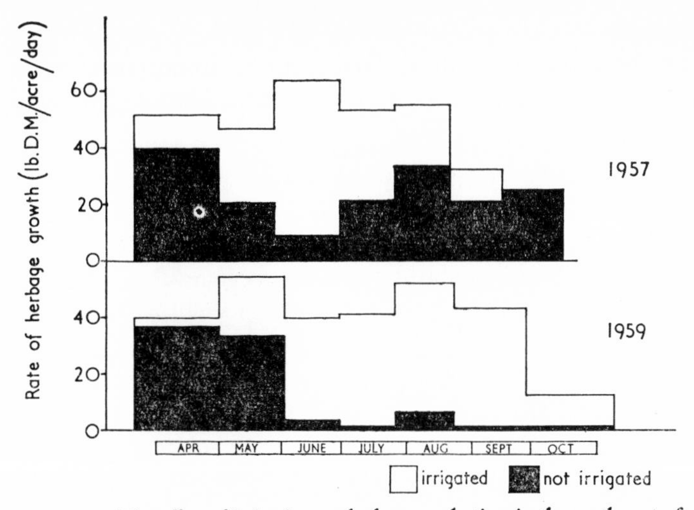

FIG. 2.4—*The effect of irrigation on herbage production in the south-east of England*

The figure shows the rate of growth, averaged between cuts at approximately monthly intervals, of a perennial ryegrass/white clover sward (S. 23/S. 100) sown in 1951, in two different years:

1957, a year of near-average summer rainfall with no severe drought;

1959, a year of severe drought.

The fertilizer applied contained sufficient P and K for a yield of 15,000 lb D.M./acre/annum but no nitrogen.

The total annual yields of D.M./acre were as follows:

	Unirrigated	*Irrigated*
1957 . .	5350	9960
1959 . .	2970	9280

These illustrate an important effect of supplementary irrigation in increasing the uniformity and therefore the predictability of herbage yield.

(Source : T. E. Williams & W. Stiles, unpublished data from experiments at the Grassland Research Institute)

further, however, it is necessary to distinguish between the effect of light on vegetative and on reproductive growth. Grasses exhibit a marked tendency to flower and set seed. This involves a period of rapid dry-matter increase, as the flowering stem elongates, but the structural tissues involved may result in a reduction in food value to the animal. The time at which flowering occurs varies with the species but is generally a response to day-length (124). The effect of light on these reproductive aspects of growth will not be considered here because it is the vegetative phase that most concerns the grazing animal. This is not, however, to belittle the importance, for grazing of understanding the implications of the reproductive phase and the factors that influence it. For sheep grazing, it might almost be stated as a major aim of management, that a pasture should be maintained in vigorous vegetative growth. With some species this can be done with relatively few well-timed defoliations; with others it is virtually impossible to eliminate flowering completely.

The important thing about light for vegetative growth is that it should be intercepted (16). Maximal crop yield cannot be obtained with a photosynthetic area that absorbs less than the optimal proportion of the incident light. Much is reflected from herbage and some is transmitted, so that the optimal leaf area is much greater than the area of the ground covered by it. Watson (122) proposed the term Leaf Area Index (L.A.I.) to describe the leaf area per unit area of soil: its optimum value varies with the light intensity (112). It is thus difficult to attempt to maintain a sward at an optimum value (see Chapter VII) but the importance of taking the photosynthetic area into account in, for example, rotational grazing, can be readily appreciated. Although the incident light cannot easily be controlled, many agronomic treatments affect the amount of light falling on each individual leaf within the sward (9). Donald (47) has argued that, even where water or nutrients impose some limit on the rate of plant growth, competition for light is still almost certain to be of importance. Warren Wilson (132) has suggested that the dispersion or spatial arrangement of the leaves would affect growth rate; the optimal arrangement would, of course, vary with different plant species. For mixed grass-clover swards it may be necessary to consider separately the L.A.I. values for the two components (111). Brougham (16, 17, 18) found that the maximum growth of New Zealand ryegrass swards occurred at a L.A.I. value of 5·5, at which stage light interception was virtually complete. As Donald (47) has pointed out, however, there are few published accounts of the existence of an optimum value, though clearly it can occur (111).

It should be emphasized that there is often a considerable difference between the amount of herbage grown and the amount harvested. The daily loss due to senescence can be substantial (140) and may amount to one-third of the 'gross aerial growth' (160).

The usefulness of the leaf area concept in pasture management has been reviewed recently by Brown & Blaser (142).

There are certainly circumstances in which L.A.I. is not the most important factor governing the growth of pasture (e.g. 136) and it is a considerable oversimplification to suppose that all leaf area operates with the same photosynthetic efficiency.

Nutrient supply

The major nutrients required by the plant from the soil are nitrogen (N), calcium (Ca), phosphorus (P), potassium (K), sulphur (S) and magnesium (Mg). The essential trace elements are iron (Fe), manganese (Mn), copper (Cu), boron (B), molybdenum (Mo) and zinc (Zn) (44, 127). In addition, other minor elements occur, including sodium (Na), chlorine (Cl), aluminium (Al), silicon (Si), selenium (Se) and cobalt (Co), which, if not essential to the plant, may be of great importance to the animal feeding on it (119). Animals require iodine (I), Cu, Zn, Fe, Mn, Co, Mo and Se and probably fluorine, bromine, barium and strontium (119). Deficiencies of Fe are unlikely under grazing conditions: a dietary deficiency of manganese has not yet been demonstrated for sheep, but deficiencies of selenium and zinc have.* Both molybdenum and selenium can occur in toxic quantities and silica is linked with two problems. The first is wear in sheep's teeth due to the presence of amorphous silica as opal phytoliths; the second is the occurrence of urinary calculi, which are frequently made of silicates, though sometimes they may be phosphates. Most of the silica present in herbage and ingested by the animal is excreted in the faeces (156, 157). Although the addition of various salts to the diet may affect the incidence of urinary calculi in wether lambs (143), more success has been achieved in some circumstances by ensuring that the calcium/phosphorus ratio in the diet is close to 3:1. The effect of Cu, Mo, Co and I on animal health will be referred to later in this chapter and in Chapter IV. Marked variations in iodine content of grasses have been found between species and seasons (135) and there are indications of differences between varieties of perennial ryegrass and white clover. The general level of iodine in herbages appears to be rather low and may be below animal requirements during pregnancy and lactation.

The main fertilizers in agricultural practice contain lime, phosphates, potash and nitrogen: impurities generally supply sufficient of the

* See SMITH, W. H., OTT, E. A., STOB, M. & BEESON, W. M. (1962). Zinc deficiency syndrome in the young lamb. *J. Anim. Sci.* *21*, 1014 (Abstr.).

other minerals (113) but should fertilizers be further purified, it may be necessary to apply minor elements also. As with animal feeding, the balance of nutrients may be important and excesses can be as serious a disadvantage as can deficiencies. Any plant nutrient can, of course, be the factor limiting growth. In Britain, the need is generally for lime where acidity has to be corrected, and for phosphates, potash and nitrogen, the so-called 'complete' fertilizer, to encourage growth.

The continued application of unnecessary quantities of an element may involve some risk (34), particularly to animal health. This is certainly the case in relation to potash, which, in excess, may lead to an increased incidence of hypomagnesaemia (a reduction in the magnesium content of the blood serum which frequently results in tetany). Excessive use of nitrogen has also been associated with metabolic disorders but the evidence here is less certain (see Chapter XIII). There is no doubt that heavy application of nitrogen can result in large increases in herbage production (21, 58, 102) and exercise a considerable influence on the distribution of growth during the season (20, 32). Reith *et al.* (102) reported responses of 14·0 and 15·7 lb of herbage dry matter per lb of fertilizer nitrogen and Castle & Reid (23) recorded a response of 15 lb. There is also no doubt that an increase in herbage nitrate content frequently follows the application of fertilizer (29, 54, 90). Griffith & Johnston (54) found this to vary with the species and variety of grass. Nowakowski (90) found that the nitrate content of grass receiving 112 lb nitrogen per acre depended upon the form of the fertilizers and the time and method of application. Nitrate contents were highest with calcium nitrate (1384 p.p.m.) or ammonium nitrate (1418 p.p.m.), lowest with ammonium sulphate (386 p.p.m.) and intermediate with urea (724 p.p.m.). Although nitrate-nitrite poisoning can occur in sheep (48, 107), it has not been recorded from a grass diet. This subject is discussed further in Chapter XIII. The relationship between animal health and nitrogen fertilization is an extremely important question agriculturally because there is no doubt that, on a soil that is adequately fertilized for crop health, the major factor in influencing yield is nitrogen. Fig. 2.5 shows the kind of response that is obtainable by the use of applied nitrogen for swards with and without clover. Clover does not generally respond to added nitrogen (30) and a mixed sward may respond ultimately by losing its clover content (15, 68, 77, 96). This may occur anywhere within the range of about 100 lb to 200 lb nitrogen per acre, after which the applied nitrogen may completely replace the contribution of nitrogen by clover, which may be as high

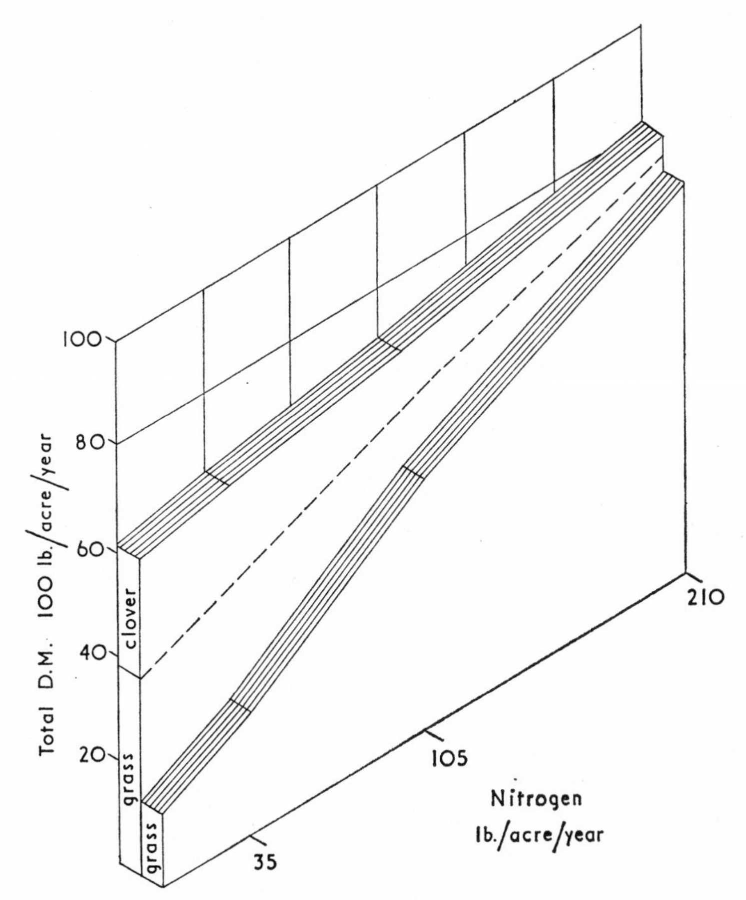

FIG. 2.5—*Mean annual yield of two 3-year leys, cut four or five times a year*

Note (1) The near-linear response to fertilizer nitrogen in both grass and grass/clover swards;

(2) Grass growing with white clover outyielded the pure grass sward at all levels of nitrogen;

(3) With increasing level of nitrogen the yield of clover was reduced, and so was the benefit conferred by clover upon the associated grass;

(4) The yield of the mixed sward (40 per cent clover) without fertilizer nitrogen is comparable with that of a pure grass sward receiving about 160 lb N/acre;

(5) The response of the mixed sward was 10 lb D.M. per lb N applied; of the grass sward almost 30 lb D.M.

(J. O. Green, from data given in Table 1 of D. W. Cowling (1961), *J. Brit. Grassld Soc.* *16* (4), 285)

as 200 lb nitrogen per acre per year; in the north of New Zealand an annual fixation of 400–500 lb nitrogen has been reported (105). Beyond this, further responses will be obtained from the almost pure grass sward. Green & Cowling (52) suggested that dependence on clover nitrogen limited production to about 6,000 lb D.M. per acre per year and that if it is desired to raise the yield of the grass crop beyond this level, in temperate regions, it is necessary to add fertilizer nitrogen. Cowling (31) found that where 210 lb nitrogen per acre was applied annually to a sward of mixed grass and white clover, the clover contributed only 5 per cent, and was altogether responsible for less than 8 per cent, of the total annual production of dry matter. These changes in grass-clover balance also alter the nature of the animal's food supply because grass and clover differ in chemical composition and may well differ considerably in food value (see page 131).

In recent years, much higher levels of nitrogenous fertilizer have been used experimentally (up to 800 or 1,000 lb N per acre) and the yield of pure grass swards has generally been found to increase in a linear manner up to about 300 lb N per acre. Since nitrogen is taken up very rapidly by the plant, it may be very inefficiently used if the defoliation interval is short, as there will have been too little time for a full response in terms of crop growth (138, 139, 147, 154, 159).

In small plot work, with ample nitrogen and water, Cooper (146) has recorded yields from S.23 ryegrass of 18,000 and 19,000 lb D.M. per acre per annum.

In New Zealand, yields of D.M. varying from 9,000 lb per acre per annum to 19–25,000 lb have been obtained (141) and a yield of 10,000 lb D.M. or more appears possible over a wide range of environments.

In south-eastern Queensland, also, yields of up to 25,000 lb D.M. per acre per annum have been obtained from several mixtures (158), using irrigation.

Grass yields tend to be very sensitive to moisture supply as well as to that of nitrogen whether derived from an associated legume or not. Lucerne, on the other hand, is little affected by drought and requires no added nitrogen. Green (152) recorded yields of between 13,000 and 14,000 lb D.M. per acre per annum from lucerne with a little meadow fescue in the mixture and similar yields have been obtained with lucerne alone. Vartha & O'Connor (163) have drawn attention to the considerable potential, for grazing in Canterbury (New Zealand), of mixtures of lucerne and ryegrass.

Animal residues

It is difficult to assess the fertilizer needs of a pasture. Deficiencies in the plant can be diagnosed and corrected but the problem remains as to the prediction of needs for future growth. This is further complicated by the return of minerals in the animal's faeces and urine (62). Table VII gives a typical analysis of sheep excreta and an estimate of the daily return made to the pasture. (The faecal output of grazing sheep can be measured by the use of collection bags and harnesses or estimated by using indigestible markers (73).) When all the herbage is cut and removed, it is possible to estimate the weight of minerals required to replace those taken away. Under grazing conditions, however, a considerable proportion of these minerals may be returned

TABLE VII

AN EXAMPLE OF THE RETURN OF FAECAL AND URINARY NITROGEN

Figures taken from a 100 lb sheep grazing high-quality pasture. The ratio of N:P:K in total excreta (faeces and urine) has been found to be 1·00:0·14:1·12 for sheep grazing a grass-clover pasture. (Wolton, K. M. (1963), *J. Brit. Grassld Soc.* *18* (3), 213)

Daily Output	g D.M.	g	% N	Amount of N returned to the Pasture		
				g/sheep/day	*per sheep per annum* g	lb
Faeces .	300		3·0	9	3,285	7·2
Urine .		4,500	0·66	30	10,950	24·1
		Total . .		39	14,235	31·3

to the soil surface (10, 11, 121). The weight of minerals removed when meat animals are slaughtered is relatively small; the amount removed by a milking cow may be much greater. Certainly, when grazing sheep take part in a mineral cycle some elements are used over and over again. Some losses occur from the system, due to leaching and as ammonia from freshly deposited dung and urine. The patchy nature of these depositions can produce local areas with a quite different, and sometimes ill-balanced, nutrient supply. The pasture may then become, spatially, a highly variable source of food (88).

Excreta may be very irregularly distributed where there are marked patterns of animal behaviour. Hilder (153), in Australia, found that sheep deposited about one-third of their total faecal output on less than 5 per cent of the area.

Nutritive Value of Herbage

While there is no great point in growing large quantities of dry matter or green matter unless it satisfies some nutritional criteria, it is not easy to say what these should be. One kind of herbage cannot be expected automatically to be equally satisfactory for animal growth, fattening and milk production and for different rates of production. Three criteria of food value can be accepted for obvious reasons.

(*a*) First, no food can be regarded as valuable if the animal will not eat it in the necessary quantities (33, 36). Many factors may be concerned in food-intake (see Chapter VII): for a given animal, however, they may be considered as attributes of the food. (*b*) Secondly, only a part of the food is digested (59) and this proportion must have a big influence on food value. This is commonly expressed as 'apparent' digestibility because it is most easily measured as the difference between food intake (I) and faecal output (F). In fact, some part of the faeces is the result of metabolic processes; 'apparent' digestibility is thus only an index, although a very useful one, of the proportion of the food consumed that is absorbed. Fortunately, an assessment of digestibility (D) can now be obtained by an *in vitro* laboratory technique (108, 116), without recourse to an animal feeding trial. It is customary to calculate apparent digestibility as a percentage, from the equation

$$D = \frac{I - F}{I} \times 100$$

In many cases, sheep consume less of the less digestible feeds, though this is more accurately visualized as a consequence of their relative 'indigestibility'. (*c*) Thirdly, the biological value of the absorbed fraction is important. This simply expresses how closely the composition of the food matches the animal's need for different components. The need is determined by the kind of tissues being built and the energy involved in all the animal's activities. The fact that something is absorbed does not necessarily mean that it is useful. In many British pastures, for example, the nitrogen content of the herbage is greater than that needed by the animal. There are exceptions but, in general, excess nitrogen has to be excreted and, if the quantity is very large, this could actually impose a burden on the excretory system. Similarly, a significant deficiency can reduce the value of the whole feed. The question of excesses and deficiencies will be further considered at the end of this chapter. The ruminant, as will emerge again in subsequent chapters, is not entirely dependent on its food, however, for the supply of all its essential nutrients. What matters chiefly about the composition of the food is that it should result, on fermentation in the rumen, in appropriate amounts of the right mixture of volatile fatty acids and other compounds for the animal's particular purposes (8, 13, 97). Again, there is some evidence to suggest that more digestible herbages result in more favourable digestion products.

At present, then, apparent digestibility provides the best guide to the nutritive value of herbage (70, 82). In fact, it may be asking too much to expect that nutritive value can ever be expressed completely in one figure. However, Blaxter (13) has summarized the position as follows: 'Digestibility coefficients are not a basis for the exact prediction of animal performance on grassland feeds, but can, if interpreted in relation to present knowledge of the overall energy exchange, allow a general appraisal of nutritive value.'

The apparent digestibility of the feed has been shown by Raymond and his co-workers (*a*) to vary with the stage of growth of the plant, (*b*) to decline somewhat as the growing season advances, (*c*) to vary from one part of the plant to another and (*d*) to differ consistently between species and even varieties of grass and clover (83, 97). An understanding of these variations and their relationship to animal performance (81) is vital to an understanding of the sheep production process at pasture. The more important variations in digestibility are illustrated in Figs. 2.6 and 2.7.

Broadly speaking, fertilizer applications do not affect digestibility (144), except indirectly. The indirect influence may be by virtue of changes in the botanical composition of the sward, resulting in a changed mean digestibility of the pasture, or changes in growth rate. A faster growing plant more rapidly reaches the stage of maturity at which digestibility declines (101). Thus at the same age it *may* be less digestible. If, on the other hand, pastures are left until they have

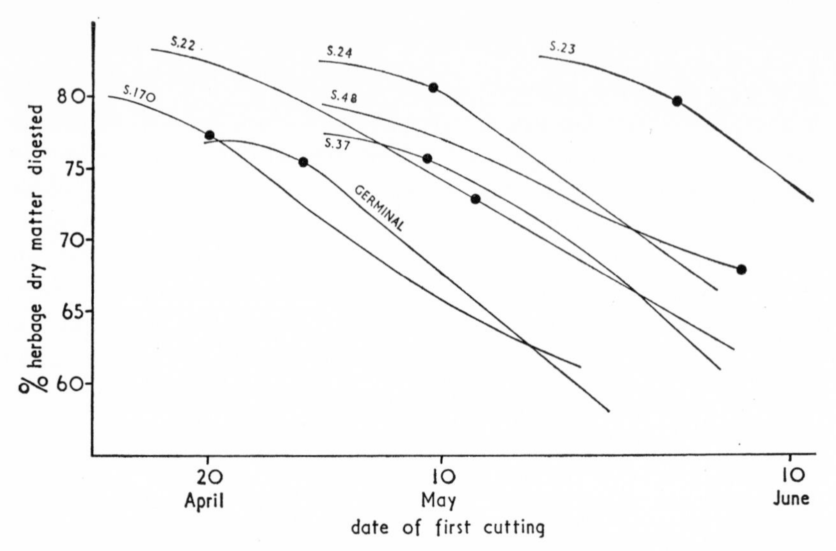

FIG. 2.6—*The digestibility of different herbage species as they mature during their first growth in the Spring*

Key:	
S. 23, S. 24	Perennial ryegrass
Germinal, S. 37	Cocksfoot
S. 22	Italian ryegrass
S. 170	Tall fescue
S. 48	Timothy
●	Date of first ear emergence

(Source : W. F. Raymond, data from the Grassland Research Institute)

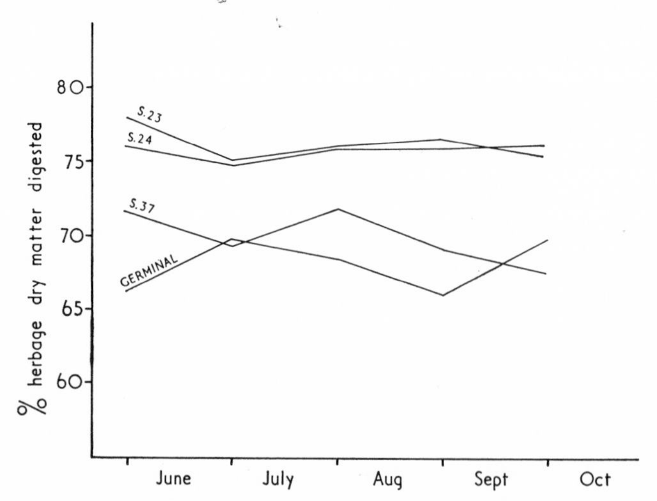

FIG. 2.7—*The digestibility of monthly regrowths, after a first cut in May, of varieties of ryegrass (S. 23, S. 24) and cocksfoot (Germinal, S. 37). Note the consistently higher digestibility of the ryegrass*

(Source : W. F. Raymond, data from experiments at the Grassland Research Institute)

produced the same *quantity*, the more heavily fertilized will have a higher digestibility because it is younger, having produced that quantity in a shorter time. Thus fertilizer applications can result in the production of a much greater total quantity of nutrients without reduction in food value as measured by digestibility. As grass plants age they tend to become less digestible, to contain a lower proportion of nitrogen and soluble carbohydrate, and to have a higher content of dry matter and structural carbohydrate. Data on the digestibility of the main British grasses and legumes have recently been summarized by Dent & Aldrich (148), in relation to herbage production and stage of growth.

The production of nutrients

Since the food value of an individual plant declines with maturity, a crop of high value must be harvested at a relatively young stage of growth. This is usually before the maximum amount of dry matter has been produced per acre. Thus a choice is often presented of a big crop of lower food value or a smaller crop of higher value. This is rarely a simple choice, however, for the sooner one crop is cut, the sooner the next can begin to grow. Furthermore, whenever the size of the crop exceeds the current animal needs, some part of it must incur the losses and cost of conservation. An important consideration in the choice of harvesting frequency is the total production of nutrients, though these must be in a suitable form. Since rate of growth varies with the season, and inadequate temperature or water supply can reduce growth unexpectedly, a harvesting programme must be flexible. It is possible for the total quantity of nutrients harvested to be the same in two cases, but for the form in which most of it occurs to be quite different. Obviously, in some circumstances, perhaps where the growing season is short, maximum growth rates must be encouraged for as long as possible. In other conditions, it might be better continually to aim at herbage of precisely the quality required.

Modifications to the pattern of herbage production

Whether expressed in terms of nutrients or dry matter, herbage growth is characterized by its non-uniform distribution in time. One of the reasons for this is the shortage of one or more plant nutrients during parts of the year: major climatic differences are responsible for the biggest variations, however. The pattern of herbage production is most readily altered by application of fertilizers, notably nitrogen, and water. This tends towards greater uniformity within the growing season but does little about the difference between the major peak in the spring and the trough in winter. Where climatic conditions are truly limiting, it may be possible to improve production by the introduction of other species. The problem for the herbage agronomist is to devise seed mixtures and managements which make the contributions of such plants additional to that already obtained during the rest of the year. The use of kale and roots is an example of the fact that, if a special crop for the winter involves the ground being out of action at some other time, then it must yield heavily.

Weeds

It is usual to regard as a weed of pasture any plant which is not required, even if it cannot be considered undesirable. There are, of course, poisonous plants, ragwort (*Senecio jacobaea*), for example, and there are plants which can become toxic by concentrating some element, such as selenium (119), to an exceptional extent. It should be noted, incidentally, that sheep are sometimes short of selenium and responses in growth and health have been obtained to selenium administration (14, 51). A number of plants have been recorded as poisonous to sheep (151) but many are not readily eaten and others have to be consumed in large quantities before they have any great effect. Common British poisonous plants are listed in Table VIII, with an indication of the circumstances in which trouble is likely. In other parts of the world, other toxic plants occur in sheep pastures. For example, oxalate-bearing plants are distributed throughout the world and one of these, *Halogeton glomeratus*, is reported to be common in the western United States (155) and extensively grazed by sheep. Many grasses, e.g. *Poa* spp., are regarded as weeds of productive leys and, within a context of high nitrogen use, clover has sometimes been so described. As mentioned earlier, however, the balance of pasture species present tends to reflect management and it is quite erroneous in most cases to attribute the poor production of a pasture simply to its weed content. The advent of effective herbicides (43, 53, 71, 114) has made it possible to measure the effect of weed eradication. The resulting increase in production is relatively small unless accompanied by a management which will encourage an increase in the desirable species (6). Often the application of this management alone would have produced the same result, if given time. Sowing and spraying achieve the desired botanical change more rapidly but are ineffective without management. It is therefore frequently the case that weeds occupy space that could be occupied by more productive species, but it is not always true that these weeds *prevent* them from doing so; they may merely colonize bare ground. Pasture receiving a lot of nitrogenous fertilizer does not normally support many broad-leaved weeds, and close grazing by

TABLE VIII

BRITISH POISONOUS PLANTS

British poisonous plants, commonly available to sheep and reported to be poisonous to them (from Forsyth, 151). It is not implied that the risk of poisoning is necessarily serious, only that it has been known to occur.

Common Name	Latin Name	Circumstances in which trouble is most likely
Horsetail	*Equisetum* spp.	when present in hay
Bracken	*Pteridium aquilinum*	when green, fresh or cut
Buttercups	*Ranunculus* spp.	eaten only when alternative food very scarce
Charlock	*Sinapis arvensis*	when pods are well formed
White mustard	*Sinapis alba*	when pods are well formed
Kale	*Brassica oleracea*	during pregnancy
Rape	*Brassica napus*	when plants contain high Nitrate content, following heavy N-fertilization
St. John's Wort	*Hypericum* spp.	all stages can cause photo-sensitization
Chickweed	*Stellaria media*	when large quantities are eaten by lambs
Water Dropwort	*Oenanthe crocata*	usually when the roots are eaten
Sheep's Sorrel	*Rumex acetosella*	especially during lactation
Oak	*Quercus* spp.	green acorns
Ragwort	*Senecio jacobaea*	all stages, if eaten

sheep, if it starts early enough in the year, will usually eliminate most of them.

In deciding whether a plant is a weed an important question is whether it has anything to contribute to the nutrition of the sheep. Quantitatively, this is unlikely. Qualitatively, the most important contribution of broad-leaved plants is their mineral content. This is normally much higher than that of grasses, and the latter are occasionally deficient in minerals, from the point of view of animal food requirement. Experiments in Britain have failed to demonstrate any improvement in animal health or performance in the presence of such plants as yarrow (*Achillea millefolium*), chicory (*Cichorium intybus*) and ribwort (*Plantago lanceolata*) (40), but this does not exclude the possibility that they may have great value at particular times and in relation to particular stages in animal production, such as at the peak of lactation.

The relationship between grassland weeds and animal productivity has been discussed by Spedding (161). It is more likely that weeds will cause a significant reduction in animal productivity in fairly efficient husbandry systems, in the sense that less efficient systems rarely utilize all the available herbage anyway.

PASTURE AND ANIMAL HEALTH

As a background to the subject of animal health the foregoing discussion needs to be focused on the known deficiencies of pasture as a feed for sheep. Table IX gives representative mineral contents of herbage and herbage products; it must be emphasized, however, that large differences exist between species and can be created by fertilizer and other agronomic treatments.

Herbs have a notably high proportion of minerals and crude protein: clovers, too, are generally higher than grasses in both of these constituents. Grasses tend to be higher than clovers in barium, silicon and molybdenum but lower in calcium, strontium and boron (119).

Grasses tend to be low, at times, in minerals of considerable importance to animal health. These are magnesium (60), sodium (75, 76, 94), copper (49) and cobalt. Low levels of these may occur in the grass at particular times, as with, for example, magnesium in the early spring (117); or they may be induced by an excess of another mineral. Excessive use of fertilizer potash may depress the magnesium content (95, 110, 133) and a high potassium/nitrogen ratio appears to reduce the content of sodium. Deficiencies may arise in the animal as a result of these low contents, or because the mineral is in a relatively non-available form, or because some other element interferes with its absorption or metabolism. Whitehead (165, 166) has recently summarized data on the mineral composition of grassland herbage and detailed descriptions are given by Ferguson (150): variation in the content of other constituents has been discussed by Waite (164). The mineral relationships of the ruminant have been discussed by Brouwer (19), who has emphasized the importance of mineral ratios. In grass tetany (hypomagnesaemia), the Dutch

TABLE IX

MINERAL CONTENTS OF HERBAGE AND HERBAGE PRODUCTS

The values are expressed on a dry matter basis.

	Perennial Ryegrass	Cocksfoot	Silage*	Clover†	Lucerne	Herbs
Ca % .	0·43–1·54	0·38–0.74	0·68	0·65–2·78	1·22–4·13	1·10–1·86
Mg % .	0·09–0·21	0·08–0·24	0·10	0·14–0·26	0·15–0·36	0·45–1·43
P % .	0·23–0.42	0·31–0·52	0·21	0·21–0·42	0·20–0·33	0·22–0·52
K % .	1·31–4·28	2·01–3·76	1·28	1·08–3·52	0·36–1·27	1·62–5·52
Na % .	0·02–0·25	0·07–0·15	0·06	—	—	0·03–0·33
Inorganic Sulphate Sulphur %	0·14–0·24	0·24–0·35	0·13	—	—	—
Cu p.p.m.‡	3·4 –11·4	3·6 –13·6	8·5	6·3 –11·9	7·0 –12·8	6·3 –15·6
Zn p.p.m.	13–63	12–156	41	11–29	12–51	—
Mo p.p.m.	0·18–3·9	0·26–2·8	1·2	0·08–0·37	0·10–1·60	—
Mn p.p.m.	56–162	39–179	59	14–64	10–109	18–75
Co p.p.m.	0·14–0·19	0·16–0·22	0·15	—	0·09–0·16	0·11–0·28
Fe p.p.m.	77–891	31–165	—	83–355	72–212	174–625
Se p.p.m.	0·6 –0·7	0·4 –0·7	0·5	—	—	—

(Sources : the values for herbs are taken from Thomas, B., Thompson, A., Oyenuga, V. A. & Armstrong, R. H. (1952), *Emp. J. exp. Agric.*, *20*, 10. All other values are from herbage grown at Hurley or an area on the Bagshot Sands. The table was compiled by E. C. Jones, 1963)

* One sample only.
† Includes both red and white clover.
‡ Parts per million.

workers have long felt that the danger was greatest when the potassium content of the pasture was high in relation to that of calcium, magnesium and sodium (56). Although hypomagnesaemia has been prevented by the use of fertilizers containing magnesium (1, 12, 92, 110), there has often appeared to be little relationship between the incidence of tetany and the magnesium content of the herbage. This has been discussed by Allcroft (1) and others (103, 115) and it has been pointed out that the availability of the herbage magnesium may differ between pastures (145). Parr (91) found no gross difference, however, in the proportion of water-soluble magnesium in pastures and faeces from hypomagnesaemic and normal situations. Rook & Wood (103) have stressed the effect that herbage intake may have on the magnesium status of the animal and 't Hart (57) has discussed the influence of meteorological conditions on the incidence of tetany. This complexity of metabolic disease problems and, incidentally, the need for co-operation between workers of different disciplines in their solution, has been emphasized by Seekles (106). Frequently, the balance of minerals in the diet is important and the obvious manifestations of deficiency may be delayed.

The use of copper provides a good example of the complication of the problem (85, 119). Swayback is a condition of lambs, characterized by progressive incoordination, particularly of the hind limbs, that can be completely prevented by administration of copper to ewes in late pregnancy. The swards on which it occurs, however, may be noteworthy, not for their low copper content, but for their high content of molybdenum or inorganic sulphate. It is arguable that grass should be produced in maximal quantity and deficiencies met by supplementation, as is commonly done with both copper and magnesium. But it is by no means certain that high production need result in grass of lower value in any respect. In relation to magnesium, even the use of heavy applications of nitrogen (see Table X) or the use of irrigation (see Table XI) may have little effect on the magnesium content of the herbage. What high production may do is (*a*) increase the performance, possibly the stress and certainly the nutritional need, of the individual animal, (*b*) allow many more animals to be kept per acre and (*c*) allow animals to go out to, or depend on, pasture much earlier in the spring.

The variation in magnesium and copper contents, over the year, of predominantly ryegrass swards is shown in Fig. 2.8. Although species such as red clover tend to have a higher magnesium content than grass it must be remembered that, in Britain, there may be very little clover in a pasture before April. By contrast, a high clover

TABLE X

THE EFFECT OF NITROGENOUS FERTILIZER APPLICATION ON THE MAGNESIUM CONTENT OF TWO GRASS SPECIES

The values given are the means of all harvests from each treatment over the period April–October (Expt. H. 9, 1957).

	N (LB PER ACRE)	Mg (% D.M.)	$\frac{K}{Ca+Mg}$ (mEq.)†
Ryegrass (S. 23)	0	0·16	1·4
	67·5	0·17	1·4
	135	0·15	1·7
	270	0·15	1·4
Cocksfoot (S. 37)	0	0·13	1·9
	67·5	0·12	1·9
	135	0·13	1·6
	270	0·13	1·6

(Source : T. E. Williams, unpublished data, Grassland Research Institute)

† milli-equivalents.

TABLE XI

THE EFFECTS OF METHOD OF DEFOLIATION AND IRRIGATION ON THE MAGNESIUM CONTENT OF SOME SPECIES OF HERBAGE

The values given are means of samples taken from April to October from several levels of nitrogen application (Expts. H. 9, H. 126, 1957).

SPECIES	Mg (% D.M.)	$\frac{K}{Ca+Mg}$ (mEq.)
Irrigated Ryegrass (S. 23)	0·17	1·5
Non-irrigated Ryegrass (S. 23)	0·18	1·4
Ryegrass (S. 23) Grazed	0·16	1·5
Cut	0·15	1·7
Cocksfoot (S. 37) Grazed	0·14	1·9
Cut	0·11	1·9
Clovers (S. 100+S. 184)	0·16	0·7

(Source : T. E. Williams, unpublished data, Grassland Research Institute)

content is sometimes regarded as suspect, as indeed is any 'lush' (meaning either low in dry-matter content, dark green or rapidly grown) pasture. This is often thought to result in scouring, bloat (50, 98) and enterotoxaemia. These and other similar questions will be specifically considered in Chapter XIII but it should be noted that enterotoxaemias can be avoided by vaccination, bloat is not a major

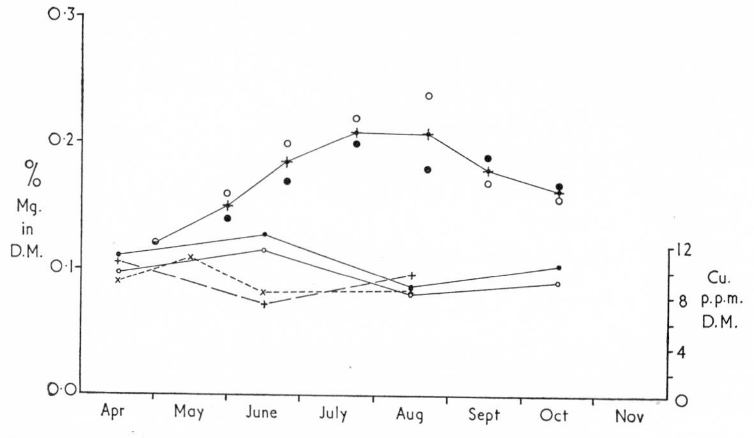

Fig. 2.8—*Seasonal variation in the content of magnesium (Mg) and Copper (Cu) in the dry matter of predominantly ryegrass pasture*

% Mg { ○ In unirrigated pasture; ● In irrigated pasture; + Mean of both treatments }

(Source: T. E. Williams, unpublished data from H. 126, (1957), Grassland Research Institute)

p.p.m. Cu {
● unirrigated pasture grazed at a high stocking rate (1961)
× ,, ,, ,, ,, ,, ,, ,, (1962)
○ ,, ,, ,, ,, low ,, ,, (1961)
+ ,, ,, ,, ,, ,, ,, ,, (1962) }

(Source: unpublished data from H. 202, Grassland Research Institute, see Spedding *et al.* (162))

problem in sheep, at least, in Britain, and scouring can be due to many other causes. Some of these causes, such as internal parasites—see Chapters IV and V—may be associated with the high stocking rates supported by highly productive grassland.

The diseases of the herbage itself have been relatively little studied. Some, such as 'rust' (infection with the fungus *Puccinia perplexans*) may reduce the food intake of sheep by rendering the herbage unattractive. 'Rust' appears to be more prevalent on older herbage and less so on actively growing, highly fertilized swards. A fungus which grows saprophytically on dead herbage (leaf 'litter') has caused considerable outbreaks of facial eczema in sheep in New Zealand. The disease (25, 79, 80, 109) is caused by a toxin ('sporidesmin') in the spores of *Pithomyces chartarum* (86). The fungus has recently been found in Britain but the climatic conditions which allow it to multiply to a dangerous level are not common in this country. Nevertheless, it would be unwise to ignore the fact that similar problems may occur in association with the methods of pasture utilization of the future. It is the essence of an ecological outlook that an awareness should be maintained of the unlooked-for consequence, which occurs as the end result of a complex chain of events. Reports of 'ryegrass staggers' in sheep (26), the possibilities of goitrogenic effects associated with a diet of white clover (22, 118) and the possible effect of nitrate content on vitamin A deficiency (63), are simply examples of the kind of consequence that might follow changes in botanical composition.

REFERENCES

(1) ALLCROFT, R. (1960). Prevention of Hypomagnesaemia. *B.V.A. Conf. on Hypomagnesaemia*, p. 102. Lond.

(2) BAKER, H. K. (1960 a). Permanent Grassland in England. *Proc. VIII int. Grassld Congr.* (Reading), 394–9.

(3) BAKER, H. K. (1960 b). The production of early spring grass. I. *J. Brit. Grassld Soc. 15* (4), 275–80.

(4) BAKER, H. K., CHARD, J. R. A. & HUGHES, G. P. (1961). The production and utilization of winter grass at various centres in England and Wales, 1954–60. I. *J. Brit. Grassld Soc. 16* (3), 185–9.

(5) BAKER, H. K., CHARD, J. R. A. & JENKINS, D. G. (1961). The production of early spring grass. II. *J. Brit. Grassld Soc. 16* (2), 146–52.

(6) BAKER, H. K. & EVANS, S. A. (1958). The control of weeds in permanent pasture by M.C.P.A. and the subsequent effect on herbage productivity. *Proc. IV Brit. Weed Control Conf.*, 16.

(7) BAKER, H. K. & GARWOOD, E. A. (1961). Studies on the root development of herbage plants. V. *J. Brit. Grassld Soc. 16* (4), 263–7.

(8) BALCH, C. C. (1960). Rumen digestion and herbage utilization. *Proc. VIII int. Grassld Congr.* (Reading), 528.

(9) BARLING, D. M. (1961). Some recent aspects of grassland research. *Agric. Progr. 36*, 1–9.

(10) BARROW, N. J. (1961). Mineralization of nitrogen and sulphur from sheep faeces. *Aust. J. agric. Res. 12* (4), 644–50.

(11) BARROW, N. J. & LAMBOURNE, L. J. (1962). Partition of excreted nitrogen, sulphur and phosphorus between the faeces and urine of sheep being fed pasture. *Aust. J. agric. Res. 13* (3), 461–71.

(12) BIRCH, J. A. & WOLTON, K. M. (1961). The influence of magnesium applications to pasture on the incidence of hypomagnesaemia. *Vet. Rec. 73* (45), 1169–73.

(13) BLAXTER, K. L. (1960). The utilization of the energy of grassland products. *Proc. VIII int. Grassld Congr.* (Reading), 479.

(14) BLAXTER, K. L. (1963). The effect of selenium administration on the growth and health of sheep on Scottish farms. *Brit. J. Nutr. 17*, 105–15.
(15) BROCKMAN, J. S. & WOLTON, K. M. (1963). The use of nitrogen on grass-white clover swards. *J. Brit. Grassld Soc. 18* (1), 7–13.
(16) BROUGHAM, R. W. (1956). Effect of intensity of defoliation on regrowth of pastures. *Aust. J. agric. Res. 7*, 377–87.
(17) BROUGHAM, R. W. (1957). Pasture growth rate studies in relation to grazing management. *Proc. N.Z. Soc. Anim. Prod. 17*, 46.
(18) BROUGHAM, R. W. (1959). The effects of frequency and intensity of grazing on the productivity of a pasture of short rotation ryegrass and red and white clover. *N.Z. J. agric. Res. 2*, 1232–48.
(19) BROUWER, E. (1961). Mineral relationships of the ruminant. In *Digestive Physiology and Nutrition of the Ruminant.* Ch. 14. London, Butterworth.
(20) BURG, P. F. J. van (1960). Nitrogen fertilization and the seasonal production of grassland herbage. *Proc. VIII int. Grassld Congr.* (Reading), 142.
(21) BURG, P. F. J. van (1962). Internal nitrogen balance, production of dry matter and ageing of herbage and grass. *Wageningen Centrum Landb. Proefschrift* (Thesis).
(22) BUTLER, G. W., FLUX, D. S., PETERSEN, G. B., WRIGHT, E. W., GLENDAY, A. C. & JOHNSON, J. M. (1957). Goitrogenic effect of white clover (*Trifolium repens* L.). II. *N.Z. J. Sci. Tech.*, Sec. A, *38* (8), 793–802.
(23) CASTLE, M. E. & REID, D. (1960). Production and use of grass. *Chem. Soc. Ind.* Monograph No. 9.
(24) CASTLE, M. E. & REID, D. (1960). Irrigation of grassland in S.W. Scotland and its influence on the utilization of fertilizer nitrogen. *Proc. VIII int. Grassld Congr.* (Reading), 146.
(25) CLARE, N. T. (1957). Recent progress in facial eczema research. *Proc. Ruakura Fmrs' Conf. N.Z.*, 3–7.
(26) CLEGG, F. G. & WATSON, W. A. (1960). Ryegrass staggers in sheep. *Vet. Rec. 72* (36), 731–3.
(27) CLEMENT, C. R. & WILLIAMS, T. E. (1962). An incubation technique for assessing the nitrogen status of soils newly ploughed from leys. *J. Soil Sci. 13* (1), 82–91.
(28) CLEMENT, C. R. (1961). Benefit of leys—structural improvement or nitrogen reserves. *J. Brit. Grassld Soc. 16* (3), 194–200.
(29) CONROY, E. (1961). Effects of heavy applications of nitrogen on the composition of herbage. *Irish J. agric. Res. 1*, 67–71.
(30) COWLING, D. W. (1961 a). The effect of nitrogenous fertilizer on an established white clover sward. *J. Brit. Grassld Soc. 16* (1), 65–8.
(31) COWLING, D. W. (1961 b). The effect of white clover and nitrogenous fertilizer on the production of a sward. I. *J. Brit. Grassld Soc. 16* (4), 281–90.
(32) COWLING, D. W. (1963). Nitrogenous fertilizer and seasonal production. *J. Brit. Grassld Soc. 18* (1), 14–17.
(33) CRAMPTON, E. W., DONEFER, E. & LLOYD, L. E. (1960). A nutritive value index for forages. *Proc. VIII int. Grassld Congr.* (Reading), 462.
(34) CROWLEY, J. P. & MURPHY, M. A. (1962). Basic slag toxicity in cattle and sheep. *Vet. Rec. 74* (44), 1177–78.
(35) CURRIE, J. A. (1962). The importance of aeration in providing the right conditions for plant growth. *J. Sci. Food Agric.* (7), 380–5.
(36) DAVIES, H. L. (1960). The feeding value of pasture plants. *Breeding and Feeding of Livestock, Fmr's & Scientist's Conf.* Univ. Western Austr. 40–2.
(37) DAVIES, W. (1951 a). Grassland development in two centuries. *Woodheads Seeds Ltd, 1801–1951.* Leeds.
(38) DAVIES, W. (1951 b). Lessons from American grasslands. *J. Brit. Grassld Soc. 6* (1), 11–17.
(39) DAVIES, W. (1951 c). The production and utilization of grass throughout the year. *Proc. Brit. Soc. Anim. Prod.*, 73–89.

(40) Davies, W. (1952). Grassland management. *Vet. Rec.* *64*, 463.
(41) Davies, W. (1953). Grassland research and the nation. *Lecture at the Roy. Dublin Soc.* March.
(42) Davies, W. (1955). Do we waste our grass? *Agric.* Lond. *62*, 153–6.
(43) Davies, W. (1958). Grassland developments in the United Kingdom in the last twenty-five years. *Emp. J. exp. Agric.* *26* (103), 195–207.
(44) Davies, W. (1960 a). *The Grass Crop.* 2nd ed. London, Spon.
(45) Davies, W. (1960 b). Temperate (and Tropical) grasslands. *Proc. VIII int. Grassld Congr.* (Reading), 1.
(46) Davis, A. G. (1955). Lucerne. *J. R. agric. Soc. of England* *116*, 50–9.
(47) Donald, C. M. (1961). Competition for light in crops and pastures. *Symp. Soc. exp. Biol. XV*, 282–313.
(48) Diven, R. H., Reed, R. E., Trautman, R. J., Pistor, W. J. & Watts, R. E. (1962). Experimentally induced nitrite poisoning in sheep. *Am. J. vet. Res.* *23* (94), 494–6.
(49) Elst, F. C. C., Hupkens van der (1962). Development of peat-land. *N.Z. J. Agric.* *105* (3), 231–4.
(50) Ferguson, W. S. & Terry, R. A. (1956). Bloat investigations. *Outlook on Agric.* *1* (2), 75–8.
(51) Gardiner, M. R., Armstrong, J., Fels, H. & Glencross, R. N. (1962). A preliminary report on selenium and animal health in Western Australia. *Aust. J. exp. Agric. Anim. Husb.* *2*, 261–9.
(52) Green, J. O. & Cowling, D. W. (1960). The nitrogen nutrition of grassland. *Proc. VIII int. Grassld Congr.* (Reading), 126–9.
(53) Green, J. O., Kydd, D. D. & Jones, L. (1958). Use of herbicides in the surface reseeding of old pasture. II. *Proc. IV Brit. Weed Control Conf.*, 239.
(54) Griffith, G. ap & Johnston, T. D. (1960). The nitrate nitrogen content of herbage. I. *J. Sci. Food Agric.* *11*, 622–6.
(55) Hamilton, R. A. (1956). Utilization of grassland. *Outlook on Agric.* *1* (1), 5–11.
(56) 'T Hart, M. L. (1957). Influence of potassium fertilizer on animal production from pasture. *Potassium Symposium, Int. Pot. Inst.*, Berne. 139–50.
(57) 'T Hart, M. L. (1960 a). The influence of meteorological conditions and fertilizer treatment on pasture in relation to hypomagnesaemia. *B.V.A. Conf. on Hypomagnesaemia*, 88, Lond.
(58) 'T Hart, M. L. (1960 b). The yield of grassland on the nitrogen experimental farms. *Stikstof* *4*, 19–25.
(59) Head, M. J. (1961). Cellulose digestibility and the rumen. In *Digestive Physiology and Nutrition of the Ruminant*, Ch. 23. London, Butterworth.
(60) Hemingway, R. G. (1961). Magnesium, potassium, sodium and calcium contents of herbage as influenced by fertilizer treatments over a three-year period. *J. Brit. Grassld Soc.* *16* (2), 106.
(61) Henry, K. M., Kon, S. K., Thompson, S. Y., McCallum, J. W. & Stewart, J. (1958). The vitamin D activity of pastures and hays. *Brit. J. Nutr.* *12* (4), 462–8.
(62) Herriott, J. B. D., Wells, D. A. & Dilnot, J. (1959). The grazing animal and sward productivity. *J. Brit. Grassld Soc.* *14* (3), 191–8.
(63) Huber, W. G. (1962). Current vitamin A problems. *Vet. Med.* *57* (3), 311–13.
(64) Hughes, G. P. (1952). The comparative seasonal output of ultra-simple and general purpose seeds mixtures. *J. agric. Sci.* *42* (4), 413–21.
(65) Hughes, G. P. (1954). The production and utilization of winter grass. *J. agric. Sci.* *45* (2), 179–201.
(66) Hughes, G. P. (1955). Tall fescue and lucerne drills for winter and summer grass. *J. Brit. Grassld Soc.* *10* (2), 135–8.
(67) Hughes, G. P. & Davis, A. G. (1951). The development of swards sown with simple mixtures at different rates of seeding under varying systems of management and manuring. *J. Brit. Grassld Soc.* *6* (3), 167–77.

(68) HUGHES, G. P. & EVANS, T. A. (1951). Response of an established sheep ley to dressings of sulphate of ammonia. II. *Emp. J. exp. Agric. 19* (74), 65–72.
(69) HUGHES, G. P., TAYLER, J. C. & EVERITT, G. C. (1952). Foggage and silage as winter feed for sheep. *Agric. 59*, 376–81.
(70) IVINS, J. D. (1960). Digestibility data and grassland evaluation. *Proc. VIII int. Grassld Congr.* (Reading), 459.
(71) JONES, L. (1962). Herbicides to aid pasture renovation. *J. Brit. Grassld Soc. 17* (1), 85–6.
(72) KERNICK, M. D. (1960). The recovery of fertilizer nitrogen from various depths below swards. *J. Brit. Grassld Soc. 15* (1), 34–40.
(73) LAMBOURNE, L. J. (1957). Measurement of feed intake of grazing sheep. II. *J. agric. Sci. 48* (4), 415–25.
(74) LANGER, R. H. M. (1959). A study of growth in swards of timothy and meadow fescue. II. *J. agric. Sci. 52* (3), 273–81.
(75) LEHR, J. J. (1960). The sodium contents of meadow grass in relation to species and fertilization. *Proc. VIII int. Grassld Congr.* (Reading), 101.
(76) LEHR, J. J. (1961). Symptoms of sodium deficiency in animals and results of too large doses of common salt. *T. Diergeneesk.*, 86.
(77) LINEHAN, P. A. & LOWE, J. (1960). Yielding capacity and grass-clover ratio of herbage swards as influenced by fertilizer treatments. *Proc. VIII int. Grassld Congr.* (Reading), 133.
(78) LOW, A. J. & ARMITAGE, E. R. (1959). Irrigation of grassland. *Outlook on Agric. 2* (5), 213–18.
(79) MCMEEKAN, C. P. (1957). Feeding of hay in facial eczema control. *Proc. Ruakura Fmrs' Conf. Wk., N.Z. 1957*, 3–7.
(80) MCMEEKAN, C. P. (1961). Facial eczema of sheep and cattle in New Zealand. *Outlook on Agric. 3* (2), 89–96.
(81) MILFORD, R. (1960 a). *Exp. Grassland Res. Inst.* (Hurley) *13*, 77.
(82) MILFORD, R. (1900 b). Nutritional value of sub-tropical pasture species under Australian conditions. *Proc. VIII int. Grassld Congr.* (Reading), 474.
(83) MINSON, D. J., RAYMOND, W. F. & HARRIS, C. E. (1960). The digestibility of grass species and varieties. *Proc. VIII int. Grassld Congr.* (Reading), 470.
(84) MITCHELL, K. J. (1956). The influence of light and temperature on the growth of pasture species. *Proc. VII int. Grassld Congr.* (Palmerston North, N.Z.), 1–12.
(85) MOLNAR, I. (1961). *A Manual of Australian Agriculture.* London, Heinemann.
(86) MORTIMER, P. H. & TAYLOR, A. (1962). The experimental intoxication of sheep with sporidesmin, a metabolic product of *Pithomyces chartarum*. I. *Res. vet. Sci. 3*, 147.
(87) MURDOCH, J. C. (1960). The effect of temperature on silage fermentation. *Proc. VIII int. Grassld Congr.* (Reading), 502.
(88) NORMAN, M. J. T. & GREEN, J. O. (1958). The local influence of cattle dung and urine upon the yield and botanical composition of permanent pasture. *J. Brit. Grassld Soc. 13* (1), 39–45.
(89) NORMAN, M. J. T. (1960). The relationship between competition and defoliation in pasture. *J. Brit. Grassld Soc. 15* (2), 145–9.
(90) NOWAKOWSKI, T. Z. (1961). The effect of different nitrogenous fertilizers, applied as solids or solutions, on the yield and nitrate-N content of established grass and newly sown ryegrass. *J. agric. Sci. 56*, 287.
(91) PARR, W. H. (1961). The proportion of soluble magnesium in pastures and faeces in relation to hypomagnesaemia in cattle. *Res. vet. Sci. 2* (4), 320–5.
(92) PARR, W. H. & ALLCROFT, R. (1957). The application of magnesium compounds to pasture for the control of hypomagnesaemia in grazing cattle: a comparison between magnesian limestone and calcined magnesite. *Vet. Rec. 69* (45), 1041–47.

(93) PENMAN, H. L. (1962). Woburn irrigation, 1951–59. II. Results for grass. *J. agric. Sci.* 58, 349.
(94) PEIRCE, A. W. (1962). Studies on salt tolerance of sheep. IV. *Aust. J. agric. Res.* 13 (3), 479–86.
(95) RAHMAN, H., McDONALD, P. & SIMPSON, K. (1960). Effects of nitrogen and potassium fertilizers on the mineral status of perennial ryegrass (*Lolium perenne*). I. Mineral content. *J. Sci. Food Agric.* (7–8), 422–32.
(96) RAYMOND, W. F. (1953). Manuring of grassland. II. Use of nitrogen. *J. Sci. Food Agric.* 9, 409–10.
(97) RAYMOND, W. F., TILLEY, J. M. A., DERIAZ, R. E. & MINSON, D. J. (1960). Herbage composition and nutritive value. *Chem. Soc. Industr. Monograph No.* 9, 181–90.
(98) REID, C. S. W. (1960). Bloat: the foam hypothesis. *Proc. VIII int. Grassld Congr.* (Reading), 668.
(99) REID, D. (1959). Studies on the cutting management of grass-clover swards. I. *J. agric. Sci.* 53 (3), 299–312.
(100) REID, D. & MACLUSKY, D. S. (1960). Studies on the cutting management of grass-clover swards. II. *J. agric. Sci.* 54 (2), 158–65.
(101) REID, J. T., KENNEDY, W. K., TURK, K. L., SLACK, S. T., TRIMBERGER, G. W. & MURPHY, R. P. (1959). Symposium on forage evaluation: I. What is forage quality from the animal standpoint ? *Agron. J.* 51, 213.
(102) REITH, J. W. S., INKSON, R. H. E., STEWART, A. B., HOLMES, W., MACLUSKY, D. S., REID, D., HEDDLE, R. G., CLOUSTON, D. & COPEMAN, G. J. F. (1961). The effects of fertilizer on herbage production. I. *J. agric. Sci.* 56, 17.
(103) ROOK, J. A. F. & WOOD, M. (1960). Mineral composition of herbage in relation to the development of hypomagnesaemia in grazing cattle. *J. Sci. Food Agric.* (3), 137–43.
(104) RUNCIE, K. V. (1961). The zero-grazing of dairy cows. *79th Ann. Congr. B.V.A.*, Oxford.
(105) SEARS, P. D. (1960). Grass-clover relationships in New Zealand. *Proc. VIII int. Grassld Congr.* (Reading), 130.
(106) SEEKLES, L. (1960). Aspects of physiological disorders in grazing livestock. *Proc. VIII int. Grassld Congr.* (Reading), 15.
(107) SETCHELL, B. P. & WILLIAMS, A. J. (1962). Plasma nitrate and nitrite concentration in chronic and acute nitrate poisoning in sheep. *Aust. vet. J.* 38 (2), 58–62.
(108) SHELTON, D. C. & REID, R. L. (1960). Measuring the nutritive value of forages using *in vitro* rumen techniques. *Proc. VIII int. Grassld Congr.* (Reading), 524.
(109) SMITH, J. D. (1962). Further progress in facial eczema research. *N.Z. J. Agric.* 105 (3), 194–200.
(110) SMYTH, P. J., CONWAY, A. & WALSH, M. J. (1958). The influence of different fertilizer treatments on the hypomagnesaemia proneness of a ryegrass sward. *Vet. Rec.* 70 (42), 846–9.
(111) STERN, W. R. & DONALD, C. M. (1962 a). Light relationships in grass-clover swards. *Aust. J. agric. Res.* 13 (4), 599–614.
(112) STERN, W. R. & DONALD, C. M. (1962 b). The influence of leaf area and radiation on the growth of clover in swards. *Aust. J. agric. Res.* 13 (4), 615–23.
(113) STOJKOVSKA, A. & COOKE, G. W. (1958). Micro-nutrients in fertilizers. *Chem. & Ind.*, 1368.
(114) TEMPLEMAN, W. G. (1956). The uses of plant growth substances. *Outlook on Agric.* 1 (1), 24–31.
(115) THOMPSON, A. (1960). Soil-plant-animal relationships and hypomagnesaemia. *B.V.A. Conf. on Hypomagnesaemia*, 75. British Veterinary Ass. Lond.
(116) TILLEY, J. M. A., DERIAZ, R. E. & TERRY, R. A. (1960). The *in vitro* measurement of herbage digestibility and assessment of nutritive value. *Proc. VIII int. Grassld Congr.* (Reading), 533.

(117) TODD, J. R. (1961). Magnesium in forage plants. Pt. I. *J. agric. Sci. 56*, 411; Pt. II. *J. agric. Sci. 57*, 35.
(118) TURNER, C. W. (1960). Administration of hormones to grazing animals. *Proc. VIII int. Grassld Congr.* (Reading), 586.
(119) UNDERWOOD, E. J. (1962). *Trace Elements in Human and Animal Nutrition.* N.Y. and London, Academic Press Inc.
(120) VOISIN, A. (1959). *Grass Productivity.* London, Crosby Lockwood.
(121) WATKIN, B. R. (1954). The animal factor and levels of nitrogen. *J. Brit. Grassld Soc. 9* (1), 35–46.
(122) WATSON, D. (1947). Comparative physiological studies on the growth of field crops. I. *Ann. Bot. Lond. 11*, 41.
(123) WATSON, S. J. & NASH, M. J. (1960). *The Conservation of Grass and Forage Crops.* Edinburgh, Oliver and Boyd.
(124) WHYTE, R. O. (1960 a). *Crop Production and Environment.* p. 193. London, Faber.
(125) WHYTE, R. O. (1960 b). Grassland in a developing world. *Proc. VIII int. Grassld Congr.* (Reading), 11.
(126) WIERINGA, G. W. (1960). Some factors affecting silage fermentation. *Proc. VIII int. Grassld Congr.* (Reading), 497.
(127) WILLIAMS, R. DORRINGTON (1951). The effects of deficiencies of several trace elements on timothy (*Phleum pratense L.*) grown in solution culture. *Plant & Soil III* (3), 257–66.
(128) WILLIAMS, R. DORRINGTON (1960). Nutrient uptake by grass roots. *Proc. VIII int. Grassld Congr.*, 283–6.
(129) WILLIAMS, R. DORRINGTON (1962). On the physiological significance of seminal roots in perennial grasses. *Ann. Bot. 26* (102), 129–36.
(130) WILLIAMS, T. E. (1960). The value of leys in crop rotation. *Esso Fmr. 12*, 16–18.
(131) WILLIAMS, T. E., CLEMENT, C. R. & HEARD, A. J. (1960). Soil nitrogen status of leys and subsequent wheat yields. *Proc. VIII int. Grassld Congr.* (Reading), 237.
(132) WILSON, J. W. (1960). Influence of spatial arrangement of foliage area on light interception and pasture growth. *Proc. VIII int. Grassld Congr.* (Reading), 275.
(133) WOLTON, K. M. (1963). Fertilizers and hypomagnesaemia. *N.A.A.S. Quarterly Rev. 14* (59), 122.
(134) ALDERMAN, G. (1968). Conserving grass. *J. Fmrs' Club*, April, 19–24.
(135) ALDERMAN, G. & JONES, D. I. H. (1967). The iodine content of pastures. *J. Sci. Fd Agric. 18*, 197–9.
(136) ANSLOW, R. C. (1965). Grass growth in midsummer. *J. Brit. Grassld Soc. 20* (1), 19–26.
(137) ANSLOW, R. C. & GREEN, J. O. (1967). The seasonal growth of pasture grasses. *J. agric. Sci., Camb. 68*, 109–22.
(138) BLAND, B. F. (1967). The effect of cutting frequency and root segregation on the yield from perennial ryegrass-white clover associations. *J. agric. Sci., Camb. 69*, 391–7.
(139) BROCKMAN, J. S. (1966). The growth rate of grass as influenced by fertilizer nitrogen and stage of defoliation. *Proc. X int. Grassld Congr.* (Helsinki), 234–40.
(140) BROUGHAM, R. W. (1962). The leaf growth of *Trifolium repens* as influenced by seasonal changes in the light environment. *J. Ecol. 50*, 449–60.
(141) BROUGHAM, R. W. (1966). Potential of present type pastures for livestock feeding. *N.Z. Agric. Sci. 2* (2), 19–22.
(142) BROWN, R. H. & BLASER, R. E. (1968). Leaf area index in pasture growth. *Herb. Abstr. 38* (1), 1–9.
(143) BUSHMAN, D. H., EMERICK, R. J. & EMBRY, L. B. (1968). Effect of various chlorides and calcium carbonate on calcium, phosphorus, sodium, potassium and chloride balance and their relationship to urinary calculi in lambs. *J. Anim. Sci. 27* (2), 490–5.

(144) CALDER, F. W. & MACLEOD, L. B. (1968). *In vitro* digestibility of forage species as affected by fertilizer application, stage of development and harvest dates. *Can. J. Pl. Sci. 48*, 17–24.
(145) CARE, A. D. (1965). Factors which affect the availability of magnesium. *Proc. Nutr. Soc. 24*, 99–105.
(146) COOPER, J. P. (1967). *W.P.B.S. Rep. for 1966*, p. 14.
(147) COWLING, D. W. (1966). The response of grass swards to nitrogenous fertilizer. *Proc. X int. Grassld Congr.* (Helsinki), 204–9.
(148) DENT, J. W. & ALDRICH, D. (1969). In *Crop Grasses and Legumes in British Agriculture* (ed. C. R. W. Spedding, C.A.B.). In press.
(149) DOUTRE, J. (1964). Vacuum compressed silage. *N.Z. Jl Agric. 109*, 365–9.
(150) FERGUSON, W. S. (1963). Ch. 28 in *Animal Health, Production and Pasture* (ed. A. N. Worden, K. C. Sellers & D. E. Tribe). London, Longmans.
(151) FORSYTH, A. A. (1968). British poisonous plants. *Min. of Agric. Fish. and Food Bull. No. 161, H.M.S.O.*
(152) GREEN, J. O. (1955). *In* Lucerne investigations, 1944–1953. *Grassld Res. Inst. Memoir No. 1.*
(153) HILDER, E. J. (1966). Distribution of excreta by sheep at pasture. *Proc. X int. Grassld Congr.* (Helsinki), 977–81.
(154) HOLLIDAY, R. & WILMAN, D. (1965). The effect of fertilizer nitrogen and frequency of defoliation on yield of grassland herbage. *J. Brit. Grassld Soc. 20* (1), 32–40.
(155) JAMES, L. F., STREET, J. C., BUTCHER, J. E. & SHUPE, J. E. (1968). Oxalate metabolism in sheep, I and II. *J. Anim. Sci. 27* (3), 718–29.
(156) JONES L. H. P. & HANDRECK, K. A. (1965). The relation between the silica content of the diet and the excretion of silica by sheep. *J. agric. Sci., Camb. 65*, 129–34.
(157) JONES, L. H. P. & HANDRECK, K. A. (1967). Silica in soils, plants and animals. *Adv.* in *Agron. 19*, 107–49.
(158) JONES, R. J., DAVIES, J. GRIFFITHS, WAITE, R. B. & FERGUS, I. F. (1968). The production and persistence of grazed irrigated pasture mixtures in south-eastern Queensland. *Aust. J. exp. Agric. Anim. Husb. 8*, 177–89.
(159) LAZENBY, A. & ROGERS, H. H. (1965). Selection criteria in grass breeding, IV and V. *J. agric. Sci., Camb. 65*, 65 and 79.
(160) MORRIS, R. M. (1967). *Pasture growth in relation to pattern of defoliation by sheep.* Ph.D. Thesis, Reading University.
(161) SPEDDING, C. R. W. (1967). Weeds and animal productivity. *Proc. 8th Br. Weed Control Conf. 3*, 854–60.
(162) SPEDDING, C. R. W., BETTS, J. E., LARGE, R. V., WILSON, I. A. N. & PENNING, P. D. (1967). Productivity and intensive sheep stocking over a five-year period. *J. agric. Sci., Camb. 69*, 47–69.
(163) VARTHA, E. W. & O'CONNOR, K. F. (1968). Prospects for high production from Canterbury plains grasses. *Proc. N.Z. Soc. Anim. Prod. 28*, 65–73.
(164) WAITE, R. (1965). The chemical composition of grasses in relation to agronomical practice. *Proc. Nutr. Soc. 24* (1), 38–46.
(165) WHITEHEAD, D. C. (1966a). Data on the mineral composition of grassland herbage from the Grassland Research Institute, Hurley, and the Welsh Plant Breeding Station, Aberystwyth. *Grassld Res. Inst. Tech. Rep. No. 4.*
(166) WHITEHEAD, D. C. (1966b). Nutrient minerals in grassland herbage. *Mimeograph No. 1.* C.A.B.
(167) WILKINS, R. J. (1967). Fodder conservation. *Proc. Occ. Symp. No. 3, Brit. Grassld Soc.*
(168) WOODFORD, E. K. (1966). The need for a fresh approach to the place and purpose of the ley. *J. Brit. Grassld Soc. 21* (2), 109–15.

III

THE ANIMAL POPULATION

Every sheep flock is really a population composed of individuals which may differ from each other in age, size and sex. It is therefore an oversimplification to visualize the utilization of grass as a process of conversion by a uniform group of animals producing one kind of product. Some confusion can be traced to the fact that the farmer must be continually concerned with the flock as a whole, whereas the agricultural scientist must often, in order to obtain the control and precision he requires, focus his attention on one part of it, perhaps for only part of the year. The integration of the results of such 'partial' studies in agricultural systems is a problem in applied ecology. It must not be forgotten, of course, that the population is in a relatively controlled environment and is itself subject to considerable control.

The agricultural purposes for which sheep are kept make them quite different from natural communities of animals: certainly the agriculturalist would hardly be satisfied with 'maintenance of the species'. Many of the more recent ecological concepts, however, notably that of 'energy flow' within ecosystems (17), have a considerable bearing on agricultural problems. Occasionally, the great value of an ecological approach is marred by the implication that man should not interfere but allow the various competing populations to work out a natural balance amongst themselves. The latter process is always fascinating and sometimes agriculturally advantageous. In general, however, it is many years too late for this attitude. Agriculture is already highly artificial and farming consists of 'interfering with nature' in the most profitable manner.

It has been noted that the criteria to be applied to the productive efficiency of sheep may be economic or biological, partly because the value of a product may not always reflect its cost of production, in whatever terms this is expressed. This chapter will be solely concerned with the biological efficiency of agricultural production, in the sense that a sheep that produces more meat per pound of food eaten may be regarded as biologically more efficient, even if, at the time it does so, the meat produced is worth less per pound. This productive efficiency is greatly influenced by the nature of the animal population, including its size and structure.

SIZE OF THE POPULATION

There are some ways in which sheer numbers (the *size* of a flock) may affect productivity. For example, the possibilities of 'mis-mothering' when ewes and lambs are moved shortly after lambing, is greater in large flocks (see Chapters VIII and IX). Apart from such aspects of the relationship between animal numbers and behaviour, the significance of population size lies in its relation to other quantities, such as the area occupied or the amount of food available. Kemp (15) has drawn attention to the effect that group size might have on the experimental assessment of production. 'It is uncertain', he wrote, 'whether the same relative results would be obtained, for example, with a basic unit of 2 animals on 1 acre as with 20 animals on 10 acres. Little is known about the validity of such assumptions.' Little is known, also, about the importance of scale in sheep farming practice. Clearly, if sheep were kept on concrete or slatted floors, and more may be kept in these or similar ways in the future (3), the actual number of animals would be relevant only to a discussion of business size, of the magnitude of an economic unit, and of labour or mechanical handling problems and such physical considerations.

The number per unit area would have much wider implications, such as the effect of overcrowding on the spread of disease and the relationship between numbers and feeding surfaces. Some of these problems of housed or yarded sheep are of importance already in countries where the winter is severe, Iceland and Scandinavia for instance, or where field fencing has been scanty, as in France; they may become more important where sheep are normally pastured in the winter (as in the United Kingdom), if sheep production becomes more intensive, particularly on the higher ground. For grazing conditions it is sometimes argued that the number of sheep per acre is the key to productivity, and in recent years much emphasis has been laid on the importance of stocking rate (12, 18, 22, 26).

Population Measurement

In discussions of this kind it is essential that the terms used should be precisely defined. It cannot be claimed that the definitions used here are generally agreed: it would avoid much confusion if a standard terminology were adopted.

Stocking rate

Stocking rate is commonly used to refer to the number of sheep per unit area (20), without considering the size of the sheep, the quantity of food available on the area, the time for which the sheep are to be kept on it or whether they are to receive any food additional to that grown on it. It is proposed, therefore, to define stocking rate simply as the number of sheep per unit area. It is thus used as a rate (rather like m.p.h., or lb per sq. in.), and different relationships between sheep, space and time may all be expressed in this way. There is no further expression in common use which takes into account the weight or size of the sheep, except the direct calculation of the total liveweight of sheep per acre, so stocking rates can only be strictly comparable for comparable sheep. This does not invalidate comparisons of the maximum stocking rates possible with two different breeds on the same kind of pasture. This is the same definition as that used by Mott (20) and is clearly the simplest. Its chief disadvantage is that it does not immediately make clear what period of time is involved. Since the agricultural significance of a number of sheep per acre depends almost entirely on the period during which this number is maintained, it is important to clarify the time factor.

In general, therefore, the period to which the stocking rate relates must always be stated. It might then be preferable to refer to the 'mean stocking rate' whenever the stocking rate had been varied during the period considered. The *Mean Stocking Rate* is therefore defined as the total number of sheep divided by the total number of acres grazed during this period. Within such a period, the number of sheep per acre at any one time might vary considerably: a further term, *Stocking Density*, is therefore useful, to describe the number per acre on the separate occasions within the whole period.

Stocking density

Stocking density is defined here as the stocking rate or the number of sheep per acre at a given point in time (5). Clearly it could be synonomous with stocking rate (in a set-stocking system of management, for example) and both could be qualified by a reference to the time for which they were maintained. Since the only way of stating the relationship between the total number of sheep and the total number of acres over a period is the mean or average stocking rate, it is proposed to reserve the term density to the number at any instant.

Thus, if 20 sheep graze rotationally 5 paddocks of 1 acre each, the mean stocking rate is always 4 per acre and the stocking density is always 20 per acre. Density is then simply the spatial distribution of the sheep at any one time. Since, in this example, the stocking density is always the same, it may seem unnecessary to elaborate further. If, however, the sheep had moved into paddocks of differing sizes, the stocking density would have varied slightly. Clearly, then, the *mean stocking density* might differ from all the other figures and would be calculated as a mean of the numbers per acre each day. It would thus take into account the number of days at any one density. Such a mean does not appear to be of very great value, however. The relationship between density and the mean stocking rate can vary greatly (Fig. 3.1).

Stocking intensity or grazing intensity

Stocking intensity and grazing intensity are terms frequently used as alternatives to 'stocking rate'; so, intensive grazing simply means the same as heavy stocking. If a special meaning is to be attached to them it might reasonably be the relationship between the stocking rate and another resource: money invested, labour employed, fertilizer applied, and so on. Exactly what the number of sheep is being related to must be stated, and only in unit terms of this relationship can stocking or grazing be considered intensive. It is permissible, then, to say that a pasture has been intensively grazed when it has received much grazing in relation to the amount of herbage it has produced; this would also be called hard grazing. It might be better to use Grazing Intensity as synonomous with Grazing Pressure, as Mott (20) has done, and to use Stocking Intensity as synonomous with Stocking Rate.

Intensive production

Intensive production is defined as a high output of products per unit of input of some one resource. It is therefore quite legitimate to

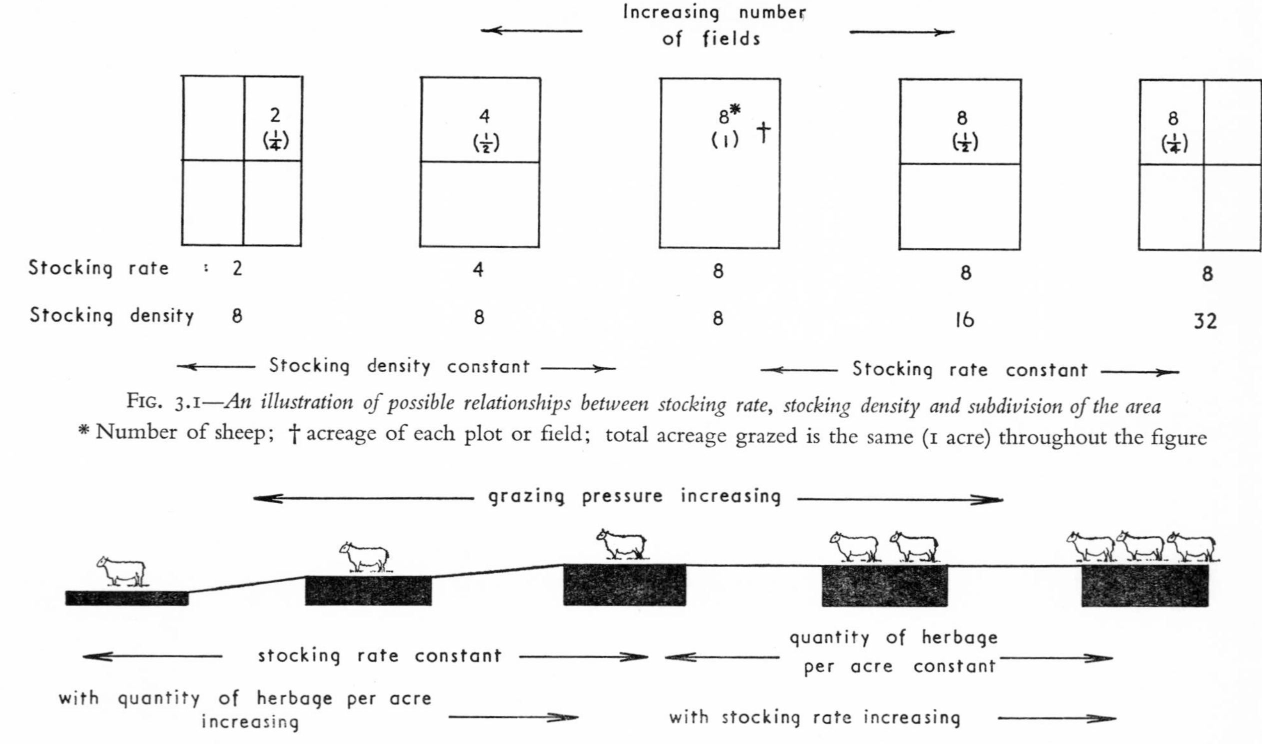

FIG. 3.1—*An illustration of possible relationships between stocking rate, stocking density and subdivision of the area*
* Number of sheep; † acreage of each plot or field; total acreage grazed is the same (1 acre) throughout the figure

FIG. 3.2—*Relationship between grazing pressure, stocking rate and quantity of herbage per acre (represented by the height of the solid blocks). Grazing pressure is increased by increasing the stocking rate or decreasing the amount of herbage per unit area of ground*

distinguish between intensity of production per acre and per ewe, without regard to such things as nutrient supply.

The term 'productivity' should simply refer to the rate of production, although, when used of the process of conversion of food to sheep products, it frequently refers specifically to the relationship of production per unit of food consumed rather than, for example, per unit of time.

Grazing pressure

Grazing pressure denotes the relationship, over any stated period, between the supply of herbage and the total animal needs: it is sometimes defined as the number of animals per unit of available forage (20, 21). At a low-grazing pressure, therefore, all animals have a surplus of herbage available, whatever the stocking rate. This is illustrated in Fig. 3.2. Some care must be taken, however, in the use of the word 'available'. Little is yet known about the relative availability of herbage to the sheep. There is probably very little above-ground herbage that a sheep cannot bite off: it is quite a different matter whether a sheep can maintain a given level of herbage intake per day when grazing on extremely short or sparse vegetation (see Chapter VII).

Significance of the stocking rate

The best way to put the significance of the stocking rate into perspective is to visualize the grass crop, or any other crop normally grazed, being cut and fed to housed or yarded animals. This process (zero-grazing), here envisaged purely to simplify the harvesting process, also has practical implications, particularly in relation to disease control. Now the amount of grass cut daily will feed a given number of sheep, no more and no less, provided that it is intended to give each animal the food it requires. Any animal production process must be based on an animal achieving some performance that can be regarded as economically satisfactory. Each animal must therefore receive its ration. To divide the available food amongst more animals is pointless unless (*a*) each animal was previously overfed or (*b*) food was previously being wasted.

Increased stocking rates have resulted in greater animal production from pasture largely because of (*b*). As noted in Chapter II, it has been estimated that, in Britain, probably as little as one-half of the herbage grown is ever consumed by animals (7).

Once this slack has been taken up, however, there may be no virtue in further increasing the stocking rate. Obviously enough sheep are needed to consume most of the herbage grown: it does not automatically follow that this number should be the maximum that can be fed. There are difficulties in having the right number at all times of the year, when the herbage production varies greatly from month to month (6) (see Chapter II). If it was possible to define 'correct stocking' in objective terms (see Chapter VII) it would probably be worth aiming at this state at all times. Since the needs of the animal population change with the season, there is theoretically a possibility of matching supply and demand. Indeed, in some environments where the climatic conditions and structure of the animal population allow it (in parts of New Zealand, for example) this can be largely achieved in practice (11). In Britain it is virtually impossible, except perhaps with an extremely low annual stocking rate. The result is that conservation of grass when it is surplus to current requirements must be a part of the utilization system, where animals are wholly (or almost wholly) fed from the acres they graze. Conservation can, of course, be completely avoided if food is purchased: this enables more sheep to be kept than could be completely supported by food produced on the farm.

Both systems really amount to a similar state of affairs in that reserves of food are maintained or available, and there is no real need for sheep to be stocked at rates which are in any sense too high or too low.

Maintaining the 'right' stocking rate, provided this can be defined, is a reasonable aim but it must also take into account the interaction between animal and plant (4, 9, 14, 19). The most important aspects of this interaction are the effect of the stocking rate, (*a*) on the amount of grass produced, (*b*) on the proportion consumed of what is grown and (*c*) on the food intake of the individual animal. These will be dealt with in more detail in Chapter VII. Here it is sufficient to note that animal production per acre usually continues to increase with increasing stocking rate beyond the point at which the production per individual animal begins to decrease. In many parts of the world, however, gross overstocking occurs and results in very poor production (8).

STRUCTURE OF THE POPULATION

Population structure is a term referring to the proportions of the total number of sheep that are male, female, old, young, breeding, fattening and so on. Rams do not usually constitute a significant item as far as food consumption is concerned. Important as is the male in sheep breeding, only one is normally required for every 20 to 60 females; 40 would be a reasonable average. Castrated males (wethers) are present in roughly equal numbers with female lambs in most fat-lamb producing enterprises. Apart from slight differences in growth rate (see Chapter V) this should not have much effect.

Sheep are kept in a variety of kinds of flock (2, 10, 13, 25), however, and this may influence age proportions particularly. With ewes in 'regular ages' (equal numbers of each age group), some are drafted out when too old for their purpose, often too old for their environment, and younger ewes, either ewe lambs, or 'gimmers' at 1 year to 18 months old, are bought every year. In many flocks based on cross-breeding no lambs are retained, both wether and ewe lambs being sent for slaughter, if fat, or sold as 'stores' for further finishing (or fattening). Other flocks, of such breeds as Clun Forest, Dorset Horn and Kerry Hill, generally breed their own replacements and only the wether lambs or the surplus females are sold. Thus it is that flocks may vary greatly in age structure, not only from flock to flock, but also during the year. This affects the pattern of demand for food, as does also the changing productive state of the ewe during pregnancy, lactation and without young. An illustration of the resulting food requirement curve is given in Fig. 3.3 for a fat-lamb producing flock of half-bred ewes. These curves are based on requirements of Scottish half-bred ewes and their single or twin lambs measured indoors. A caution about averages is relevant here. A food supply adequate during pregnancy for a flock in which each ewe, on average, carried one foetus, might be catastrophically inadequate for the few ewes carrying 3 or 4 lambs each. This kind of consideration is most important in relation to the lambing percentage.

Lambing percentage

Lambing percentage is variously calculated as the number of lambs, born alive, or dead, tailed, weaned, per 100 ewes, mated, marked, or put to the ram. Each method of calculation has its proponents; in fact, the value of the method depends entirely on what it is intended to express and all such calculations may be useful and appropriate for different purposes. For example, the number of lambs born per ewe, better termed 'litter size', whether dead or alive, is a better measure of fertility but a poorer measure of agricultural productivity than the number of lambs successfully reared to a particular age.

A standard terminology has been proposed by Desvignes (27) and

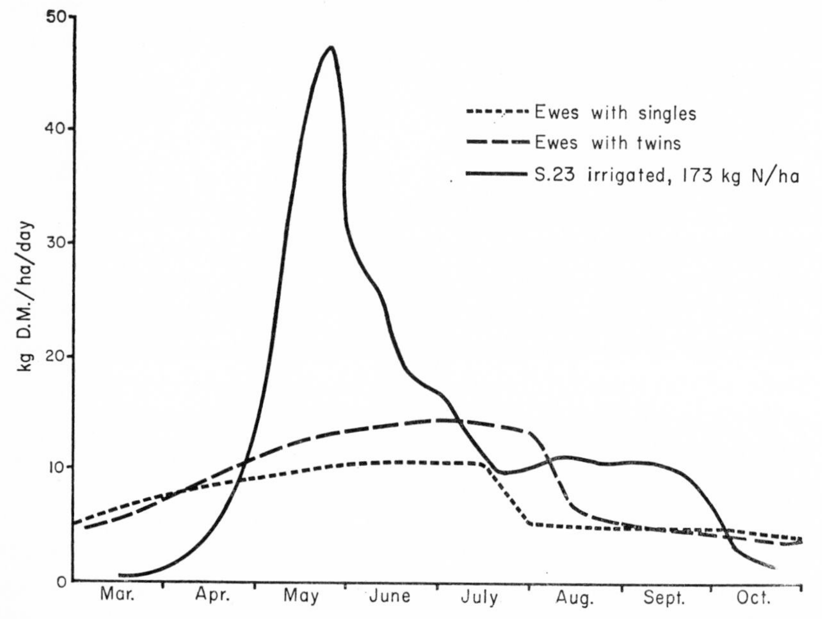

FIG. 3.3—*Herbage supply and sheep requirement for ewes with singles and ewes with twins at 3·34 ewes/ha*

this would be of great value, particularly in clarifying results expressed in different languages.

A comprehensive list of terms, based on the proposal by Desvignes, is given in Table XII.

Lambing percentages vary enormously between breeds and a good deal within breeds. Management at or before mating ('tupping'), can influence the ovulation rate of the ewe (see Chapter IV), and disease may further affect the number of lambs born alive. In Britain, mountain breeds, which are usually of small mature size, have lambing percentages around 100, i.e. number of lambs born per 100 ewes put to the ram, but they are often capable of much higher figures under suitable conditions. On the hills of Scotland and Wales, however, in years when the weather is severe in winter, the figure may drop to 40 per cent or even less. It is thus essential to distinguish between what a breed is capable of, and what it is allowed to do. Lowland breeds vary from some in which lambing percentages do not differ greatly from 100 per cent to those which regularly achieve 160 to 170 per cent and may, under excellent management, reach 230 per cent.

There are breeds, the Romanov and East Friesian (24), for example, which have even larger litters, and hormone treatments can induce similar litter sizes in other breeds (see Chapter XV).

Significance of the lambing percentage

In this chapter it has been stressed that the sheep population usually consists of different kinds of animals engaged in a variety of biological activities and that these differences are important in the conversion of grass. It is equally important to be able to attach a meaning to any one of the different categories and to appreciate, for example, the full significance of the fact that one ewe is lactating and another barren. Some of these problems will be considered in subsequent chapters: at this stage the following should be noted. A given lambing percentage does not necessarily have the same significance in two breeds. One lamb may be as big a burden to a small hill ewe as are two to a big lowland ewe. The intake of food by the lamb relative to that of the ewe may be of importance both nutritionally (see Chapter VI) and parasitologically (see Chapter IX). These ratios might differ greatly between breeds.

Higher lambing percentages always mean higher stocking rates of young susceptible lambs for any given stocking rate of ewes, unless there is also a higher proportion of older, non-breeding stock (such as wool-producing shearling wethers).

Probably the most important single aspect of the lambing percentage is the proportion of lambs which are born and reared as singles, twins or triplets. The very great differences, in nutrition, growth and parasitism, between such lambs are described in Chapter V. In many circumstances they may almost be regarded as different kinds of sheep. The proportions of singles and twins present in relation to lambing percentage are shown in Fig. 3.4. A simple case is illustrated where all ewes either have one lamb or two. Few flocks, in practice, have neither barren ewes nor triplet lambs but, in general, the pro-

TABLE XII. TERMINOLOGY OF REPRODUCTION

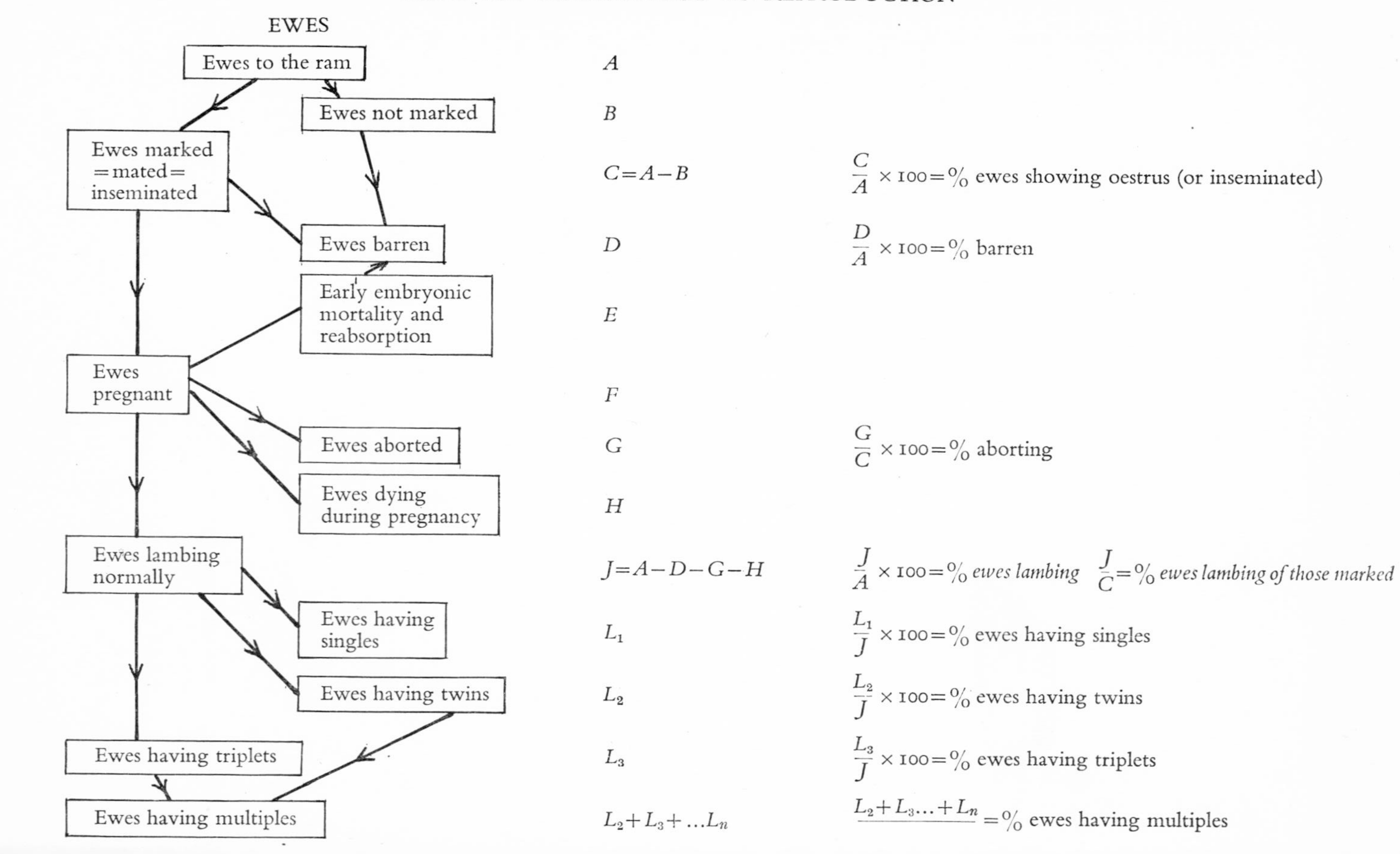

Ewes	Symbol	Calculation
Ewes to the ram	A	
Ewes not marked	B	
Ewes marked = mated = inseminated	$C=A-B$	$\frac{C}{A}\times 100=\%$ ewes showing oestrus (or inseminated)
Ewes barren	D	$\frac{D}{A}\times 100=\%$ barren
Early embryonic mortality and reabsorption	E	
Ewes pregnant	F	
Ewes aborted	G	$\frac{G}{C}\times 100=\%$ aborting
Ewes dying during pregnancy	H	
Ewes lambing normally	$J=A-D-G-H$	$\frac{J}{A}\times 100=\%$ *ewes lambing* $\frac{J}{C}=\%$ *ewes lambing of those marked*
Ewes having singles	L_1	$\frac{L_1}{J}\times 100=\%$ ewes having singles
Ewes having twins	L_2	$\frac{L_2}{J}\times 100=\%$ ewes having twins
Ewes having triplets	L_3	$\frac{L_3}{J}\times 100=\%$ ewes having triplets
Ewes having multiples	$L_2+L_3+\ldots L_n$	$\frac{L_2+L_3\ldots+L_n}{}=\%$ ewes having multiples

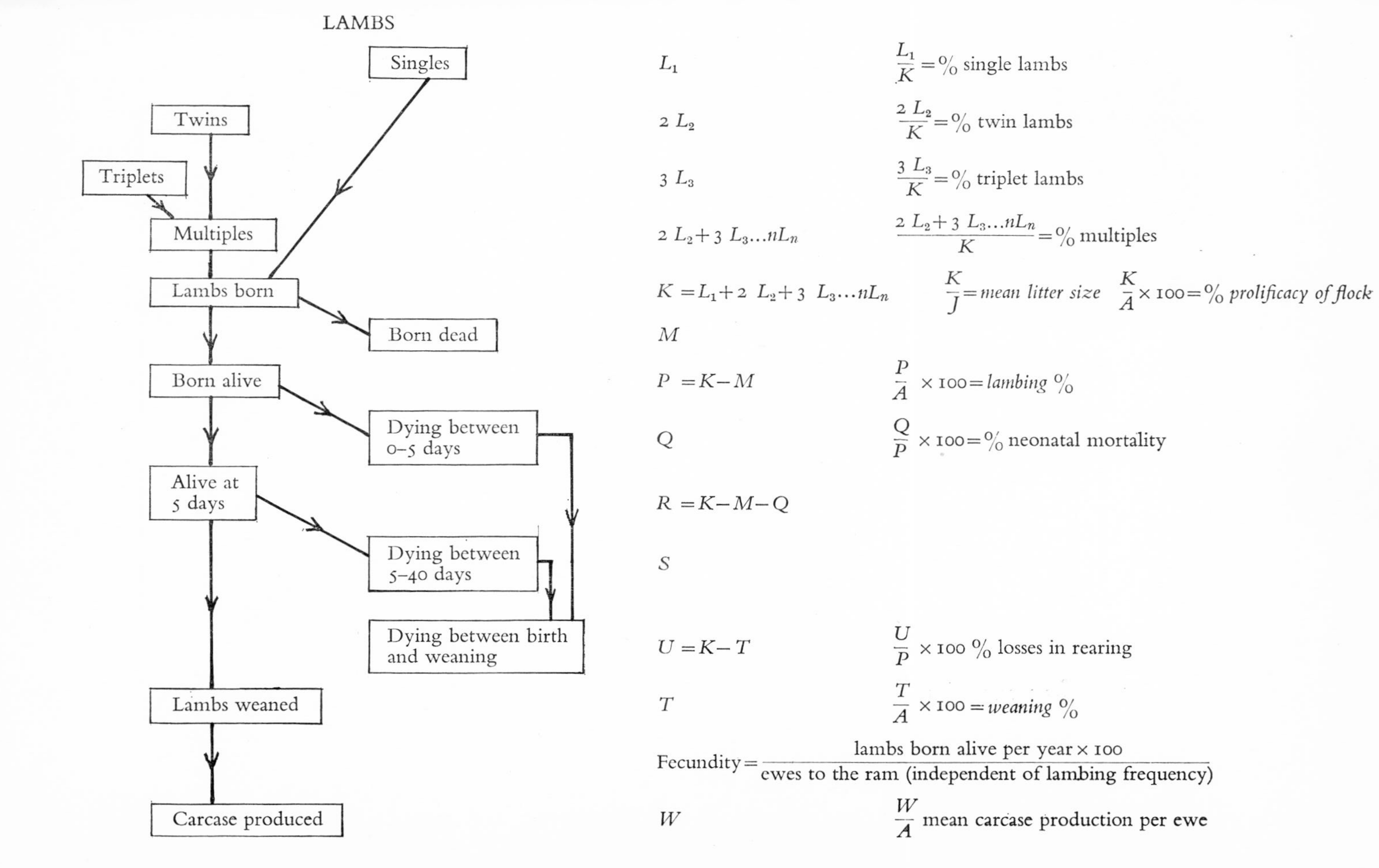
LAMBS
Singles
Twins
Triplets
Multiples
Lambs born
Born dead
Born alive
Dying between 0–5 days
Alive at 5 days
Dying between 5–40 days
Dying between birth and weaning
Lambs weaned
Carcase produced
L_1 $\frac{L_1}{K}$ = % single lambs
$2\ L_2$ $\frac{2\ L_2}{K}$ = % twin lambs
$3\ L_3$ $\frac{3\ L_3}{K}$ = % triplet lambs
$2\ L_2 + 3\ L_3 \ldots nL_n$ $\frac{2\ L_2 + 3\ L_3 \ldots nL_n}{K}$ = % multiples
$K = L_1 + 2\ L_2 + 3\ L_3 \ldots nL_n$ $\frac{K}{J}$ = *mean litter size* $\frac{K}{A} \times 100$ = % *prolificacy of flock*
M
$P = K - M$ $\frac{P}{A} \times 100$ = *lambing* %
Q $\frac{Q}{P} \times 100$ = % neonatal mortality
$R = K - M - Q$
S
$U = K - T$ $\frac{U}{P} \times 100$ % losses in rearing
T $\frac{T}{A} \times 100$ = *weaning* %
Fecundity = $\frac{\text{lambs born alive per year} \times 100}{\text{ewes to the ram (independent of lambing frequency)}}$
W $\frac{W}{A}$ mean carcase production per ewe

portions shown would not be far wrong. It is particularly worth pointing out that with a change of lambing percentage from 120 to 150 (both these figures are within the range commonly found within a lowland breed) the proportion of lambs changes from 2 singles for every twin to 2 twins for every single. Such changes can have enormous implications for management as well as for potential productivity

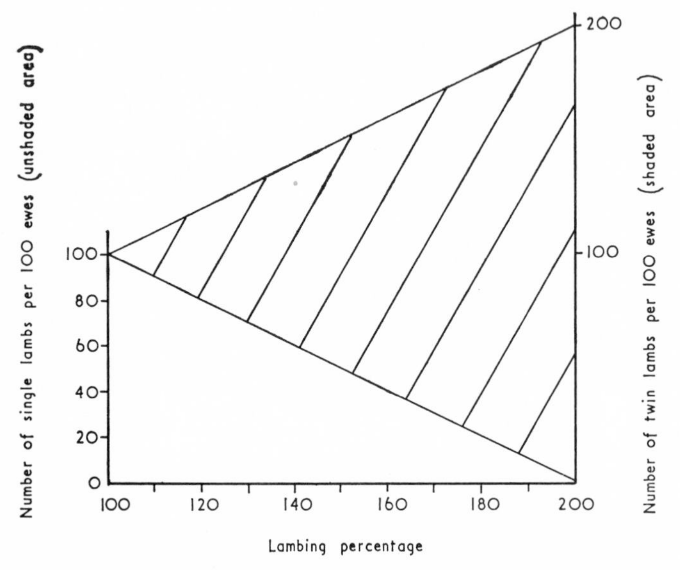

FIG. 3.4—*Proportion of single and twin lambs at different lambing percentages, assuming the simple case with no barren ewes and no triplets, etc.*

(Source : C. R. W. Spedding, T. H. Brown & R. V. Large (1960), *Proc. VIII int. Grassl. Congr.* (Reading), 718)

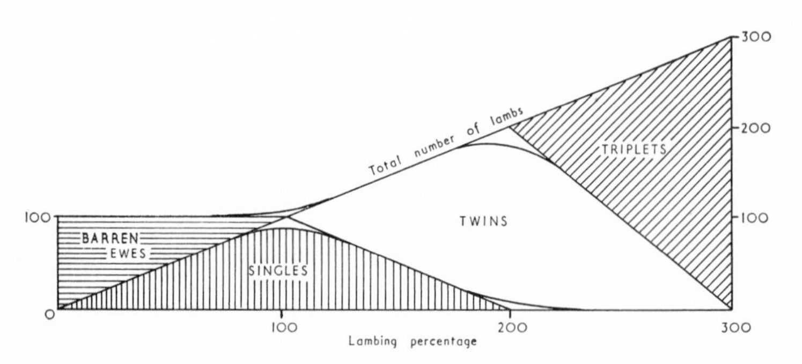

FIG. 3.5—*Proportions of barren ewes and single, twin and triplet lambs at lambing percentages from 0–300 per cent*

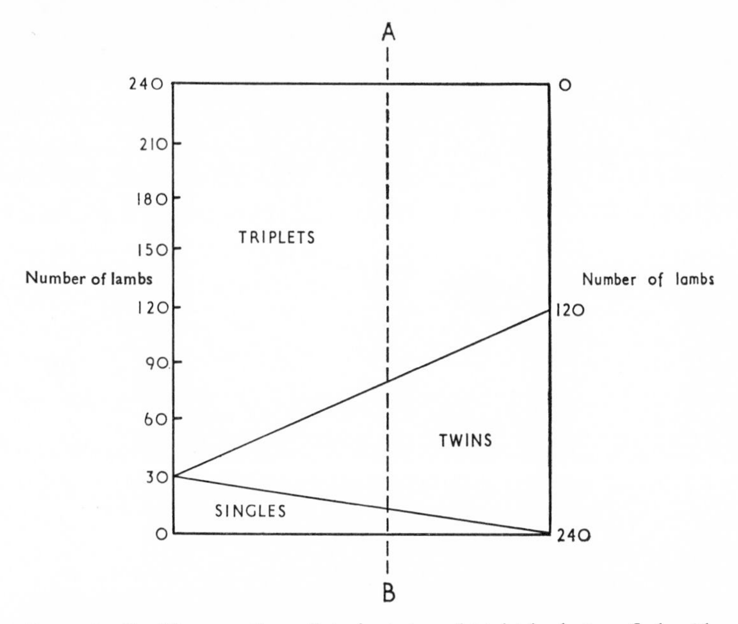

FIG. 3.6—*Possible proportions of single, twin and triplet lambs in a flock with a lambing percentage of 240. A similar figure could, of course, be constructed for every other lambing percentage. The possibilities are represented by the vertical lines (e.g. A–B) that can be drawn: the numbers of lambs of each kind are then represented by the length of vertical cut off by the sloping lines*

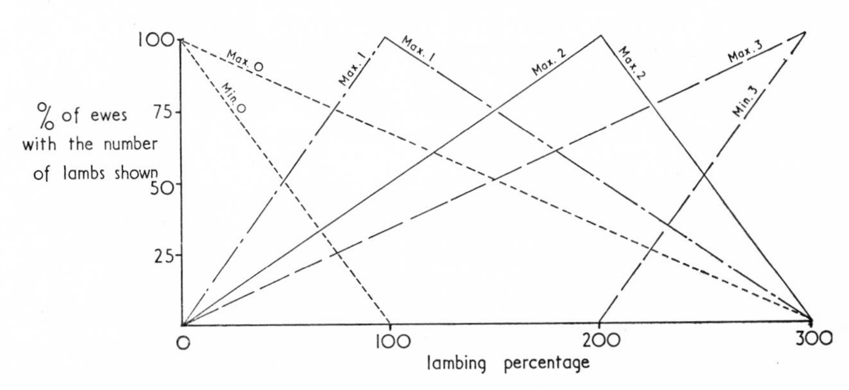

FIG. 3.7—*Maximum (max.) and minimum (min.) number of ewes having 0, 1, 2 or 3 lambs at any lambing percentage from 0 to 300. At each lambing percentage, the actual percentage of ewes which are barren or have singles, twins or triplets (quads are excluded for simplicity), must lie between the lines giving maximum and minimum values. The minimum for ewes with 1 or 2 is always 0; the minimum for ewes with 3 is 0 up to a lambing percentage of 200; the minimum for barren ewes is 0 beyond 100 per cent (based on calculations by H. L. Back, Grassland Research Institute, Hurley)*

(1, 13, 16, 23). Where triplets are involved the situation may be far more complex. To take an extreme example, a lambing percentage of 200 could describe two flocks, one with 100 ewes each with a pair of twins and the other with 100 ewes, 50 of which had singles and 50 with sets of triplets. The needs and problems of these two flocks would be entirely different. Fig. 3.5 is an extension of Fig. 3.4 to include both higher and lower lambing percentages. The straight lines represent the simplest case where no barren ewes occur beyond 100 per cent, no twins occur before 100 per cent is achieved and no triplets before 200 per cent. The curved lines are an estimate of the more likely situation in practice.

It will be clear that the lambing percentage may give a very imperfect picture of the flock structure. Fig. 3.6 illustrates the range of possible structures associated with one high lambing percentage. Fig. 3.7 shows the possible number of ewes with each type of lamb, at lambing percentages from 0 to 300.

Thus it is that the pattern of food requirement of the sheep population may vary seasonally and be very inadequately stated in terms of numbers of sheep. It may be quite unlike that of any individual sheep, yet it must be based on an understanding of all these individual needs.

REFERENCES

(1) BARBER, D. (1961). *Sheep and Sums.* N.S.B.A. Yearbook, 6–10.

(2) CHERRY, G. (1962). *Farming Facts.* Ed. Cherry, G. London, The Graham Cherry Organization.

(3) COLBURN, O. (1961). Progress in sheep. *Jl R. agric. Soc. 122,* 54.

(4) COOPER, M. M. (1961). Grazing systems. *J. King's Coll. agric. Soc. 15,* 5.

(5) COWLISHAW, S. J. (1962). The effect of stocking density on the productivity of yearling female sheep. *J. Brit. Grassld Soc. 17* (1), 52–8.

(6) DAVIES, WM. (1957). The use and misuse of grassland. *Potassium Symposium,* 41–57. Ed. Int. Pot. Inst., Berne.

(7) DAVIES, WM. (1960 a). Temperate (and tropical) grasslands. *Proc. VIII int. Grassld Congr.* (Reading), 6.

(8) DAVIES, WM. (1960 b). Pastoral systems in relation to world food supplies. *The Advancement of Science,* No. 67.

(9) DAVIES, WM. (1961). The British Grassland Society in the world of tomorrow. *J. Brit. Grassld Soc. 16* (2), 83–8.

(10) FRASER, A. (1951). *Sheep Husbandry.* Chap III. London, Crosby Lockwood.

(11) HAMILTON, R. A. (1956). Utilization of grassland. *Outlook on Agriculture 1,* 5–11.

(12) IVINS, J. D., DILNOT, J. & DAVISON, J. (1958). The interpretation of data of grassland evaluation in relation to the varying potential outputs of grassland and livestock. *J. Brit. Grassld Soc. 13* (1), 23–8.

(13) James, P. G. (1960). Sheep on the arable farm. *Rep. Fm. Econ. Camb. Univ.* *49*, 33.

(14) Kammlade, W. G. & Esplin, A. L. (1962). Sheep are efficient users of forage. Chap. 61 in *Forages*, 2nd ed. Ed. Hughes, H. D., Heath, M. E. & Metcalfe, D. S. Iowa St. Univ. Press.

(15) Kemp, C. D. (1960). The need for a dynamic approach to grassland experimentation. *Proc. VIII int. Grassld Congr.* (Reading), 728.

(16) Luxton, H. W. B. & Loadman, M. (1959). Fat sheep production in Devon, 1957/58. *Rep. Dep. Agric. Univ. Bristol* *115*, 23.

(17) Macfadyen, A. (1957). *Animal Ecology*. London, Pitman.

(18) McMeekan, C. P. (1956). Grazing management and animal production. *Proc. VII int. Grassld Congr.* (Palmerston North, N.Z.), 146–55.

(19) McMeekan, C. P. (1960). Grazing management. *Proc. VIII int. Grassld Congr.* (Reading), 21.

(20) Mott, G. O. (1960). Grazing pressure and the measurement of pasture production. *Proc. VIII int. Grassl Congr.* (Reading), 606.

(21) Mott, G. O. (1962). Evaluating forage production. Chap. 10 in *Forages*, 2nd ed. Ed. Hughes, H. D., Heath, M. E. & Metcalfe, D. S. Iowa St. Univ. Press.

(22) Mott, G. O. & Lucas, H. L. (1952). *Proc. VI int. Grassld Congr.* (Pennsylvania), 1380–85.

(23) Simpson, I. G. (1961). The ewe flock on the lowland farm. Interim report on a Yorkshire survey. *Fmrs' Rep. Univ. Leeds*, p. 150.

(24) Terrill, C. E. (1962). The reproduction of sheep. Chap. 14 in *Reproduction in Farm Animals*, ed. Hafez, E.S.E. London, Baillière, Tindall & Cox.

(25) Thomas, J. F. H. (1955). *Sheep*, p. 42. London, Faber and Faber.

(26) Wheeler, J. L. (1962). Experimentation in grazing management. *Herb. Abstr.* *32*, 1.

(27) Desvignes, A. (1968). Proposition de définition de critères zootechniques et économiques en matière d'élevage ovin. *E.A.A.P. Commission mtg.* (Dublin).

IV

THE EWE

The contributions of the ewe to sheep productivity are many and varied. Some are direct, such as ewe mutton, wool, new-born lambs and milk; others are indirect and include important influences on both the pasture and the parasite populations available to the growing lamb.

Food requirement of the ewe (Fig. 4.1)
Ewes vary, according to breed, size, age, fertility and so on, in their potential contribution to the production process but the ability of any ewe to express her potential depends on adequate nutrition. Largely because her activity varies with the season, her needs are highly seasonal (93), and, at present, they are rather ill-defined. Clearly, the pregnant animal has different requirements from those of the ewe in lactation, and the needs of each may vary according to the number of lambs carried or suckled. The nutrient requirements of sheep have been summarized in a specially prepared publication (135), for energy, protein and minerals. Energy requirements are given as metabolizable energy (M.E.), which can be calculated for grass from the per cent organic matter digestibility (per cent. O.M.D.), as follows:

$$\text{M.E. (k cal/g D.M. of forage)} = 0{\cdot}037\%\ \text{O.M.D.} + 0{\cdot}122 \pm 0{\cdot}083$$

Protein needs are given as 'available protein' and an additional allowance should be added, for metabolic losses of nitrogen in faeces, of approximately 16·8 g for each kg of dry matter eaten.

The minimum requirements given for protein (in g/day) for maintenance of sheep weighing from 5 to 70 kg live weight, vary from 17 to 30 g/day. For a growth rate of 400 g/day, the requirements rise from 100 to 110 g/day of available protein over the same weight range. Requirements for maintenance and wool production vary from 16 to 25 g/day for hill breeds weighing from 30 to 70 kg live weight: over the same weight range, the requirements for lowland longwools is given as 25 to 34 g/day. During the last 3 months of pregnancy, a supplementary allowance of 3 g/day rising to 50 g/day for ewes with singles (6 kg birth weight), compares with 9 g/day rising to 70 g/day for ewes with twins (10 kg total birth weight).

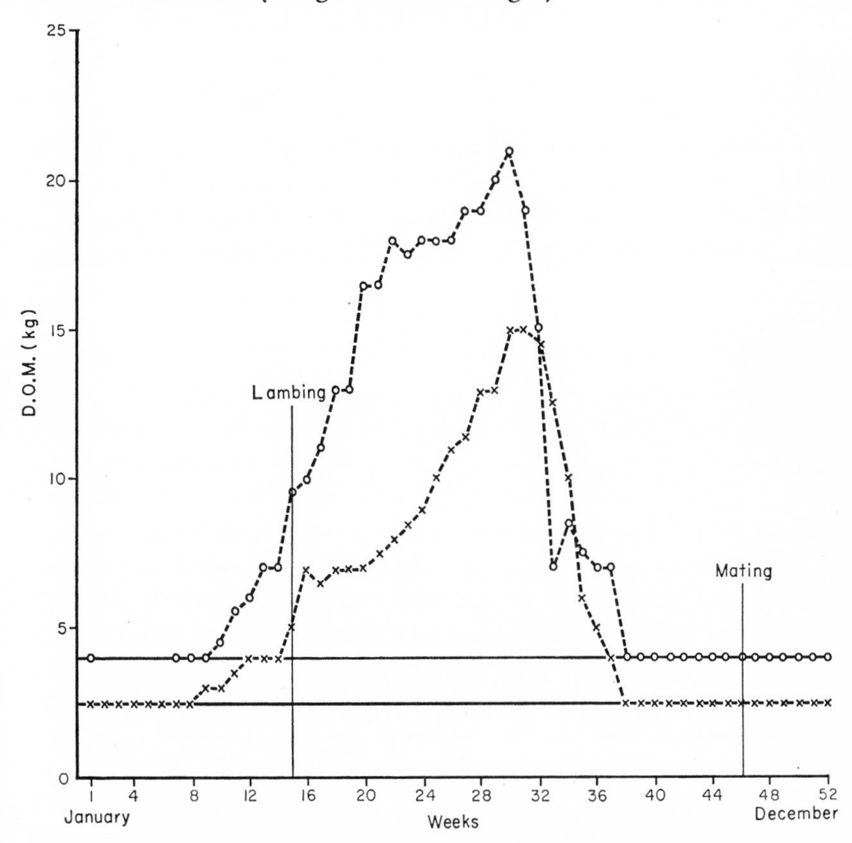

FIG. 4.1—*Distribution of the annual food requirement of the ewe, showing kg D.O.M. required weekly for maintenance (———) and the total for a ewe with twins (- - - -), for two breeds (Kerry Hill weighing approx. 55kg liveweight, –o–; Welsh Mountain weighing 33 kg, –x–)*

The energy requirement of a dry ewe (70 kg live weight and walking 2 miles per day) is given as about 2·0 Mcal metabolizable energy per day, though this varies with the diet; at the peak of lactation, the figure is of the order of 6·0 Mcal.

A detailed treatment of the subject can be found in a recent account by McDonald, Edwards & Greenhalgh (201) and detailed studies on the effect of dietary protein in gestation on ewe and lamb performance have been reported by Forbes & Robinson (174) and Robinson & Forbes (229, 230).

WOOL

Most sheep produce wool: in Britain, the Wiltshire Horn is exceptional in that its wool production is of negligible importance.

The importance of wool varies enormously, however. Australian sheep farmers may derive half their total income from the wool clip. Within the United Kingdom the proportion of the total farm income attributable to sheep that is provided by wool may vary from 15 per cent in fat-lamb producing lowland flocks to 40 per cent in the hill flocks of Scotland and Wales (82).

Quantity and Quality of Wool

The quantity of wool produced per head varies greatly with the breed, not only with the obvious weight and size differences but also between breeds with a similar mature weight. Table XIII shows a small selection of fleece weights that may be regarded as typical for ewes of the breeds listed, under the conditions of the region in which they were measured. Typical liveweights are also given for mature ewes under the same conditions. Wool yield is in part determined by size but clearly there exist great differences between breeds producing similar fleece weights, in the liveweight of the mature ewe. The subject is dealt with fully by Barnard (138) and by Ryder & Stephenson (233) and some of the values (references (f) and (g)) are taken from their books.

Wool quality varies between and within breeds but also from one part of the fleece to another (49). In general the finest wool is grown on the shoulder and the poorest towards the extremities—head, tail and legs. The value of a fleece is affected by the variation of type within it, the length of the staple, the diameter of the fibres, freedom from extraneous matter, which varies from marking fluids to weed seeds, by characteristics such as crimp (see Plate II) and by freedom from faults. Some of the last named are nutritional in origin, the most notable being the occurrence of a 'break' or point of weakness in the fibres which results from a period of nutritional stress: it is often associated with poor nutrition of the ewe during a severe winter.

TABLE XIII

FLEECE WEIGHT OF SHEEP IN RELATION TO BREED
(and average liveweight of the breed)

Breed	Typical Fleece Wt. lb	Typical Liveweight lb	Source
Clun Forest	5·4	124	(a)
Ideal (Polwarth)	6·5–7·7	65–76	(d)
Cheviot (in Eire)	7·0	147	(b)
Greyface	7·0	154	(b)
Half-bred (Border Leicester × Cheviot)	7·5	160	(b)
Corriedale (in Brazil)	7·7–8·8	75–85	(d)
Romney Marsh (in Brazil)	8·5–9.0	100–110	(d)
Galway	9·0	188	(b)
German Merino	10·0	150–180	(c)
German White-headed Mutton Sheep	13·0–14·0	220	(c)
Devon Longwool	17·5	170–190	(e)
Welsh Mountain	2·0–3·0	70	(g)
New Zealand Romney	9·5–11·0	140	(g)
Polwarth (Australia)	11·0		(g)
Merino (Australia)	8·7–12·8	110	(f)
Fine wool (Merino)	12·0	110	(h)
Corriedale (housed)	16·2	130	(h)
Lincoln (housed)	19·2		(h)

Compiled from the following sources :

(a) Personal communication from M. Bichard and B. C. Yalcin (25) ;
(b) Personal communication from A. Conway (35) ;
(c) *Livestock breeding in the Federal Republic of Germany*, 1961 (75) ;
(d) Personal communication from Prof. Geraldo Velloso Nunes Veira, Brazil (122)
(e) Luxton, H. W. B. & Loadman, M., 1957–8 (271).
(f) Barnard, 1962 (138)
(g) Ryder and Stephenson, 1968 (233)
(h) Daly and Carter, 1955 (153)

Wool growth rate

Wool does not grow at a uniform rate throughout the year. In some parts of the world sheep are shorn more than once a year and there is some variation in the time of shearing. In the United Kingdom ewes are normally shorn in late spring (May or June), just before the normal temperatures may cause discomfort to a sheep with a heavy fleece. With spring lambing this is also a time when the inevitable disturbance of shearing is unlikely to lead to mismothering of lambs.

The rate at which wool is grown is influenced by the plane of nutrition but this is not the only reason for the seasonal growth rhythm (51). The environmental temperature may be responsible (23) and photoperiodicity has been implicated (22). Whatever the reasons, the amount of wool grown per unit area of skin varies from one part of the year to another. These variations are greater in some breeds and in the Scottish Blackface they are thought to be greater than for many others. Doney & Smith (48) found that 80 per cent of the year's total was grown between July and November; and the growth during February and March amounted to only 16 per cent of that grown in August and September.

Stresses due to pregnancy and lactation (158, 159), poor nutritional supply, worm-infestation or disease may affect wool growth (87) and, because of the seasonal rhythm, such effects may be more or less serious according to when they occur. It cannot be assumed that significant changes in lambing percentage or stocking rate will leave wool growth unaffected. Total fleece yield is, of course, the resultant of the wool produced per unit area and the total area of skin. The latter may be greatly increased in some breeds, e.g. the Merino, by enormous folds.

Full details of all aspects of wool growth can be found in the recent and comprehensive book by Ryder & Stephenson (233). A development of great potential importance (169) is the use of formaldehyde treatment of dietary proteins to protect them from degradation in the rumen. This enables proteins to pass unchanged into the abomasum, resulting in much more efficient use of them and greatly increased wool growth.

The sheep ked, *Melophagus ovinus*, has been shown to reduce both liveweight gain of lambs and wool growth in ewes (208). In one investigation, it was found that ked-free ewes produced 11 per cent more wool than infested ewes.

Wool shedding

The wool shorn may not represent all that has been grown during the preceding year. Shedding of fibres is a common feature in some breeds and total loss of the fleece can occur as a result of extreme stress (Plate III (a)). Austwick (16) has pointed out that actual loss of wool over great areas of the skin can result from mycotic dermatitis ('lumpy wool'). This is caused by a minute fungus-like organism (*Dermatophilus dermatonomus*) which attacks the skin, causing great irritation. Attacks follow heavy rain and are thus more severe in wet years. Wool quality and value may be adversely affected even when no shedding occurs (17). In addition to the effect of the plane of nutrition there are certain specific nutrients especially required for wool growth. Included amongst these are inorganic sulphates and copper; insufficient of the latter results in a reduction of wool quality due to loss of crimp and the growth of so-called 'steely' wool.

Protective Value of the Fleece

Thus far wool has been considered as a product, something that the ewe contributes to the productivity of the system: this will be further discussed in Chapter VI. Wool also functions as a protective coat and in certain conditions can influence the efficiency with which food is converted. This efficiency is reduced when food has to be used in order to maintain body temperature. Although the rectal temperature of the sheep may vary within the range 38° to 40° C (26), in cold conditions energy must be used to maintain it. This is required when the environmental temperature drops below a 'critical' level. The critical temperature is that below which an animal has to increase its heat production in order to prevent its body temperature falling. For sheep, the critical temperature varies with the plane of nutrition and the insulating capacity of the fleece. Blaxter (28) has calculated values for the critical temperature ranging from −3° C, for a sheep at maintenance with a fleece length of 100 mm, to 31° C, for a fasting sheep with a fleece length of only 5 mm. The insulating properties of the fleece naturally vary with the wind speed and whether the fleece is wet or not. Less obviously but most importantly, they vary with the structure of the fleece. In many breeds an undercoat of dense, fine wool is over-laid by an overcoat of long, coarse fibres so arranged that they shed water most efficiently. Other breeds lack this topcoat or have different kinds of fibres in different proportions: in some, the fine and coarse fibres are mixed up and not arranged in layers. Obviously, the appropriateness of a breed to a particular environment may be governed in part by the suitability of its fleece (see Chapter XII).

Studies on Blackface sheep (157, 206) and on Welsh Mountain ewes (254) have demonstrated that winter exposure, especially combinations of wet and windy conditions, may reduce body weight or increase food intake. The length of fleece is important, however.

Claims have been made for the value of hessian coats, both in terms of ewe survival and of wool growth, but Peart (215) found no benefit.

Apart from shelter belts and full-scale housing, Cresswell (151) has revived the idea of a circular, partly-roofed stell. Such a shelter (700 sq. ft, enough for 100 hoggs) can be produced relatively cheaply and sited where required.

REPRODUCTION

As indicated in the previous chapter, the number of lambs born per ewe is of importance both in determining the kind of animal population that has to be fed and in influencing productivity per acre and per ewe. In relation to the balance between food supply and the needs of the flock, a major limiting feature is the breeding season of the ewe.

The Breeding Season

The Ram

There are differences in breeding activity between rams of different breeds and between individuals within a breed. Lees (192) found that Kerry rams were active during the 'out-of-season' period when Suffolk and Hampshire rams were not. It was also observed that activity by the ram within 24 hours of parturition could interfere with lamb suckling (191).

Within a Merino flock in Australia, considerable differences have been noted in the number of services per day, varying from 8 to 38 (198), and this has been found to vary with season (217). Seasonal changes in semen characteristics have also been found (133). Semen examination (164) may have especial relevance, therefore, in out-of-season breeding. It is not known what effects nutrition may have on ram fertility (245).

Where artificial insemination (A.I.) is used, it is possible to pool semen and thus eliminate the great variation between rams. An example of the latter is given by Robinson & Lamond (231): conception rates, at the first synchronized oestrus after progestagen sponge withdrawal from the ewes, ranged from 0·4 per cent to 60 per cent. They concluded that the normally accepted figure of $100–150 \times 10^6$ spermatozoa for A.I. needs to be greatly exceeded in the ewe inseminated at this first synchronized oestrus (a normal ejaculate contains 4,000 million sperm).

It is generally considered that a normal, mature ram can maintain 20–40 services a day without reducing the content of spermatozoa in the ejaculate below that normally required: it does not follow that this number of ewes could necessarily be satisfactorily mated daily. It appears, indeed, that ewes served more than once have a better chance of becoming pregnant (199). The ratio of rams to ewes that is satisfactory will vary with circumstances. In many normal circumstances, one ram per 100 ewes is adequate, but much higher ratios may be needed where breeding is synchronized (possibly exceeding 5 rams per 100 ewes).

In general, rams reach puberty in 4–5 months and give good service up to 8 years of age, although libido may decrease after 2–3 years (237).

The Ewe

Most British breeds of sheep have a restricted breeding season, largely determined by natural variation in daylength. The onset of breeding activity may be affected by the presence or absence of rams, ambient temperature and date of previous lambing (193), age of the ewe (240), time of shearing (194) and many other factors, but the most important influence is seasonal change of daylength (259).

It is generally assumed that the increased incidence of oestrus associated with decreased hours of daylight is due to stimulation of sexual activity by the reduced daylength: it is possible, however, that the important effect may be one of active inhibition of ovarian activity by increasing daylight (220, 241).

If ewes are subjected to a reversal of their normal seasonal variation in daylength, as by moving from one hemisphere to another, their breeding season may be almost completely reversed (247). Under equatorial lighting, the normal pattern may be lost after a year (247) or persist for $2\frac{1}{2}$ years (258).

Merinos at the equator have been reported to show oestrus throughout the year (119): variation often occurs in the reproductive performance of Merinos in Australia (21, 67, 112, 114, 132). Different breeds differ in the length of their breeding seasons and, under Australian conditions, Kelley & Shaw (69) demonstrated a difference between strains within the Merino breed.

The Awassi fat-tailed ewe is reported (134) to have an extended breeding season, as has the Dorset Horn among British breeds (211). Tropical sheep may be relatively insensitive to the influence of light: Symington & Oliver (244) found this to be so for the German Merino, Persian Blackhead and indigenous 'native' ewes, all reared in Rhodesia.

Within the breeding season ewes come on heat at intervals of about 17 days and each oestrus lasts for 1–3 days. Jones (185), working with Corriedales in Australia, confirmed the findings of Hulet *et al.* (182) that ewes allowed rams to mount during a period of about 19 hours: periods of up to 3 days have been recorded, however. In breeds with restricted breeding seasons, peak activity generally occurs shortly before the shortest day: the ovulation rate is highest at this time and progressively lower thereafter.

Breeding from ewe lambs

It should be emphasized that age and season also affect ovulation rate. Ewe lambs, less than 1 year old at mating, have a somewhat later breeding season and a lower ovulation rate. Old ewes, over about 8 years, also exhibit reduced ovulation rates. Breeding from ewe lambs is increasingly practised because (*a*) the results can serve as an early guide to breeding value, (*b*) life-time performance may be greater thereby and (*c*) there is evidence of improved subsequent performance. Since the possibility of breeding from a ewe lamb is dependent on an interaction between physiological development and a sufficient period of shortening days, it is necessary either that ewe lambs should not be born late in the season or that their rate of growth should be high.

Lactation anoestrus

There is a widespread feeling that lactation is one cause of the anoestrus state in sheep and Hunter (183) concluded that, at the present time, it may be helpful to wean lambs before remating their ewes. Since lactation often coincides with periods when many ewes are outside their normal breeding season, however, and since variation in the plane of nutrition may be important, the issue is rarely clear-cut and cannot be regarded as proven experimentally.

Smith (239) and Pretorius (218) found significant effects of suckling on the interval between lambing and first oestrus but Mauleon & Dauzier (200) concluded that lactation did not greatly suppress ovulation in Ile-de-France ewes which had lambed in the autumn.

It may be that *post-partum* heats are short and of low intensity and that this is particularly so during lactation.

Synchronization of Oestrus

Effect of the ram

The introduction of a ram shortly before the start of the breeding season has been claimed by many workers to synchronize the season's first oestrus in a high proportion of ewes (183).

The general pattern appears to involve an initial 'silent' ovulation within a few days of the ram being introduced, followed by the first oestrus of the season about one oestrus cycle length later (184). This results in a peak of oestrus activity in the ewes about 16 days after the introduction of the ram, lasting for about one week.

In general, rams have been used successfully to induce an earlier onset to the breeding season and any synchronization has been a by-product of this.

It has been reported from New Zealand (165) that the presence of the ram influenced the time of onset of the breeding season of both young and mature Romney ewes, that fertility was unaffected and that teasers (vasectomized rams) were less effective than entire rams. It has been claimed that vaginal pessaries can be more effective than rams (256) and result in closer synchronization.

It should also be noted that one effect of synchronization is a need for more rams (209) and the importance of the male in controlled breeding, especially with A.I., needs greater emphasis (231).

Effect of hormones

In the last few years, a good deal of work has been done using vaginal pessaries (152, 209, 227) impregnated with progestagenic compounds, as described by Robinson (226). Injections and oral administration have also been used (221).

Results have varied greatly: probably differences in the precise timing in relation to the natural breeding season have been of great importance.

Synchronization is obtained because the absorbed progestagen suppresses oestrus during the normal cycle; when the supply is stopped, as in sponge withdrawal, oestrus occurs within a few days. If all ewes are treated together, a high proportion (up to 95 per cent) may be served within about 5 days of the removal of the pessaries.

Conception rates at the first oestrus after pessary removal are often low, however, and it appears preferable to mate ewes at the second oestrus (251). The degree of synchronization still found at this time may vary but, on occasions, can be as good as that obtained at the first oestrus (257). Conception rates at the second oestrus are generally about normal: Robinson & Lamond (231) reported conception rates up to 76 per cent with A.I.

Artificial insemination is both more practicable, more advantageous and, perhaps, more necessary where large numbers of ewes are synchronized. In the case just quoted, where sponges were withdrawn over a three-day period, 90 per cent of the ewes were in oestrus over a five-day period commencing 16 days later.

Effect of light control

When out-of-season breeding (see next section) is induced by control of daylength, some degree of synchronization may result. Present indications, however, are that individual variation is considerable and that no significant effect of synchronization can be expected from the methods of light control currently being explored.

Out-of-Season Breeding

Hormones have been mentioned in relation to both litter size and compact lambing; in the latter case, they are closely related to out-of-season breeding (269). Before going further into this, it should be pointed out that the control of daylight, which housing could allow, is an alternative way of influencing the breeding season. It could prove inexpensive, particularly if, in a context of intensive production, the housing did not have to be provided especially for this purpose.

It will be noted (Chapter X) that housing may confer substantial advantages in reducing losses. The environment may exert a major influence on the feed requirements of sheep and housing could thus also influence the efficiency of production by modifying maintenance needs.

Control of daylength

Since Yeates (259) demonstrated the importance of daylight in controlling the onset of ovarian activity in ewes, several methods of harnessing this information to practical use have been tried.

The reduction of daylength generally acts as a stimulus to ewes and causes the onset of oestrus. It is not yet clear whether this reduction must be gradual or whether it can be an abrupt change. Certainly, abrupt changes may influence oestrus, but they may not result in conception.

In general, it appears that light control is capable of producing a breeding season at almost any time of the year, if other factors are not limiting (e.g. lactation, nutrition). There is, however, a substantial interval between the onset of treatment and the onset of induced sexual activity: this 'reaction interval' is variable but, in temperate regions, is from 80 to 125 days (180).

At certain times of the year, it is possible to achieve the desired patterns of daylength change simply by 'adding' light. In winter, for example, supplementary light can be provided to give longer days (up to 24 hours, if necessary) and then steadily decreased. The advantage lies in the fact that it is cheaper to add light, which does not require housing, than to increase the 'dark' hours. There is no information yet as to the importance of light intensity in these treatments.

The use of hormones

Out-of-season breeding can be induced by the use of progesterone followed by the injection of P.M.S.

The most successful treatments so far have involved about 750 i.u. of P.M.S. injected subcutaneously some 1–2 days after the last use of progesterone. The latter can be injected daily, or every other day, for 10–14 days, or can be introduced by the use of a vaginal sponge or pessary, as described for synchronization of oestrus (80). Several of these pessaries are now available commercially, arising out of the work of Robinson and his co-workers (227, 228) in Australia.

These sponges are impregnated with a progestagen and inserted into the parous ewe's vagina with the aid of a special implement. This can be done by a shepherd, provided that due care is taken and no unnecessary force used. Sponges are left in position for about 14 days and then withdrawn, using the string left protruding during the entire period.

The results achieved so far, however, have been generally disappointing (143) and not more than 20 per cent of ewes treated have actually lambed, although more than 75 per cent may have shown oestrus and been marked by a ram.

Better results have been obtained with the Dorset Horn (211) but this is to be expected with any breed exhibiting an extended breeding season. Results may also be much better close to the onset of the natural breeding season and the method can then be used to induce an earlier-than-normal onset to breeding.

Lambing frequency

Since gestation occupies anything up to 155 days and, in practical circumstances, conception is unlikely to be achieved again within 3–4 weeks of parturition, twice a year is about as frequent as lambing is ever likely to be.

Hunter (183), in a recent comprehensive review of breeding frequency in sheep, came to the conclusion that 'although published data are scarce, there are breeds which, when given the opportunity, are capable of lambing regularly at 6–8 month intervals without the need for stimulating *post-partum* oestrus by using exogenous hormones'. He considered the best examples so far reported to be the East African Blackheaded sheep (234) and Hampshire and Suffolk crossbred ewes (150).

In breeds where the natural breeding season is long, especially where it exceeds 6 months, it is possible, by careful timing, to achieve lambing intervals of about 6 months. The crucial factor is the interval between lambing and the first oestrus. This interval may be influenced by nutrition and by the presence and activity of the rams used; the age of the ewe does not appear to be of great significance, however.

It is possible that individuals or strains within breeds may exhibit a substantially longer-than-average breeding season and that such animals could be identified and selected for this characteristic. Little knowledge as yet exists of the genetic variation or heritability of the ability to lamb frequently, however (195). The identification of such useful attributes is only one of the many purposes for which organized recording is required.

Recording

This is perhaps too obvious to require mention. It is difficult to imagine any of these developments succeeding without full recording and the proper means of identification and measurement that are an indispensable part of it. As Cooper (264) pointed out, in a discussion of the potential for expansion in animal production, 'there is an almost complete absence of measurement of performance in sheep, and selection based largely on eye appraisal has not been very effective in improving economic traits'.

In 1961, a committee set up in Britain to examine sheep recording and progeny testing, concluded (204) that recording should be encouraged in commercial flocks, mainly as an aid to management, but also to provide more accurate bases for culling and selection of ewes for future breeding. They regarded recording as essential to constructive ram breeding and they were in favour of progeny testing to allow improvement of characters that are both economically important and of low heritability.

These general conclusions and recommendations were endorsed by the sheep panel at a National Livestock Breeding Conference (207) and a national recording scheme for Britain is now (1968) under consideration by the Meat and Livestock Commission.

The main objections to recording are the labour involved and the difficulty of ensuring that the records taken are usefully employed. The main answer is to make sure that only those features are recorded that are known to be important; in other words, to know what use is going to be made of the information *before* it is collected.

The Number of Lambs Born

The number of lambs born to each ewe is chiefly determined by three things: first, the number of eggs shed at the time of conception; secondly, the number of fertilized eggs which ultimately establish themselves as viable foetuses; and thirdly, adequate nutrition and freedom from disease during late pregnancy.

Breeding

The number of lambs born has been regarded as a characteristic associated with breed. Certainly, some breeds are normally more prolific than others: these include the East Friesian (mean per cent of lambs born of ewes lambing, given by Thomson & Aitken (246) as 296), Romanov (195–231), Wensleydale (182), Wrizosowka (185) and the Finnish Landrace (200–400). The Romanov is reputed to be a small sheep, the ewes weighing between 30–45 kg, with rather coarse fleeces. The mean litter size (for live lambs per ewe lambing) is normally more than 2 and may approach that figure even for ewe lambs (155, 243). Seljarnin (236) refers to a 'lambing rate' of 250–320 lambs on some farms. Lambs are often very small and survival poor. Individual Finnish sheep may not only produce large litters, the record being 9 lambs of which 4 were stillborn, but combine this with a long breeding life. Since Finnish sheep were imported into Britain, they have been used in a variety of crossing programmes and have generally raised the prolificacy of their progeny to a level intermediate between that of the parent breeds. Donald & Read (156) found that Finnish Landrace ewe lambs would generally mate successfully at 6–10 months of age and produce lambing percentages from 150 to 200 per cent. Lambing percentages of 300 per cent are characteristic of older ewes but the early loss of lambs is high.

Cross-breeding is thus one fairly rapid way in which prolificacy can be improved. The point of doing this is to confer higher litter size on sheep that are particularly valuable in other directions, without serious reduction in the value of these attributes. If the more prolific breeds were completely satisfactory in these respects also, they could be used without further breeding.

There are also possibilities of improving the prolificacy of existing breeds by selection within them. Turner *et al.* (249) suggested that even with the Merino, in certain circumstances, selection for multiple births may be successful. Heritability estimates of litter size usually vary from 0–25 per cent and selection programmes may only achieve annual increases of the order of 4 per cent in the number of lambs born per ewe lambing.

In a great many British breeds, however, individual ewes frequently produce litters well above the average and it is probable that selection of such individuals would result in much faster progress than is usually predicted for increase in lambing percentage on a whole-flock basis.

PLATE I

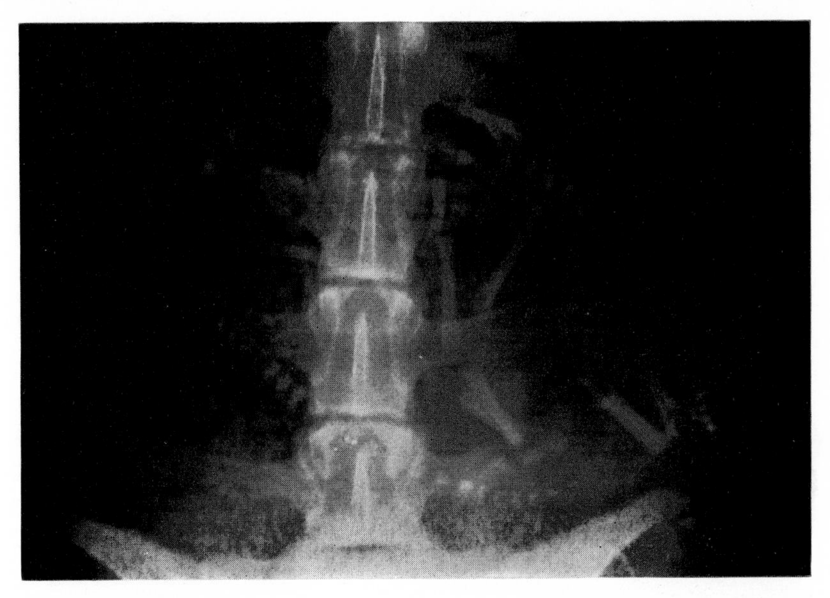

(a)

X-ray photograph (5.2.1963) of a pregnant Half-bred ewe, 98 days after conception (viewed ventrally). Two foetuses were diagnosed and two lambs were subsequently born.

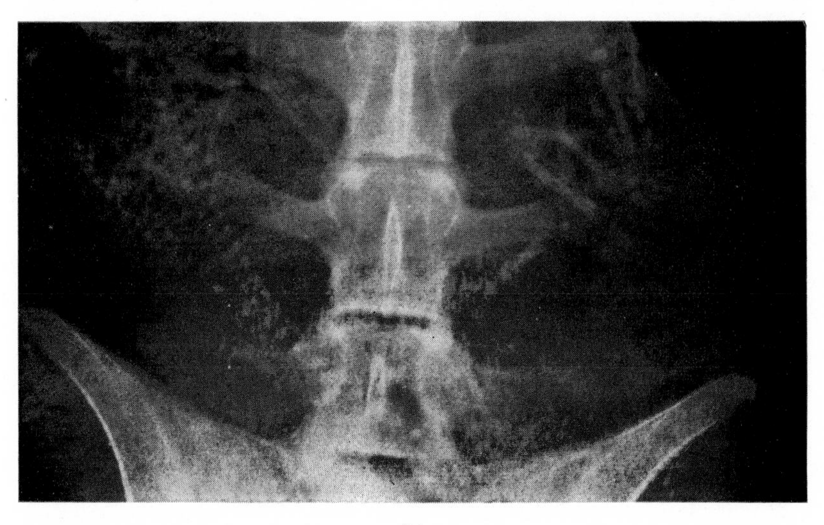

(b)

X-ray photograph (30.1.1963) of a pregnant Half-bred ewe, 85 days after conception (viewed ventrally). Four foetuses were diagnosed and four lambs were subsequently born.

PLATE II

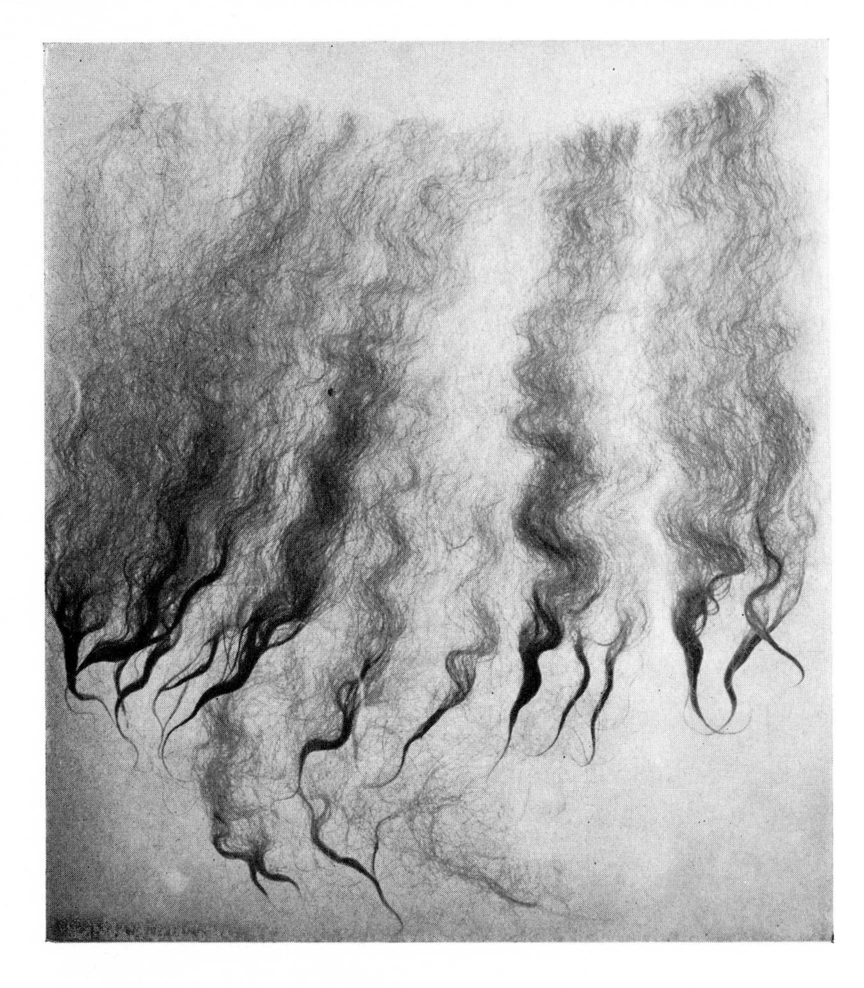

A sample of Half-bred wool showing ' crimp ' (photograph taken with back-lighting) (see page 72).

Since the plane of nutrition and the condition of the ewe (see 'Flushing' below) may influence the number of lambs born, the occurrence of a particular number does not necessarily indicate any genetic basis for the level of prolificacy expressed. This applies to singles, twins and, to a lesser extent, triplets in breeds like the Scottish Half-bred and the Kerry Hill: but when individual ewes of these breeds produce litters of 4, 5 or 6, it is improbable that management has had anything to do with it—except in the negative sense that it has not prevented it. No amount of juggling with nutrition has ever produced litter sizes of this order: yet ewes regularly do this and some may even do it repeatedly. Whether very large litters are repeated or not, however, it is likely that the ewes involved are genetically more prolific and could be used to establish a more prolific strain.

Nutrition

The number of lambs born can be affected by the nutrition of the ewe during pregnancy, in several ways.

Severe undernutrition in early pregnancy can affect placental development and foetal growth, especially in maiden ewes (140, 167). Under-nutrition can also affect ewes in late pregnancy and may lead to loss of lambs. Undernutrition is commonly represented by a low energy intake but other components of the diet may be involved (see Tassell (245) for a recent review).

Of major importance is the plane of nutrition during and just before mating.

Flushing

The ovulation rate is influenced by the state of the ewe and the plane of nutrition: the traditional view has been that the ewe should be in a fit but not fat condition and that the plane of nutrition should be rising before and during mating. Considerable trouble has often been taken to provide an improvement in the food supply to ensure this 'flushing' and very often much effort has gone into a preparatory period during which fat sheep have been kept at a very high stocking rate in order to reduce their condition. Little information has normally been available on the farm as to the weight changes of the ewe, the nutritive value of the feed or how much each ewe has in fact eaten. Great variation usually exists in the condition of individual ewes in the flock and in the length of time for which each ewe has been 'flushed' before mating. Since, in addition, the success or failure of the procedure has had to be judged by the number of lambs born (which, although of obvious importance, is influenced by other factors, abortion, for instance (125)), it is hardly surprising that the situation has not been entirely clear.

In recent years, however, it has been demonstrated, under both British and Australian conditions, that ovulation rates are higher in fat ewes than in thin ones (14, 83, 120, 124). Allen & Lamming (14) found that flushing increased the ovulation rate of ewes in store condition but did not lead to rates higher than in ewes maintained in fat condition. Fat ewes fed submaintenance diets showed no reduction in ovulation rate as long as they still had good body reserves. It is possible that the relationship between fatness and fertility may differ between breeds, particularly between those which characteristically lay down fat reserves in different parts of the body. At present, however, the evidence points to the need for good condition, erring on the fat side, to obtain high ovulation rates.

Flushing is considered not to influence the proportion of the eggs shed which are fertilized. Everitt (168) has considered the role of *components* of body weight, especially fat, in the body weight-fertility relationship. He found that differences in skeletal frame size were unimportant and concluded that differences in amount of soft body tissue must be chiefly concerned.

The most recent evidence (148, 149) suggests that the effects of flushing are twofold:

(*a*) a static effect of liveweight, such that lambing percentage rises by about 5–10 per cent for a 10 lb increase in liveweight;

(*b*) a dynamic effect due to flushing itself (i.e. due to nutritional improvement immediately prior to and during mating), amounting in practice to 6–8 per cent.

Thus about half the total effect of flushing appears to be due to liveweight and, where this is adequate, the response to flushing may be quite small (148).

Age of the ewe

The natural fertility of the ewe tends to increase for the first 2–4 years and remain fairly constant up to about 8 years of age (146) (see also Chapter XI).

The use of hormones

Hormones have been used increasingly in the last ten years. The early work of Robinson (225) showed that pregnant mare's serum (P.M.S.) could be used to increase the ovulation rate. The subject was reviewed by Gordon (178) and more recently by Robinson and co-workers (227) and Laing (188).

Provided that pregnancy nutrition is adequate, and this depends upon the application of nutritional knowledge to accurately diagnosed situations, there is no doubt that litter size can be safely increased in many breeds. The lambs will be smaller at birth but are certainly rearable, artificially if not by the ewe.

Newton & Betts (210), using a range of P.M.S. injections, on the 12th or 13th day of the oestrous cycle during the natural breeding season, found that the mean litter size (M.L.S.) was increased significantly up to 1,250 to 1,500 international units (i.u.) of P.M.S. The optimum dose level for the Scottish Half-bred was about 1,500 i.u. P.M.S. (see Table XIV). They also showed that the ram could significantly increase the mean litter size of superovulated ewes (186). The

TABLE XIV

THE EFFECT OF P.M.S. INJECTION ON MEAN LITTER SIZE (M.L.S.) IN THE SCOTTISH HALF-BRED

Ref.	P.M.S. level (i.u.)	No. of ewes lambing	No. of lambs born	M.L.S.
(a)	0	17	33	1·94
	500	10	20	2·00
	750	11	26	2·36
	1000	12	32	2·67
	1250	10	31	3·10
	1500	10	32	3·20
(b)	1250	31	81	2·61
	1500	27	76	2·81
	2000	27	66	2·44

(Source: Newton & Betts [a] (210), [b] (186))

TABLE XV

NEONATAL LOSS OF LAMBS FROM SUFFOLK × AND SOUTHDOWN × SCOTTISH HALF-BRED EWES

Litter size	No. of lambs born	% Mortality	No. of lambs alive at 1 day per ewe lambing	Mean birth wt. (kg)
1	79	6	0·94	5·70
2	380	8	1·84	4·72
3	276	10	2·69	3·72
4	132	27	2·94	3·18
5	60	13	4·33	2·91
6	12	67	2·00	1·82

(Source: compiled by J. E. Newton from Newton & Betts (210), and Expt. H.857, Grassland Research Institute)

TABLE XVI

RESPONSE TO LEVEL OF P.M.S. IN THREE BREEDS OF SHEEP

Breed	P.M.S. level (I.U.)	No. of ewes lambing	No. of lambs born	Mean litter size
Devon Longwool .	0	18	27	1·45
	500	19	26	1·37
	1000	20	38	1·90
	1500	17	32	1·88
Kerry Hill . .	0	16	28	1·75
	500	15	30	2·00
	1000	15	26	1·73
	1500	18	42	2·33
Welsh Mountain .	0	16	16	1·00
	500	13	18	1·38
	1000	13	30	2·31
	1500	11	22	2·00

(Source: compiled by J. E. Newton from Experiments at Grassland Research Institute)

kind of neonatal mortality experienced with Suffolk- and Southdown-cross Scottish Half-bred lambs is shown in Table XV, together with the mean birth weight of lambs produced.

Early results with other breeds are given in Table XVI.

It is possible that the optimum level of P.M.S. will be related to the body weight of the ewe but at present this value varies from 13 i.u. per kg body weight to 29 i.u. per kg. Natural prolificacy seems to be just as important, however, and the best current guide to obtain maximum response would be to give 1,000 i.u. P.M.S. to breeds of low natural prolificacy (M.L.S. $<$ 1·7) and to give 1,500 i.u. P.M.S. to breeds of high natural prolificacy (M.L.S. $>$ 1·7).

Effect of Litter Size on Lamb Birth-Weight

Assuming nutrition to be adequate, the optimum litter size must certainly be such that the lambs born are viable. Whether it is likely that a whole litter of lambs, from an adequately fed ewe, would be non-viable because of the size of the litter, may be doubted. The most important factor is almost certain to be uniformity. Presumably, there must be a minimum individual size or weight of lamb at birth that can be tolerated and the optimum litter size might represent that at which each lamb just exceeded this weight.

Dickinson and others (266) have discussed the influences on the size of lambs at birth and reported a study involving egg transfer. They found that a ewe was able to respond to progressively greater demands made by lambs of progressively larger, combined, potential size, but in accordance with the law of diminishing returns. They concluded that the upper limit of maternal accommodation must be considerable. Possibilities obviously exist for the transfer of several eggs of a breed with a small birthweight to a recipient ewe of a large breed. Dickinson *et al.* (266) constructed a model from which the mean birthweight of lambs can be predicted. This can be expressed in the formula

$$W = 0\cdot181\, D^{0\cdot83}\left[1 - 10^{-\frac{1\cdot1}{M}\left(\frac{R}{D}\right)^{0\cdot83}}\right]$$

where the lamb's birthweight (W) is envisaged as depending on the mature weight of the donor ewe (D lb) and of the recipient ewe (R lb) and on the multiplicity of birth (M).

Thus, in their own maternal environment, twins are expected to be

78 per cent and triplets 62 per cent of the mean weight of singles. About any predicted mean, however, there remains considerable variation in lamb size.

Effect of diet on reproductive rate

Quite apart from the influence of the level of nutrition (e.g. in 'Flushing'), there are constituents of some pasture plants that have a particular effect on the reproductive capacity of the ewe (246).

Legumes

The most important of these is the group of oestrogenic substances found in some legumes. In Australia, Bennetts, Underwood & Shier (141) first suggested that the observed infertility in sheep was associated with oestrogenic potency in subterranean clover (*Trifolium subterraneum*). Since that time 'clover disease', as the infertility syndrome was called, has been much studied in Australia and a great deal of research carried out. Barrett *et al.* (139) reported that lambing percentages declined progressively in ewes grazing subterranean clover over a period of 5 years and short-term exposure to oestrogens during mating has also been shown to cause reproductive disturbance (203).

Many plants have now been found to contain oestrogenic substances, in many parts of the world (see Bickoff (144) for a recent review).

It is not yet clear what the precise relationship is between the several oestrogenic compounds found in these plants and the reproductive physiology of the animal. It is possible that important changes take place in the rumen and that the compounds in the plant are pro-oestrogens (145).

What is clear is that they may have a deleterious effect on the reproductive rate of the ewe and may cause permanent damage to the reproductive tract.

In Britain, the subject has been little studied so far and there is no strong evidence yet of damaging effects. The only experiment reported did, however, show a reduced twinning rate in ewes grazed on red clover (212).

It is not possible, at the present time, to use a simple chemical analysis to determine whether a legume is likely to affect sheep or not. Nor are bio-assays on mice sufficiently reliable for this purpose. At the moment, tests have to be carried out using sheep and large numbers of ewes would be required over substantial periods of time to verify whether the Australian findings apply in Britain. There is no reason to suppose that they do not, however.

The simplest relevant assay is to allow wethers to graze on the legume to be tested and to measure the length of their teats. The latter lengthen substantially and rapidly if the sward is oestrogenically active. Quite commonly, teat length doubles within two weeks. Several common legumes have been tested in this way by Newton and his colleagues (213) over the last two years (1967 and 1968). Their findings can be summarized as follows:

1. Oestrogenic activity has been detected at significant levels in the following legumes:

Red clover . . .	S.123
	Altaswede
	Broad Red
White clover . . .	S.100
	S.184
Lucerne	Eynesford
	Provence

2. No activity has been detected in sainfoin (Cotswold Common and English Giant).

3. When activity has been detected it has been so during the entire grazing season (April to November for red clover; June to November for white clover and lucerne).

It would be unwise either to overemphasize the importance of these results or to ignore them. White clover is the most widespread of the legumes mentioned but is not usually grown as a pure stand for grazing; not a great deal of red clover is grown in Britain.

It may be a simple matter to breed varieties without these undesirable properties and, in any event, whatever risk there is, applies only to the breeding animal: growth of lambs tends to be better on these legumes than on pure grass swards.

At present, then, it would seem wise to entertain some reservations about using those varieties known to be oestrogenically active for grazing breeding ewes.

Kale

Breeding ewes fed on marrow-stem kale have been shown to suffer from anaemia and loss of appetite and to exhibit reduced duration of oestrus (252).

Food Addit ves and Supplements

This is a convenient point at which to refer to the use of food additives and supplements and the administration of hormones, directed to an increase in rate of growth (268, 277). It is probable that increases in productivity will require a more precise formulation of diets. Where this is not, or cannot be, done, deficiencies of minor constituents may be more likely. More attention may need to be given to mineral supplementation (260), vitamins (265) and, possibly, antibiotics (272, 273, 280). Assuming adequate nutrition, the role of hormones in the acceleration of growth rate may be expected to increase (267). Oestrogens, such as hexoestrol and stilboestrol, have been shown to increase the growth rate of both suckling and weaned lambs (261, 262, 274, 275). Burgess & Lamming (263) found that treatment with stilboestrol resulted in a 33 per cent increase in rate of gain in hoggets, but no significant effect was found in suckling lambs. Shepherd (278) also failed to obtain a response to hexoestrol in suckling lambs. Lamming, Stokes & Horspool (270) considered that the optimum time for implanting hoggets with hexoestrol and diethylstilboestrol was approximately 100 days prior to slaughter. Preston, Greenhalgh & MacLeod (276) have reported on the responses of ram, wether and female lambs to hexoestrol. Implantation with androgens has not influenced growth in female hoggets from 5–7 months of age (279). Whether any form of hormone treatment can be expected to increase the growth rate of lambs born in large litters must depend on their viability and an adequate supply of food. Before considering this further, it would be as well to discuss what developments are possible in the direction of increased litter size and the methods by which they might be achieved.

Gestation

Establishment of foetus

Little is known yet of the effect of such things as fatness and plane of nutrition on the phase during which the fertilized egg establishes itself as a developing foetus. There are indications, notably from work on pigs, that overfatness may be a disadvantage and result in foetal atrophy. This failure to develop could be an important source of loss and might be particularly so where the object was to produce larger litters. Since there is also evidence that poor feeding at this time can lead to foetal loss there would appear to be grounds for the generally held view that body weight should be maintained or slightly increased during early pregnancy, i.e. for the first 3 months (93).

Little is known of the extent to which embryos are lost shortly after fertilization. Edey (163) considered that losses before day 11 are unlikely to be distinguished from fertilization failure. Mattner & Braden (199) have pointed out that ewes that lose *all* embryos and return to service are not necessarily of great significance. The important losses, mainly one embryo out of two, they considered to be substantial, however (up to 20 per cent). Pre-natal losses can be still higher (205) and increase with increasing degrees of heat stress (248).

Pregnancy diagnosis

Many methods have been employed but few have yet reached the stage at which they can make a contribution in practice.

External palpation was used in Germany (224) and internal palpation, using a laparotomy incision, has been used successfully in ewes 4–8 weeks pregnant (189).

The use of a vaginal smear (181) was successful at 40–50 days of gestation in Canada. In the U.S.A., foetal electrocardiograms have been tried (190).

Radiography has been by far the most satisfactory method of diagnosing pregnancy in sheep and of determining the number of foetuses (136, 142, 161, 175, 187, 222, 250, 255). The use of X-rays involves some risks, however, the necessary equipment is both expensive and cumbersome, and diagnosis depends upon the development of foetal bones. This generally means that a higher proportion of correct diagnoses is possible later in pregnancy, depending on the number of foetuses present, the size of the ewe and the nature of its fleece covering.

For research purposes, radiography has proved of tremendous value, allowing ewes to be fed during pregnancy in relation to their foetal burden. At the Grassland Research Institute, I. A. N. Wilson has recorded the success rate over a wide variety of conditions (see Table XVII). For practical purposes, however, the disadvantages, and especially the cost, tend to be prohibitive.

More recently (176, 196, 253), ultrasonics have been used in sheep. The advantages are that the equipment is cheaper, portable and safe, and does not require a skilled operator in the sense that X-radiography

TABLE XVII

RADIOGRAPHIC PREGNANCY AND LITTER-SIZE DIAGNOSIS IN THE EWE AT DIFFERENT STAGES OF GESTATION, 1967–1968

Results are expressed as the ratio: number of correct diagnoses/number of exposures made, under each heading

Stage of Gestation in days	Pregnant	Ewes carrying						Totals at all litter sizes
		Singles	Twins	Triplets	Quads	Quins	>5	
<60	1/1	—	1/1	—	—	—	—	1/1
61–70	3/9	0/2	1/7	—	—	—	—	1/9
71–80	27/31 87·1%	18/20 90·0%	6/10 60·0%	— —	—	—	—	24/30 80·0%
81–90	298/300 99·3%	129/132 97·7%	141/157 89·8%	9/12 75·0%	3/7 42·8%	—	—	282/308 91·5%
91–100	355/357 99·4%	130/138 94·2%	155/165 93·9%	42/48 87·5%	10/10 100%	2/2	0/1	339/364 93·1%
>100	160/162 98·8%	48/49 97·9%	94/96 97·9%	21/25 84·0%	0/1	0/1	0/1	163/173 94·2%
Totals	844/860 98·1%	325/341 95·3%	398/436 91·3%	72/85 84·7%	13/18 72·2%	2/3 66·7%	0/2	810/885 91·5%

Diagnosis of barrenness: 204/207 (98·5%)

(Source: I. A. N. Wilson)

does. At the present time (1968), the accuracy of pregnancy diagnosis can be as high as 100 per cent after the eightieth day of gestation (X-rays giving approximately this result after 90 days) and the operation takes little time (about 1 minute): the determination of foetal number is, however, much less successful by this method.

Duration of Gestation

The duration of gestation in sheep may be influenced by nutrition (1) and by the number of lambs carried, but is normally about 147 days (119). It is well known that the growth of the foetus is greatly accelerated during the second half of pregnancy, but it should not be forgotten that structures such as the placenta, essential to foetal development, are being grown somewhat earlier (167). Although the foetus increases greatly in weight during the last 4 to 6 weeks of pregnancy, a very high proportion of this increase is in water, since water content of the body tends to vary inversely with fat content. It is possible that the size of the foetal burden is of greater consequence than its weight, for it has been suggested that the space occupied by the foetuses can significantly reduce that available for the alimentary tract and its contents. In this way, the very ewes with greatest need could be most limited in their nutrient intake from a bulky feed. The reasons for this may be of metabolic rather than digestive origin, however (138).

The duration of gestation has been found to vary with breed (242) and, although individual variation occurs within a breed, the range is usually within 8–9 days.

Forbes (171) has also reported differences between breeds and an effect of lamb sex: single males were carried for 0·9 day longer than single females. Pregnancy duration decreased with age of ewe up to the fifth parity and then increased. There were significant effects of litter size and the average gestation lengths for singles, twins and triplets were, respectively, 147·3, 146·7 and 145·6 days.

Pregnancy nutrition

The extent to which the ewe requires an increased nutrient intake as pregnancy advances is ill-defined. The need has often to be satisfied against a background of a decreasing food supply and, for this reason alone, supplementary feeding may be required (54, 73, 102, 108, 109). Often, too, the quality of the feed available in late pregnancy is poor and bulky and the greatest need may therefore be for a more concentrated diet.

Inadequate nutrition during the last 6 weeks of pregnancy not only risks the loss of the lambs (78) or a reduction in their size and vitality but risks as well the total loss of the ewe, due to pregnancy toxaemia, also called 'twin-lamb disease' (63, 93, 101).

Unfortunately, adequate nutrition is difficult to specify, even when the number of lambs carried is known. Detailed studies of the nutritional needs of the pregnant ewe have been reported by Reid and his collaborators (102, 103, 104, 105) and by Ford (56); the nutrition of sheep in general has been discussed by Phillipson (95). In addition to the normal energy and other needs of an adult sheep, the pregnant ewe requires particularly calcium, phosphorus, certain proteins, iron, copper and cobalt in order to build the tissues of the developing lambs. A deficiency of some of these elements may be more serious during gestation; with others the consequences are not evident until after lambing. Thus, in areas where lambs suffer from cobalt deficiency, administration of cobalt during pregnancy may improve the growth of lambs during lactation. Similarly, administration of copper during pregnancy can prevent the occurrence of swayback in suckling lambs (12, 19, 33, 34). Ruminants are independent of an exogenous supply of vitamins of the B complex and of vitamin K (28, 31, 98), due to the activities of the microbes in the rumen (45). The rumen flora, however, is also susceptible to changes in the diet (81). Errors in the feeding of sheep have recently been discussed by King (70) and the role of trace elements in sheep nutrition recently reviewed by Wilson (129).

One major problem in achieving adequate nutrition of the ewe in late pregnancy may be simply ensuring that enough is eaten of what is offered. Forbes (170) has demonstrated that the enlargement of the abdomen during pregnancy soon reaches a limit, after which the rumen becomes compressed. He considered that this progressive reduction in the space available for the rumen would result in a decrease in the voluntary intake of fibrous foods and advocated the use of high-quality materials to overcome this.

Silage varies greatly in quality and Forbes, Rees & Boaz (173) have suggested that silage for pregnant ewes should have at least 20 per cent D.M. and 14 per cent crude protein in the D.M., but not more than 32 per cent crude fibre (in the D.M.).

When intake is inadequate, reserves are mobilized (232) and this may lead to pregnancy toxaemia.

Nutrition in relation to large litters

In the past, very little attention was paid to the nutritional status of the ewe during pregnancy and ewes with 4 or more foetuses were often treated similarly to those with one or none. It is hardly surprising that neonatal losses of lambs born were high, mostly in the first three days after birth, and correlated positively with litter size. It is obvious that a valid assessment of the usefulness of larger litters, or any techniques employed to produce them, cannot be made unless the nutritional status of the ewe is adequate. Unfortunately, there is little information on what an adequate level of nutrition is in relation to normal litter sizes, quite apart from foetal burdens in excess of the normal. A good indication, however, can be obtained from the information given in Table XVIII.

TABLE XVIII

FOOD REQUIREMENT OF EWES IN RELATION TO FOETAL NUMBER

Dry-matter intakes are given in g/day

	No. of foetuses	Total lamb birth wt. (kg)	No. of* ewes	Number of days before lambing					Mean
				43–47	36–42	22–35	15–21	8–14	
(a)	1	6·4	2	790	950	1030	1060	1120	990
	2	11·4	6	1120	1240	1380	1540	1750	1410
	3	12·9	5	1120	1360	1510	1700	1850	1510
(b)	1	6·6	4	1370	1360	1360	1380	1490	1390
	2	9·7	6	1560	1570	1650	1820	1970	1710
	3	12·2	9	1610	1740	1920	2100	2320	1940

(Source: unpublished data from two experiments with Scottish Half-bred ewes:
(*a*) (T. T. Treacher, 1969). D.M. required (as mixture of 75% maize/linseed concentrate and 25% ryegrass hay) to prevent use of body fat;
(*b*) (R. V. Large, 1966). D.M. required (as dried lucerne pellets) to gain during pregnancy approximately the same liveweight as is lost at parturition).)

* Mean liveweight of ewes, 47 days before lambing, varied between 71 kg and 79 kg.

Similar arguments apply to the results of lamb growth and mortality after birth in relation to the nutrition of the lamb itself. If a ewe, to everyone's surprise, produces five lambs, it may be doubted whether their chances of survival or their performance can be fairly assessed unless, by chance, the ewe was adequately fed during pregnancy for this extra burden. Even if this was so, however, there is no reason to suppose that her milk output would be adequate for five lambs. Indeed, it is doubtful in many cases, particularly where lambing percentages have been artificially increased, whether it is either satisfactory or relevant to leave a litter of this size on the ewe.

Litter size and lambing frequency

Not a great deal of work has yet been done on frequency of lambing: certainly insufficient data exist to relate it to litter size. Since the relationships between nutrition during pregnancy and litter size are poorly understood, it may be premature to discuss the subject. It is certainly true for smaller litters, however, that the stress on the ewe is far greater during lactation than it is during pregnancy. If it can be assumed that pregnancy, granted adequate nutrition, is unlikely to prove an undue strain to the ewe carrying a large litter, then the frequency of lambing that can be achieved may be independent of the size of the gestation burden. It is, however, unlikely to be independent of the ewe's status during lactation. It might be that once a ewe had reached her maximum milk yield, increases in the number of lambs using it would not affect her. In general, however, it is to be expected that increased litter size would lead not only to maximum milk output but also to the maximum period of lactation. Similar nutritional considerations might apply to lactation, to those discussed in relation to pregnancy. The relationships between nutrition, milk yield and weight loss, however, are not sufficiently understood to predict the outcome with any confidence. Since lambs can be reared independently of the ewe, it is worth discussing frequency of lambing as if litter size did not affect it.

Pregnancy toxaemia

Pregnancy toxaemia is incompletely understood and there is as yet no specific remedy. Prevention should therefore be the aim and this has to be based on empirical evidence. It is generally agreed that stress in late pregnancy due to a low level of nutrition in relation to the burden of lambs carried is the critical factor. Feeding should therefore be related to need. In practice, ewes are similarly fed whether they are carrying one, two or three lambs and, since it is the flock that is fed, individuals may be in stages of pregnancy differing by as much as 6 weeks.

For all ewes, therefore, the daily intake of nutrients should increase during late pregnancy. This involves the provision of food in sufficient quantity and in a form which will satisfy the needs of those ewes whose intakes are limited by, for example, foetal burden or, possibly, internal fat.

It is generally considered that fat ewes are more susceptible to pregnancy toxaemia and that this is one good reason for starting pregnancy in store condition. In many practical situations there are many factors which can come between a ewe and its feed supply in late pregnancy, including bad weather, bad feet (57), mud and insufficient feeding space.

Abortion

Apart from those associated with defective nutrition, other diseases also play their part in influencing the number of lambs born. Chief among these are the various forms of abortion (125). The loss of foetuses may pass unnoticed if it takes place early, since they are then quite small. Ewes may then successfully mate again and the result is later rather than fewer lambs. Later loss, or the birth of dead lambs, may be due to contagious or non-contagious abortion. Productivity is severely reduced by any disease which interferes with reproduction, particularly if the opportunities for breeding are limited to a short period of the year. The prevention of abortion is largely a question of management: the avoidance of rough handling or violent exercise for non-contagious abortion, and hygiene during lambing time for contagious forms. Vaccination can also be used to prevent, for example, enzootic virus abortion.

Birth Weight of Lambs

The significance of the number of lambs born depends upon the size, vitality and growth potential of each one. These features are difficult to describe but, within a breed, they are most readily summarized in the liveweight at birth. This is chiefly influenced by the number of lambs in the litter (see Chapter V, Table XXIV; the larger the litter, the smaller the lambs) and the weight of the dam (46).

A lamb must be sufficiently developed at birth to be able to struggle to its feet and suck: the rapidity with which it can do this in bad weather may make the difference between life and death. Survival, however, is not simply a matter of size or weight. In practice it is usually held that lambs can be too large and heavy, that such lambs are 'lazy' and slow to get on their feet. Recent research has tended to confirm the view that the survival rate is higher for lambs of average birth weight. Purser & Young (100) and Gunn & Robinson (65) found that lamb mortality was greater in those that were markedly heavier or lighter than the average weight at birth. Prud'hon, Denoy & Desvignes (219), in a study of lamb mortality in 'Merino d'arles' ewes, reported losses of more than 50 per cent where the birth weight was less than 2·5 kg. Mortality decreased with increasing birth weight, up to 5·5 kg. Dawes & Parry (154) also reported that survival was extremely poor in lambs with birth weights below about 2·5 kg (2·25 for females and 2·75 for males). Such critical figures must be expected to vary with breed, however.

Vitality at birth is obvious enough when observed but difficult to measure or describe. It also is usually associated with average rather than extreme birth weights. The relationship between weight and surface area could be of major importance in lamb survival immediately after birth, since the latter is much affected by heat loss and the lamb's ability to withstand it. Only recently have precise measurements been made of heat loss at parturition (30) but the need for an animal to maintain body temperature has long been recognized as one of its most fundamental requirements for survival. A valuable series of papers by Alexander and others (5, 7, 11) has stressed the importance of temperature regulation in the new-born lamb and the influence on this of birth-coat (3). The effect of deprivation of milk, even for a short time after parturition, can be very marked on lamb mortality (2, 6, 9, 10) and the onset of lactation can be influenced by the late-pregnancy nutrition of the ewe (77).

An increase in neonatal mortality of lambs has been attributed to a diet (of kale during pregnancy of the ewe) deficient in iodine (119), though other substances may be involved (129).

Probably the greatest influence of the birth weight of the lamb is on the milk production of the ewe: this will be discussed later in this

(a)

Suffolk × Half-bred lambs (age 10 months) showing shedding of the fleece due to poor nutrition combined with worm infestation (see page 75).

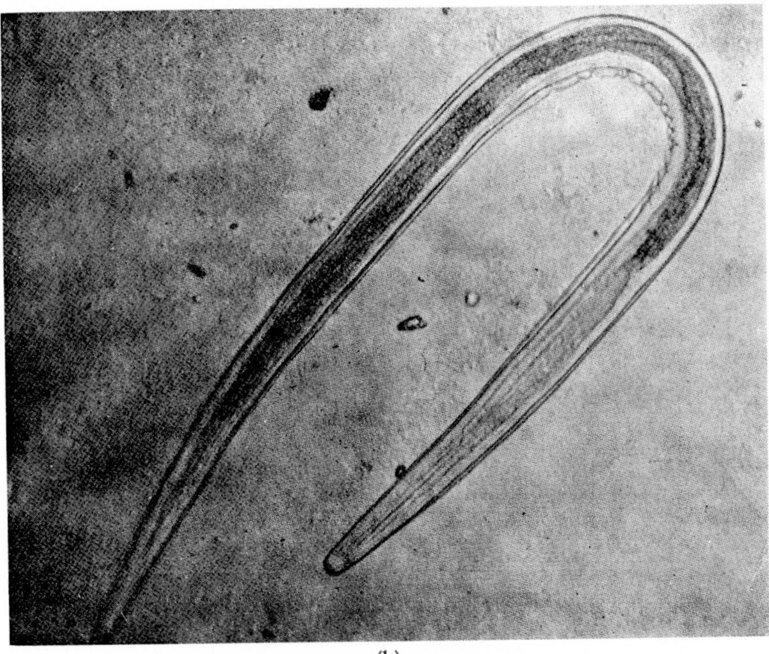

(b)

Infective larva of a small trichostrongyle showing the protective sheath (actual length 800 microns) (see page 117).

Artificially reared 10-day-old Dorset Horn lambs at feeding-time.

chapter. Apart from this, the exact weight, within quite a wide range, appears to have little influence on subsequent performance.

Nutrition during late pregnancy can influence lamb birth weight, although the ewe is capable of using her own fat reserves for foetal growth if current intake is inadequate. Poor feeding often leads to smaller lambs but the size of the ewe appears to have a limiting effect on the upper size of the lamb, however good the food supply. For this reason, lambs born at normal term do not generally prove too large for a ewe to expel without assistance. It is possible, however, for heavy lambs to be too big in certain respects. This is often true of the head, a problem which can be accentuated in horned breeds.

Maternal Care of Lambs

For a considerable time after birth the lamb is dependent on the ewe for milk, without which it would not survive. Quite rapidly, it is able to find its dam and virtually help itself whenever it wishes: this may be quite frequently up to 30 times a day at 2 weeks of age. Very often the ewe appears to call its lambs by bleating but at other times the lambs themselves appear to decide when to suck. Ewes often terminate a suckling period by walking away but rarely refuse to allow suckling at all, unless they are suffering from damaged teats. For most of the time, maternal care may operate simply to enable the lamb to receive its milk at reasonable intervals. In mountain flocks much more may be involved, particularly in guiding the lamb to exploit a difficult or harsh environment. Avoidance of danger, protection from some enemies, warmth at night and shade during the day all come within the sphere of maternal care and occur to a greater or lesser extent in both upland and lowland situations. Recent work by Lindsay & Fletcher (197), suggests that sight is the main sense by which ewes recognize their lambs but that is normally supplemented by other sensory cues as well.

It is during the first few hours, rather than weeks, after birth, however, that maternal care plays its most vital part (27). From the time the lamb is born to the time when it receives its first drink of milk may be regarded as one of the most critical periods in its life and certainly the time when it is most in need of assistance (8). The licking, cleaning and massage carried out by the ewe as soon as the lamb is born plays an essential part in this process. Failure to do this may be due to a long and difficult lambing (4), to interference from other ewes about to lamb or with dead lambs, or to confusion of ownership between two or more ewes.

A small proportion of ewes exhibit aberrant maternal behaviour, varying from erratic, or even deliberate movement of the ewe, preventing suckling, to desertion or butting. Apart from this, most trouble occurs with young ewes breeding in their first or second year. The shock of lambing may be greater on this first occasion and lambs are more likely to be abandoned. The loss of lambs immediately after parturition often passes unnoticed under farming conditions. The total loss of a lamb may mean that a ewe has been fed for a year for no more output than her fleece. It is therefore extremely important to the production process that this loss be kept as small as possible.

LACTATION

The milk yield of the ewe has a dominant influence on lamb growth rate (see Chapter V). An example of a lactation curve is shown in Fig. 4.2. This is a mean for Dorset Horn ewes milked twice daily by machine. In fact, there is great variability between individuals, within breeds, and between breeds for all aspects of lactation. It is hardly possible to construct a 'typical' curve which does not give a greater impression of uniformity than is justified. The onset of lactation can be influenced by nutrition during late pregnancy (77) but milk output normally rises very rapidly from parturition, reaching a peak after 10 to 20 days (37, 72). The first milk or colostrum contains antibodies which have a considerable protective value to the lamb (96) and the chemical composition of the milk (Table XIX) changes quite markedly during the first week or so (88, 94). Fat is an important constituent and the percentage of fat tends to rise as lactation proceeds (see Fig. 4.3). Scales (235) found little variation in milk composition between Romney, Corriedale and Merino ewes, except that the Romneys had a higher fat content during the 6–12 week period of lactation. The peak yield varies with the breed, but little is known of the relationship between total yield, length of lactation and size of ewe. At the peak of lactation, yields from 1·7 (29) to 4·9 lb (90) per ewe per day have been reported. It is noteworthy that the ewe normally has a high proportion of 'cistern' milk in the udder (127), however. Ricordeau & Denamur (106) measured a maximum yield of 1·25 litres from Préalpes ewes 25 days after parturition, but found that milking only obtained 60 per cent to 80 per cent of the milk normally removed by the lamb. Table XX gives the yield over the whole lactation for some lesser known breeds and further details are given in Chapter VI.

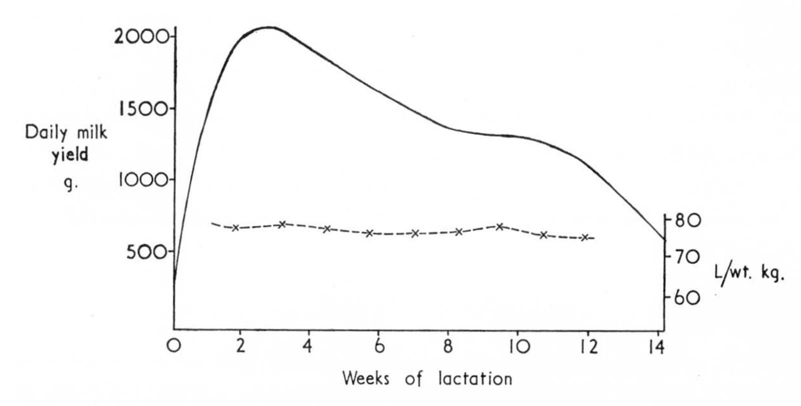

FIG. 4.2—*Lactation curve of Dorset Horn ewes machine-milked twice daily (continuous line) and their mean liveweight (broken line). The milk yields are expressed as a curve of the daily means for each week. Mean daily D.M. intake (of cocksfoot grass nuts) was 2·72 kg per ewe*

(Source : T. T. Treacher, from Expt. H. 367, Grassland Research Institute, Hurley)

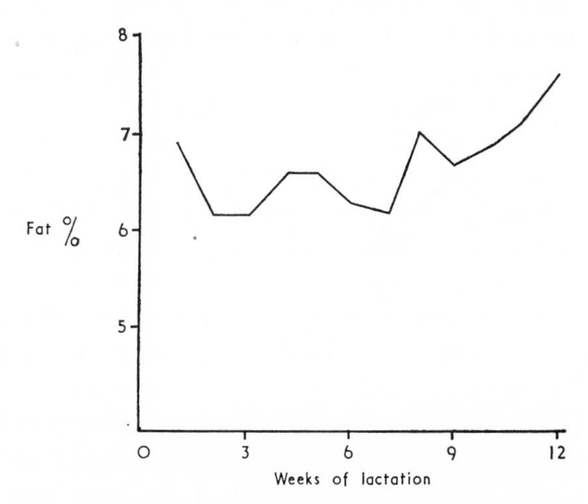

FIG. 4.3—*Changes in the fat content (mean weekly fat percentage) of ewe's milk. Means of the values for the same 4 Dorset Horn ewes as in Fig. 4.2*

TABLE XIX

COMPOSITION OF EWE'S MILK

Mean values from 24 Dorset Horn ewes (machine milked twice daily) for the first 12 weeks of lactation, sampled at morning and evening milking once a week.

	MEAN %	RANGE
Moisture	82·2	80·5– 8·37
Ash	0·825	0·76– 0·88
Fat	6·96	5·80– 8·30
Lactose	4·82	4·58– 5·18
N × 6·38 *	5·18	4·48– 5·88
Total energy†	111·1 kcal/100 g	100·1–122·7

(Source: T. T. Treacher, unpublished data from Experiment H. 367 at the Grassland Research Institute)

* Crude protein.

† Calculated from the equation: $E = 9{\cdot}11\,F + 5{\cdot}54\,(6{\cdot}38 \times T.N.) + 3{\cdot}95\,L$

Where $E \equiv$ Total energy
$F \equiv$ Fat (per cent)
$L \equiv$ Lactose (per cent)
$T.N. \equiv$ Total nitrogen (per cent)

From Perrin, D. (1958). *J. Dairy Res.* 25, 215–220.

TABLE XX

SOME MILK YIELDS REPORTED FOR NON-BRITISH BREEDS

BREED	MILK OUTPUT (kg)	PERIOD OVER WHICH LACTATION WAS MEASURED (DAYS)
Awassi	100–150	160
East Friesian	500–700	240–260
Ossimi	48	84
Préalpes du Sud	123	175
Rahmani	53	84
Sakiz	200	210

(Compiled from the literature (58, 71, 106, 113, 115): for a recent review of milk yields in a wide range of breeds, see J. G. Boyazoglu (1963), *Ann. Zootech.* 12 (4), 237–296)

Factors Influencing Milk Yield

As has been mentioned, the number of lambs suckled can influence milk yield. Thus a ewe will give more milk if suckling twins than if suckling only a single lamb, independently of the number of lambs that she actually produced at birth (47). The number of foetuses carried during gestation may also have some effect by interacting with the plane of nutrition, and it has been suggested that the level of feeding during pregnancy may affect the development of udder tissue. It has also been suggested that nitrogen intake may limit milk yield (179).

By far the most important influence on milk yield, however, has been considered to be the level of feeding during lactation (20, 38, 84) and lactating ewes may voluntarily consume 26 per cent more forage, under grazing conditions, than dry ewes (36). It has been pointed out at the beginning of this chapter that ewes commonly lose weight during the first 4–6 weeks of lactation. Their ability to do so without a reduction in milk yield must be related to their weight and condition immediately after lambing and this in turn may reflect nutrition during pregnancy. Little is known of these relationships between milk output, nutrition, body weight and weight loss, but the relationship between milk yield and body-weight change is likely to depend upon whether yield is limited genetically, nutritionally or by the amount being withdrawn, e.g. by the number of lambs suckled.

Measurement of milk output

In discussing the influence of nutrition during lactation it is important to bear in mind the methods available for measuring milk output (47, 50, 76, 85, 90, 97, 106). Machine milking is common in some countries (France and Israel are among them) where cheese is often produced, and is used in others for experimental purposes. This provides a standard method of milk withdrawal, but it can never be certain that suckling lambs would have obtained similar amounts. Hand-milking suffers from the same disadvantage and may be a less uniform procedure. Covering the udder for a period and then weighing lambs before and after suckling has proved a satisfactory method of assessing how much milk such lambs have obtained but the procedure cannot be repeated too often without risk of interference with the very situation that is being assessed. Many workers have found a correlation between milk output and lamb growth over the first 4 or 6 weeks: the weight gained by lambs during this period has therefore been used as an index of milk yield (44, 47, 123). As measured in these various ways it is clear that poor feeding during lactation may reduce output, in spite of the ewe's capacity for mobilizing reserves (216).

Water intake and D.M. content of the food

The great need of the lactating ewe for water should not be forgotten: under certain environmental conditions, water intake can be affected by temperature (18). Water consumption by ewes depends also on the nature of the diet and the quantity eaten (see Table XXI).

Little is yet known about the effects of food quality on milk quantity or quality, particularly in relation to grazing conditions. For ruminants in general, heat production during lactation may be 100 per cent greater than the non-lactating level (31). This implies a large increase in nutritional need and this must require an increase in intake, which may only be possible on foods of high quality and, perhaps, high dry-matter content.

A recent study by Forbes (172) suggests that in early lactation (the first 4 weeks) ewes may require more water than might be expected in relation to dry-matter intake (D.M.I.) and milk yield. The A.R.C. Summary (135) suggested that lactating ewes required 50 per cent more water than dry ewes.

During pregnancy, the water requirement varied with foetal number, especially from the 14th week of gestation onwards (172). The total water intakes of pregnant Scottish Half-bred ewes in late pregnancy were 3·1, 3·7 and 4·6 kg per kg of D.M.I., for ewes with singles, twins and triplets, respectively.

In general, water intake was related to D.M.I. In pregnant ewes carrying twins, the total water intake, in kg per kg of D.M.I., rose from about 2·0 in early pregnancy to nearly 6·0 close to lambing. Ewes carrying singles rose to a figure of approximately 3·5 kg per kg D.M.I.

Diseases and Disorders

As in all productive processes, disease can greatly reduce efficiency. This, of course, applies to all kinds of disease; parasitic worms, for example, have been shown to be capable of reducing milk yield (64).

Of particular concern here, however, are those metabolic disorders which are closely associated with lactation and, in some sense, caused by it.

The two most serious are hypomagnesaemia, characterized by low blood magnesium, and hypocalcaemia, low blood calcium. Both can occur at other times (55), late pregnancy for instance, but are most common at about the peak of lactation. They occur in all kinds of sheep but are frequently associated with mountain breeds moved to

TABLE XXI

DAILY INTAKE OF WATER AND FOOD

Groups	Pregnancy		Lactation	
	Water intake	*Food intake*	*Water intake*	*Food intake**
1	7·3	2·40	9·4	2·65
2	7·4	2·40	9·0	2·46
3	7·7	2·68	7·7	2·06
4	7·8	2·64	6·6	1·54

Mean daily intake of water (litres) and cocksfoot nuts (kg D.M.) by 4 groups of 6 Dorset Horn ewes during the last 4 weeks of pregnancy and the first 12 weeks of lactation (with twice-daily machine milking).

(Source : T. T. Treacher, unpublished data from Experiment H. 367 at the Grassland Research Institute)

* Intake of the nuts (D.M. content = 89 per cent) was controlled at four levels during lactation.

more productive, lowland-type pasture (15, 110). Considerable individual variability exists (68, 107) and considerable variation occurs in the values for serum magnesium that appear normal for different individuals kept under similar conditions. Death often results but can be prevented by injection of magnesium or calcium compounds, as appropriate, provided that the characteristic symptoms (126) are observed in time. Onset of visible symptoms of hypomagnesaemia, however, and indeed the onset of the commonly observed form of the disease (tetany), may only occur following sudden stress. Fright, due to handling or the presence of a dog, cold winds and movement (by lorry, for example) have been found to be responsible for the onset of a clinical condition (111).

Close observation and careful handling are part of shepherding skill but the aim must be to prevent the animal reaching a point where it is actually deficient in something.

There is no doubt that at the peak of lactation current intake of nutrients may be inadequate to supply all the needs of the ewe. The weight losses which can occur at this time demonstrate that this may be true for gross fractions, such as energy. If body reserves cannot supply the need for a particular substance, such as magnesium, or cannot make good the difference between intake and output, a deficiency must arise. This kind of situation is greatly exacerbated if, as may happen, intake is reduced (91) or the required substance is present in the feed in a non-available form. Field, McCallum & Butler (53) calculated tentatively the amount of magnesium, contained in herbage, required by two Cheviot wethers, as 800 and 950 mg/day for maintenance. The efficiencies with which these sheep utilized the magnesium present in the herbage were 13 per cent and 26 per cent respectively. More recently, Field (52) reported estimates of the percentage absorption of the magnesium in grass nuts, from 3·2 to 11·3 per cent, and an estimate of 12 per cent for the availability of the magnesium in grass nuts to all the sheep used. For the grazing sheep the supply of magnesium can be improved by the application of magnesium-rich fertilizers to the pasture or by the incorporation of calcined magnesite in supplementary feeds (13, 66, 79). Mineral licks may provide the necessary elements but depend for their effectiveness upon the use made of them. Pellets containing water-soluble magnesium have also been used, in the rumen, to prevent hypomagnesaemia (130). Clovers and other pasture constituents of high mineral content are not always available in quantity when the ewe is at the peak of lactation.

These metabolic disorders raise in acute form the question whether the aim should be to make pasture a complete feed at all times for all classes of ruminant simply because it is nearly so to start with (see Chapter II). Conversely, increased production of herbage (131), particularly as an extension of the normal grazing season, may lead to adverse effects on animal health precisely because the herbage grown may be deficient in some constituent. The desirability of maintaining animal health may not be a good reason for keeping pasture output at a low level. The desirability of increasing pasture output, however, is not a good reason for ignoring the possibility that adverse effects on animal health may result. On the contrary, it is an argument for more thorough investigation of the full consequences of any change in the pattern of feed supply.

Although disease is a cause of gross inefficiency in animal production the justification for large-scale measures of prevention must take into account the magnitude of the problem. It is quite noticeable for many of the disorders mentioned that, however serious they may be for the individual, only a small proportion of the flock is normally affected. Furthermore, it is sometimes the same few individuals that are affected on successive occasions, even year after year. Selection with health and disease in mind could have a big effect wherever preventive measures have to be taken on a flock scale in the interests of a few susceptible animals.

Longevity

Productivity over a short period of an animal's life has always to be balanced against the length of the productive life. In the breeding ewe, there is a considerable investment of food and other resources before reproduction begins to add its return to the wool production obtained all the time. It would be misleading to regard the first year or two as even relatively unproductive; but rather, production at any subsequent time should be viewed in relation to the biological cost of the breeding animal over its life up to that point. Little is known for sheep about the relationship between peak of milk yield and length of lactation, between rate of early growth and longevity, or between litter size and total number of litters produced.

Longevity, in this context, is not simply a question of the rate at which a ewe ages, whether this is judged physiologically or physically, in relation to the number of teeth still present for example (24), but also involves susceptibility to disease.

Since the object of considering longevity is to improve the assessment of productive efficiency, food intake must also be considered. The same rate of production per unit time might require a longer life in a heavier animal in order to obtain the same life-time efficiency of food conversion as in a lighter ewe. This will be further considered in Chapter VI.

EFFECT OF THE EWE ON THE PASTURE SUPPLY TO THE LAMB

The effect of the ewe on the pasture she grazes is one of the indirect effects she has on the lamb. The interaction of sheep and pasture is fully discussed in Chapter VII: here we are concerned simply with the effect on the lamb of the ewes grazing with it. The general level of stocking in relation to the herbage available, i.e. grazing pressure, affects ewes and lambs somewhat differently, because their needs are different. They not only need different *quantities* of food, because of their difference in size, but their need may differ in the proportion of their maximum voluntary intake which must be supplied at any one time. The ewe at a very high grazing pressure may receive less food than she requires for maintenance. The ewe therefore loses weight, but this may not matter unless it occurs during lactation or late pregnancy. The lamb, however, requires more than a maintenance diet, not only for production but for survival. This is another way of saying that a lamb cannot simply be 'maintained', it will continue to grow its skeleton even while it is losing fat or muscle. There is therefore an important difference between the ewe, with fat reserves that can be safely depleted and replaced at some later time, and the lamb, whose requirements for growth must be continuously satisfied if a satisfactory product is to be obtained. In a sense, then, the stocking rate can be too high for lambs even when it is not so for ewes grazing with them. In such circumstances lambs are in competition with the ewes for the available food and are bound to be adversely affected by this. Creep-grazing managements have as part of their object the elimination of this competition at high stocking rates (see Chapter IX).

The effect of ewe grazing on the pasture may also reduce the amount of herbage available to the lamb by decreasing the quantity grown. Such influences on pasture growth may be short or long term and may be due to under-stocking or over-stocking. Over-stocking may reduce herbage production quite quickly; under-stocking generally takes longer to exert any effect since, by definition, stock are in a position of surplus for some time. Under-stocking is much more likely, however, to lead to a decline in the quality of herbage available. This may be unimportant whilst enough material of high nutritive value can be selected, but as soon as animals are forced to consume the mature parts of the sward, the lamb is placed at a disadvantage.

How much effect the presence of ewes has by soiling, i.e. contamination of pasture with faeces and urine, is not known. At high stocking rates the quantity of herbage contaminated must be considerable and it is conceivable that this might affect the lamb's food intake.

Long-term influences on botanical composition or sward structure are not specifically the result of ewe grazing. Alterations in the botanical composition of the herbage on offer, however, can be a result of highly selective grazing by the ewe. In a remarkably short time, for example, the clover component can be selectively removed from a grass/clover mixture. This kind of selection can have extremely beneficial effects as, for example, in the control of broad-leaved weeds, such as thistles and docks, by heavy grazing of ewes in early spring. Even quite tough and unpalatable plants may be eaten out if the stocking rate is high and the plant is never given the opportunity to mature.

EFFECT OF THE EWE ON THE POPULATION OF PARASITES

Only those parasites which are of importance, either in the growth of the lamb or to the performance of the ewe, and on which the ewe has a significant influence will be considered here. There are, however, two ways in which the ewe exerts such influence. First, grazing may modify the height and density of the sward and, in this way, affect both the survival of parasitic stages and the lamb's intake of them. Secondly, the ewe may directly influence the parasite population by depositing faeces containing their eggs or larvae, or by ingesting parasites at the infective stages with the herbage on which they are found. These effects are of the greatest importance where numbers of parasites matter most. To some extent, the number of parasites is always important but the number ingested may not be. Thus ingestion of bacteria, which can multiply within the body, may result in a population the size of which bears no relation to the number ingested. In these circumstances the ewe may act as a carrier of infection for the lamb, but may have no great influence on the degree of infection which results.

With parasites such as the nematodes, which reproduce but do not multiply within their host, the number ingested has a far greater significance. Although the adults present in the sheep must have been individually taken in with food or water, and the number of adults cannot exceed the number ingested, there is not necessarily a good correlation between them. This is because sheep become resistant, by virtue of age (32) or previous experience of the parasite, and one manifestation of this is a reduction in the proportion of those ingested that finally establish themselves as adults in the body. The development of immunity is clearly of immense importance to the efficient combination of animal and pasture in the presence of parasites (64). This will be further discussed in Chapter IX. Even though there is no simple relationship between the damage due to parasites and the numbers on the herbage, nevertheless herbage infestation is of enormous importance (40, 80). This has been discussed in detail by Michel & Ollerenshaw (202) and by Gibson (177). Michel has found that, in general, the number of adult nematode parasites in a host animal is not determined by a steady accumulation process but is related to the rate of intake of infective larvae. The interactions with stocking rate, management and the amount of herbage grown are complex and control measures are best considered in relation to whole grazing systems.

All this is best illustrated by the parasitic worms, or helminths. These include the flatworms, i.e. flukes and tapeworms, and the roundworms, or nematodes; between them these account for the major parasites of lambs. The life cycles of these worms are often complicated and sometimes require intermediate hosts. Typical examples are shown in Figs. 4.4–4.7 Such life cycles, of course, grossly oversimplify the ecology of both host and parasite but must be borne in mind when considering the effect on the ewe. Clearly, for example, where there is no intermediate host in which the parasites may further multiply, the number of eggs or larvae deposited by the ewe is of greater significance.

Nematode Egg Output of Ewes

Much attention has therefore been given to the number of nematode eggs per unit of faeces passed by the ewe at different times of the year

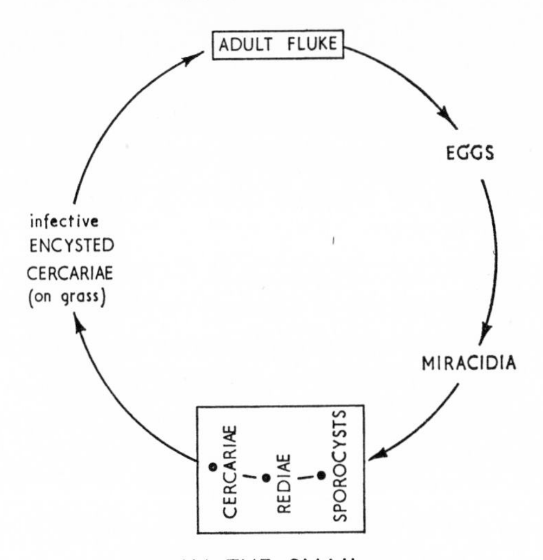

FIG. 4.4—*Life cycle of a liver fluke* (Fasciola hepatica)

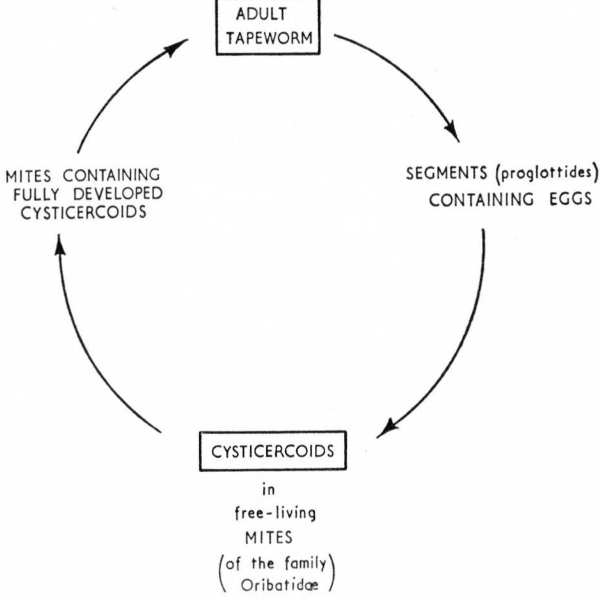

FIG. 4.5—*Life cycle of a tapeworm* (Moniezia expansa)

(usually expressed in eggs per gramme and abbreviated to e.p.g.). Fig. 4.8 shows a typical curve of egg output (in e.p.g.) during the year for a mature ewe. Enormous variations have been found in the faecal egg output of ewes within and between days (117), between sheep and between years, and from one environment to another.

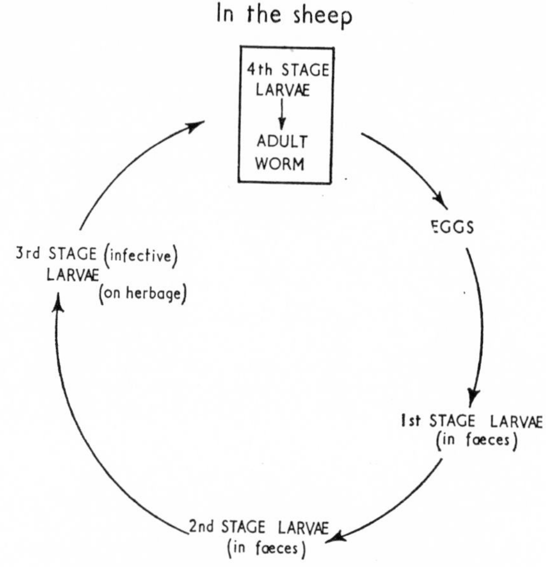

FIG. 4.6—*Life cycle of* Trichostrongylus axei

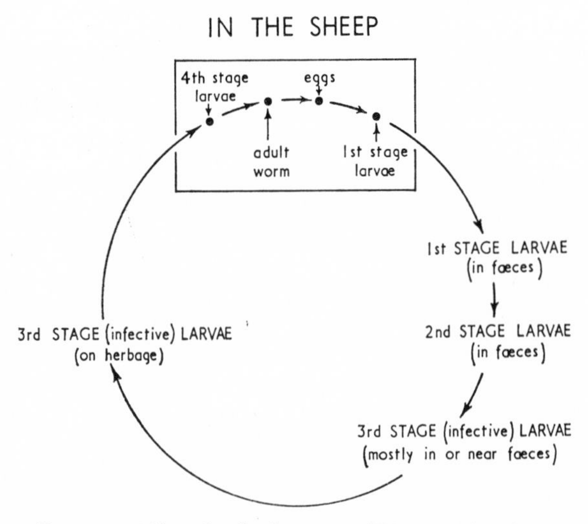

FIG. 4.7—*Life cycle of a lungworm* (Dictyocaulus filaria)

Egg counts must therefore be interpreted with caution. They are usually a better guide to the rate at which a pasture is being infected, however, than they are of the number of parasites in the animal. The level of nutrition has often been thought to influence the number of eggs in the faeces. There is no doubt that nutrition can have an effect but there is also evidence from experiments where no difference was found in the egg counts of ewes maintained at stocking rates differing by 50 per cent, where the level of nutrition was markedly lower at

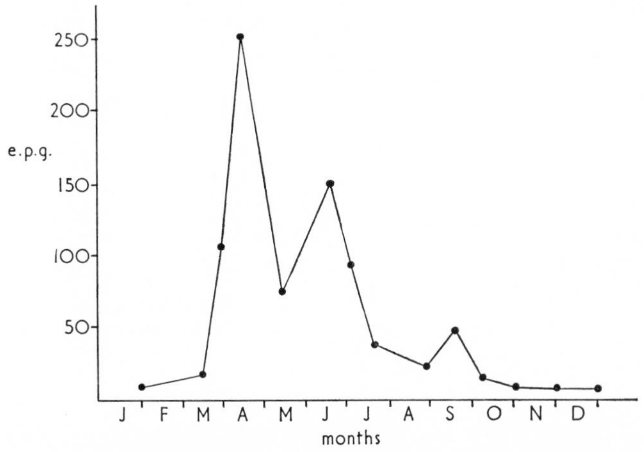

FIG. 4.8—*Seasonal variation in the egg-output of a ewe, expressed as the number of trichostrongyle eggs per gramme of faeces (e.p.g.)*
(Source : C. R. W. Spedding (1956), *Outlook on Agriculture*, 1 (3), 101)

the higher stocking rate. There remains much to be learned about the relationship between the faecal egg count and such things as the ewe's milk yield, weight and food intake.

Two features of the curve in Fig. 4.8 are important: the generally lower level over most of the year and the peak output or 'spring rise' (42) during the period immediately after lambing. The coincidence of this output with the peak of lactation is less often emphasized than the fact that it follows lambing (92). The reasons for the 'spring rise' have been discussed by Soulsby (116) and recently Connan (147) has reported studies in which reinfection contributed to the egg-output and in which a close association was shown with lactation. Crofton (42) found that an 'autumn rise' was associated with lambing in Dorset Horn ewes from October to December and Downey (160) has recorded a post-parturient rise following hormone-induced ovulation. Dunsmore

(162) considered that the 'stress' of parturition was unlikely to be the cause of the sudden burst of egg-producing activity in *Ostertagia* spp. The consensus of opinion is that the 'spring rise' is chiefly the result of the full development of numerous larvae which, until this time, were 'inhibited', lying dormant in the gut wall of the host. The magnitude of the 'spring rise' may thus be influenced by events in the winter and autumn, when those larvae may have been ingested. (For a recent discussion of this topic, see Soulsby (116).) It can also be influenced by anthelmintic treatment, such as low-level, daily feeding of powdered phenothiazine (59, 74). It should be remembered that the number of eggs per gramme imperfectly reflects the number of eggs deposited daily since this is also determined by the faecal output. The latter varies chiefly with intake, although obviously affected by the digestibility of the feed, and intake is often greatly increased during lactation. The number of eggs deposited by the ewe during the 'spring rise' is normally, therefore, enormously above the level characteristic of the rest of the year. Naturally, the level of egg output by the ewe can be influenced by anthelmintic treatment (60, 61): the latter has recently been reviewed by Gibson (62). The role of anthelmintics in delaying heavy infestation (41) until lambs have been fattened, is particularly worth noting.

Significance of the ewe's egg output

The significance of the ewe's contribution to the pasture infection (39) depends primarily on two things: (*a*) whether the lamb is going to be exposed to it and (*b*) whether it is all that the lamb is exposed to. The first question is partly a matter of management and weather conditions (see Chapter IX), and partly a matter of the source of the lamb's diet (see Chapter V). If the lamb is never going to meet the consequences of the 'spring rise', the size of it is of little moment. The second question simply draws a distinction between infection derived from the ewe whilst it is grazing with the lamb, in the same grazing season, and that derived from previous grazings, from sheep or cattle during the preceding year. Where there are heavy infestations of *Nematodirus* spp., for example, the eggs deposited by the ewe are generally of little significance. This is a particularly good example of a nematode that is chiefly of importance to the lamb and where the infestation of one lamb crop is, in the main, derived from eggs deposited by the previous lamb crop (see Chapter XI).

By contrast, where the pasture has not been grazed during the previous year, or grazed only by adult cattle, or where the pasture is a newly sown ley, the initial infection of the lamb must be derived from the ewe. In these circumstances a reduction in the size of the 'spring rise' may reduce significantly the pasture infection to which lambs are exposed.

Survival of Parasites on the Pasture

Between the time when eggs reach the ground, in the faeces, and when infective larvae (Plate III (b)) are ingested, on the grass, there is a variable period during which there are large losses by death of eggs and larvae. The survival of larvae and the magnitude of these losses depend on the interactions of temperature and humidity, the length of time, the amount of wind and sunshine, the management of the pasture and the activities of other populations, including earthworms and various insects. The ewe can modify the effect of nearly all these factors to a greater or lesser extent. Even influences which superficially appear trivial may be important in certain circumstances. An example of this is the possible effect of the ewe lying down at night on the temperature of faeces on the ground underneath her. In the early spring, when ewes may remain relatively undisturbed with their very young lambs and when night temperatures could be a factor limiting development of larvae, this could influence the parasite population. Whether this influence was large enough to matter might depend upon the stocking density and where the ewes spent the night, in relation to faecal concentration on the ground and subsequent grazing behaviour. It is difficult to steer a middle course between exaggerated complications and the too easy dismissal of possibilities which may be more important than is obvious or become so in changed circumstances. It will be stressed in subsequent chapters how great these changes may be in the way that sheep are kept in the future.

Effect of climate

The climatic factors concerned here are those pertaining to the microclimate where the parasites are: the position of the latter is influenced by the activities of the larvae themselves and by the pasture structure (139). The height and density of the sward are frequently regarded as determinants of larval survival. Clearly they can be, but pastures are themselves dynamic populations and do not remain static, even structurally, for any length of time. It is extremely difficult to dissociate, for example, the effect of shortness of grass from the effects of whatever means are employed to keep it short. Ewes do not have many specific effects on pasture structure. The most likely would be shortness combined with density: a high proportion of many-tillered grass plants with many of these tillers in a prostrate condition. This kind of effect might occur only with heavy stocking of sheep, such as could be practised with ewes at a non-productive time of year or in systems involving creep-grazing.

Certainly low-density plant populations which are cut short will expose larvae and eggs to climatic effects from which denser or taller pasture might protect them. If the climate is hot and dry, and so adverse to parasites, sward structure could probably have a very big influence on larval numbers. On the other hand, on short swards there is probably a greater risk of heavy infestation per unit of herbage in warm, humid weather. A large quantity of grass per unit area may have the effect of diluting the infestation and reducing the number of larvae per unit weight of grass. Larval populations, however, are not uniformly distributed (the faeces from which they are derived are, after all, not uniformly deposited either and infective larvae do not travel far (see Chapter IX)) and grazing may not be uniform either. This is particularly the case where large acreages are involved, or where the grazing pressure is low, since there is then a surplus of herbage. On the hills, for example, the stocking rate is an average that obscures the contrast between neglected areas and those on which sheep concentrate. This is also true of arid parts of the world where water-holes and riversides may offer more attractive vegetation as well as the water itself.

Effect of Grazing by Ewes

Finally, ewes may destroy infective larvae in the sense that only a small proportion of those ingested establish themselves in the alimentary tract of a mature, resistant sheep. Older sheep have sometimes been thought of as 'vacuum cleaners' because they ingest greater numbers of larvae than they contribute to the grazing system. This concept of a 'favourable exchange rate' is not as simple as it may at first appear.

It implies that more larvae are ingested daily than result from the eggs deposited daily (118).

If this were true without further qualification ewes would, of course, eventually become worm free. A clear understanding of these relationships is essential if the problems implicit in the possibilities of separation of ewes and lambs are to be resolved. It is therefore worth considering this question in some detail. The first qualification must concern the time interval between the numbers being related. An environment, or system of management, that prevented or favoured the development of larvae derived from the ewe's faecal egg output could alter the total result in either direction. Yet the concept must be considered as a dynamic one, involving the passage of time, since the removal by the ewe of grass and larvae cannot alter either the infection on the rest of the sward or the parasite intake of the lamb at that time.

The second qualification concerns the obvious fact that, even if, for the purpose of discussion, larval development is regarded as uniform, the 'exchange rate' can only be 'favourable' if the number of larvae on the ingested herbage is relatively high or the number of eggs deposited is relatively low. Frequently, neither is the case. Indeed, in the early spring, faecal egg output may be at its peak and pasture infestation at its lowest. Later in the season, when the ewe's egg output is low, the higher pasture infestation may be chiefly due, directly or indirectly, to the ewe's spring output.

Furthermore, even when pasture infestation is high and egg output low, the lamb will only benefit if the resulting lower infective population per acre leads to a lower number of infective larvae per unit of herbage. This must be greatly influenced by the rate of grass growth and the rate at which it is removed. For the lamb to be exposed to fewer larvae, the presence of the ewe must result in a reduction of the number of larvae per unit of herbage *eaten by the lamb*. This is a complex situation and the several effects of the ewe's presence are not easily separated. Absence of the ewe quickly demonstrates one major effect; the grass growth has to be controlled. If this is done by increasing the stocking rate of lambs more parasites could, in certain circumstances, result. It is probably reasonable to regard the ewe's role in parasite control as that of enabling a low stocking rate of lambs to be carried, without adding appreciably to the infection, and whilst controlling pasture growth.

Of course, this can be done in other ways. Grazing by other classes of stock, cattle for instance, may have the same effect. Harvesting grass by machine is considerably more efficient from this point of view, since it removes the larvae on the harvested crop without adding anything to the infection on the remainder. It is doubtful, therefore, if this 'decontamination' can be regarded as a reason for having the ewe there but rather as a possibly beneficial aspect of her presence. The preceding discussion suffers from lack of information on many important points. It also ignores the fact that management can be directed towards employing the ewe in a particular way. This has been done successfully in systems of creep-grazing. The idea is simply to use the ewe to defoliate, severely and frequently, the area receiving or carrying the heaviest infestation and to allow the lamb to graze where this is least. Differential treatment of ewes and lambs (see Chapter IX) can safely exploit the ewe's ability to destroy larvae and to keep pasture very short because the lamb is protected from the consequent risks. All this is particularly relevant to the nematodes inhabiting the alimentary tract, nearly all of which have direct life cycles (see Table XXVII, Chapter V), and to some of the lungworms.

Influence on other parasites

Similar considerations apply to the ewe's influence on the survival of tapeworm and fluke stages, except that the relevant interactions include the intermediate hosts. The ewe may still affect the number of parasites available to the lamb by virtue of the eggs she contributes and by her influence on the pasture. In addition, the ewe may affect, directly or indirectly, the population of intermediate hosts. By and large, however, the influence of the ewe is much less in these cases and the influence of management potentially greater. The control of fascioliasis, for example, depends as much on the use of molluscicides for pasture treatment, drainage of land and dosing of stock, as on grazing management of the ewe. This subject has been discussed by Ollerenshaw (89), in relation to the forecasting of the expected incidence of heavy fluke infestation and the associated Black disease (86, 128). The practical value of acting on such forecasts has recently been described by Edwards (166) and the possibility of extending the approach to forecasting the incidence of nematodiriasis has been discussed by Ollerenshaw & Smith (214).

REFERENCES

(1) ALEXANDER, G. (1956 a). The influence of nutrition upon duration of gestation in sheep. *Nature (Lond.) 178*, 1058–59.

(2) ALEXANDER, G. (1956 b). Changes in weight of lambs during the first thirty-six hours of life. *Aust. vet. J. 32* (12), 321–4.

(3) ALEXANDER, G. (1958). Heat production of new-born lambs in relation to type of birth coat. *Proc. Aust. Soc. anim. Prod. 2nd Bienn. Conf. 2*, 10–14.

(4) ALEXANDER, G. (1960). Maternal behaviour in the Merino ewe. *Proc. Aust. Soc. anim. Prod. 3*, 105–14.

(5) ALEXANDER, G. (1961). Temperature regulation in the new-born lamb. III. *Aust. J. agric. Res. 12* (6), 1152–74.

(6) ALEXANDER, G. (1962). Energy metabolism in the starved new-born lamb. *Aust. J. agric. Res. 13* (1), 144–64.

(7) ALEXANDER, G. & MCCANCE, I. (1958). Temperature regulation in the new-born lamb. I. *Aust. J. agric. Res. 9* (3), 339–47.

(8) ALEXANDER, G., MCCANCE, I. & WATSON, R. H. (1956). The relation of maternal nutrition to neonatal mortality in Merino lambs. *Proc. III int. Congr. Anim. Reprod. Sect.* I, 5–7.

(9) ALEXANDER, G. & PETERSON, J. E. (1961). Neonatal mortality in lambs. *Aust. vet. J. 37* (10), 371–81.

(10) ALEXANDER, G., PETERSON, J. E. & WATSON, R. H. (1959). Neonatal mortality in lambs: intensive observations during lambing in a Corriedale flock with a history of high lamb mortality. *Aust. vet. J. 35* (11), 433–41.

(11) ALEXANDER, G. & WILLIAMS, D. (1962). Temperature regulation in the new-born lamb. VI. *Aust. J. agric. Res. 13* (1), 122–43.

(12) ALLCROFT, R. (1957). Treatment of copper deficiency in cattle and sheep by intramuscular injection of copper glycine (copper amino-acetate). *Vet. Rec. 69* (33), 785.

(13) ALLCROFT, R. (1960). Prevention of hypomagnesaemia. *Proc. B.V.A. Conf. on Hypomagnesæmia*, p. 102, London.

(14) ALLEN, D. M. & LAMMING, G. E. (1961). Nutrition and reproduction in the ewe. *J. agric. Sci. 56*, 69–79.

(15) ANNISON, E. F., LEWIS, D. & LINDSAY, D. B. (1959). The metabolic changes which occur in sheep transferred to lush spring grass. I and II. *J. agric. Sci. 53* (1), 34–41; 42–5.

(16) AUSTWICK, P. K. C. (1960). Mycotic dermatitis and the down-grading of wool. *N.S.B.A. Yearbk.*, 1–4.

(17) AUSTWICK, P. K. C. & DAVIES, E. T. (1958). Mycotic dermatitis in Great Britain, 1954–58. *Vet. Rec. 70* (49), 1081–86.

(18) BAILEY, C. B., HIRONAKA, R. & SLEN, S. B. (1962). Effects of the temperature of the environment and the drinking water on the body temperature and water consumption of sheep. *Can. J. Anim. Sci. 42*, 1–8.

(19) BARLOW, R. M., PURVES, D., BUTLER, E. J. & MACINTYRE, I. J. (1960). Sway-back in south-east Scotland. I. *J. comp. Path. 70* (4), 396–410.

(20) BARNICOAT, C. R., LOGAN, A. G. & GRANT, A. I. (1949). Milk secretion studies with New Zealand Romney ewes. *J. agric. Sci. 39*, 44, 237.

(21) BARRETT, J. F., REARDON, T. F. & LAMBOURNE, L. J. (1962). Seasonal variation in reproductive performance of Merino ewes in northern New South Wales. *Aust. J. exp. Agric. Anim. Husb. 2*, 69–74.

(22) BENNETT, J. W., HUTCHINSON, J. C. D. & WODZICKA-TOMASZEWSKA, M. (1962 a). Climate and wool growth. *Proc. Aust. Soc. Anim. Prod. 4*, 32–3.

(23) BENNETT, J. W., HUTCHINSON, J. C. D. & WODZICKA-TOMASZEWSKA, M. (1962 b). Annual rhythm of wool growth. *Nature (Lond.) 194* (4829), 651–2.

(24) BENZIE, D. & CRESSWELL, E. (1962). Studies of the dentition of sheep. IV. *Res. vet. Sci. 3* (4), 416–28.

(25) BICHARD, M. & YALCIN, B. C. (1963). Personal communication.

(26) BLAXTER, K. L. (1958). Nutrition and climatic stress in farm animals. *Proc. Nutr. Soc. 17*, 191–7.

(27) BLAXTER, K. L. (1961). Lactation and the Growth of the Young. Ch. 19 in *Milk*, Ch. 9 in *Digestive Physiology and Nutrition of the Ruminant*. Ed. Lewis, D.

(28) BLAXTER, K. L. (1962). *The Energy Metabolism of Ruminants*. London, Hutchinson.

(29) BONSMA, F. N. (1939). Factors influencing the growth and development of lambs, with special reference to cross-breeding of Merino sheep for fat lamb production. *Univ. Pretoria Agric.*, No. 48. South Africa.

(30) BROCKWAY, J. M. & PULLAR, J. D. (1961). Heat loss at parturition in ewes. *J. Physiol. (Lond.) 159*, 10–11.

(31) BRODY, S. (1945). *Bioenergetics and Growth*. New York, Reinhold Pub. Co.

(32) BRUNSDON, R. V. (1962). Age of resistance of sheep to infestation with the nematodes, *Nematodirus filicollis* and *N. spathiger*. *N.Z. vet. J. 10*, 1–6.

(33) BUTLER, E. J. & BARLOW, R. M. (1963). Factors influencing the blood and plasma copper levels of sheep in Swayback flocks. *J. comp. Path. 73* (2), 107–18.

(34) BUTLER, E. J. & BARLOW, R. M. (1963). Copper deficiency in relation to swayback in sheep. I. *J. comp. Path. 73* (2), 208–13.

(35) CONWAY, A. (1963). Personal communication.

(36) COOK, C. W., MATTOX, J. E. & HARRIS, L. E. (1961). Comparative daily consumption and digestibility of summer range forage by wet and dry ewes. *J. Anim. Sci. 20* (4), 866–70.

(37) COOMBE, J. B., WARDROP, I. D. & TRIBE, D. E. (1960). A study of milk production of the grazing ewe, with emphasis on the experimental technique employed. *J. agric. Sci. 54*, 353.

(38) COOP, I. E. (1950). The effect of level of nutrition during pregnancy and during lactation on lamb and wool production of grazing sheep. *J. agric. Sci. 40*, 311–40.

(39) CROFTON, H. D. (1954 a). Nematode parasite populations in sheep on lowland farms. I. Worm egg counts in ewes. *Parasitology 44* (3 and 4), 465–77.

(40) CROFTON, H. D. (1954 b). The ecology of the immature phases of trichostrongyle parasites. V. The estimation of pasture infestation. *Parasitology 44* (3 and 4), 313–24.

(41) CROFTON, H. D. (1958 a). Nematode parasite populations in sheep on lowland farms. IV. The effects of anthelmintic treatment. *Parasitology 48* (3 and 4), 235–242.

(42) CROFTON, H. D. (1958 b). Nematode parasite populations in sheep on lowland farms. V. Further observations on the post-parturient rise and a discussion on its significance. *Parasitology 48* (3 and 4), 243–50.

(43) CURNOW, D. H., ROBINSON, T. J. & UNDERWOOD, E. J. (1948). Oestrogenic action of extracts of subterranean clover. *Aust. J. exp. Biol. med. Sci. 26*, 171–9.

(44) DALTON, D. C. (1962). Characters of economic importance in Welsh Mountain sheep. *Anim. Prod. 4* (2), 269–78.

(45) DAVEY, L. A. & BRIGGS, C. A.E. (1961). Descriptive microbiology of the rumen. Ch. 9 in *Digestive Physiology and Nutrition of the Ruminant*. Ed. Lewis, D. London, Butterworth.

(46) DONALD, H. P. (1958). Crossbred lamb production from cast-for-age Blackface ewes. *Proc. Brit. Soc. Anim. Prod.*, 77–93.

(47) DONEY, J. M. & MUNRO, J. (1962). The effect of suckling, management and season of sheep milk production as estimated by lamb growth. *Anim. Prod. 4* (2), 215–20.

(48) DONEY, J. M. & SMITH, W. F. (1961). The fleece of the Scottish Blackface sheep. I. Seasonal changes in wool production and fleece structure. *J. agric. Sci. 56*, 365.

(49) DONEY, J. M. & SMITH, W. F. (1961). The fleece of the Scottish Blackface sheep. II. Variation in fleece components over the body of the sheep. *J. agric. Sci.* *56*, 375.
(50) DUCOULOMBIEU. M. (1957). La production lainière. *Bull. tech. Ing. Serv. agric.* *125*, 667.
(51) FERGUSON, K. A., WALLACE, A. L. C. & LINDNER, H. R. (1960). The hormonal regulation of wool growth. I. *Int. Congr. Endocrinology, Copenhagen.* Session X, No. 510.
(52) FIELD, A. C. (1962). Studies on magnesium in ruminant nutrition. IV. *Brit. J. Nutr.* *16*, 99.
(53) FIELD, A. C., MCCALLUM, J. W. & BUTLER, E. J. (1958). Studies on magnesium in ruminant nutrition. *Brit. J. Nutr.* *12* (4), 433.
(54) FORBES, T. J. (1962). General observations on the husbandry and nutrition of the pregnant ewe. *Anim. Prod.* *4* (2), 299.
(55) FORD, E. J. H. (1958). Metabolic changes in cattle and sheep at parturition. *Vet. Rev. Annot.* *4* (2), 119–31.
(56) FORD, E. J. H. (1962). The effect of dietary restriction on some liver constituents of sheep during late pregnancy and early lactation. *J. agric. Sci.* *59*, 67.
(57) FORSYTH, B. A. (1958). Footrot in sheep. *Outlook on Agric.* *2* (2), 86–91.
(58) FRASER, A. & STAMP, J. T. (1961). *Sheep Husbandry & Diseases.* London, Crosby Lockwood.
(59) GIBSON, T. E. (1950). Observations on the value of small daily doses of phenothiazine for the control of trichostrongylosis in sheep. *J. comp. Path.* *60* (2), 117–32.
(60) GIBSON, T. E. (1959). The therapy of parasitic gastro-enteritis in sheep and cattle. *Vet. Rec.* *71* (45), 1014–23.
(61) GIBSON, T. E. (1962 a). Worming cattle and sheep. *Agriculture*, London, *69* (8), 382–5.
(62) GIBSON, T. E. (1962 b). Veterinary anthelmintic medication. *Tech. Comm. No.* *33* of the Commonw. Bur. Helm.
(63) GILL, J. C. & THOMSON, W. (1954). Some aspects of the nutrition of the in-lamb ewe. *Proc. Brit. Soc. Anim. Prod.*, 35–44.
(64) GORDON, H. MCL. (1957). Helminthic Diseases. Ch. 4 in *Advances in Veterinary Science, III.*
(65) GUNN, R. G. & ROBINSON, J. F. (1963). Lamb mortality in Scottish hill flocks. *Anim. Prod.* *5* (1), 67–76.
(66) HEMINGWAY, R. G., INGLIS, J. S. S. & RITCHIE, N. S. (1960). Factors involved in hypomagnesaemia in sheep. *B.V.A. Conf. on Hypomagnesaemia*, 58.
(67) HUNTER, G. L. (1961). Some effects of plane of nutrition on the occurrence of œstrus in Merino ewes. *Proc. IV int. Congr. Anim. Reprod.*, The Hague, *2*, 197–201.
(68) INGLIS, J. S. S., WEIPERS, M. & PEARCE, P. J. (1959). Hypomagnesaemia in sheep. *Vet. Rec.* *71* (36), 755–63.
(69) KELLEY, R. B. & SHAW, H. E. B. (1943). Fertility in sheep. *Comm. Aust. C.S.I.R.O. Bull.*, 166.
(70) KING, J. O. L. (1961). *Veterinary Dietetics.* London, Baillière, Tindall & Cox.
(71) KÖSEOĞLU, H. & AYTUĞ, C. N. (1961). Studies on the milk production of Awassi sheep which are breeding at Cukurova Stock Farm. *Lalahan zootek. Araşt. Enst. Derg.* *1* (10), 100–110.
(72) LAMBOURNE, L. J. (1955). Recent developments in the field study of sheep nutrition. *Proc. N.Z. Soc. Anim. Prod.* *15*, 36.
(73) LARGE, R. V., ALDER, F. E. & SPEDDING, C. R. W. (1959). Winter feeding of the in-lamb ewe. *J. agric. Sci.* *53* (1), 102–16.
(74) LEIPER, J. W. G. (1951). A new approach to phenothiazine therapy in sheep. *Vet. Rec.* *63*, 885.

(75) *Livestock Breeding in the Federal Republic of Germany* (1961). Compiled by Arbeitsgemeinschaft Deutscher Tierzüchter e. v., Bonn.

(76) MAZERAM, P. (1953). Orientation et amélioration de la production du lait de brebis. *Bull. tech. Ing. Serv. agric.* 76, 27.

(77) MCCANCE, I. & ALEXANDER, G. (1959). The onset of lactation in the Merino ewe and its modification by nutritional factors. *Aust. J. agric. Res.* 10 (5), 699–719.

(78) MCDONALD, I. W. (1962). Ewe fertility and neonatal lamb mortality. *N.Z. vet. J.* 10, 45–52.

(79) MICHAEL, D. T. (1962). Manurial treatment in relation to calcium and magnesium serum levels in sheep. *Vet. Rec.* 74 (6), 163–6.

(80) MICHEL, J. F. (1963). The phenomena of host resistance and the course of infection of *Ostertagia ostertagi* in calves. *Parasitology* 53, 63–84.

(81) MOIR, R. J. & HARRIS, L. E. (1962). Ruminal flora studies in the sheep. X. *J. Nutr.* 77 (3), 285–98.

(82) MORRIS, I. (1961). *Sheep Recording and Progeny Testing.* London, H.M.S.O.

(83) MOULE, G. R. (1962). Field trials in retrospect: flushing. *Proc. Aust. Soc. Anim. Prod.* 4, 195–200.

(84) MOULE, G. R. & YOUNG, R. B. (1961). Field observations on the daily milk intake of Merino lambs in semi-arid tropical Queensland. *Qd J. agric. Sci.* 18, 221–9.

(85) MUNRO, J. (1962). A study of the milk yield of three strains of Scottish Blackface ewes in two environments. *Anim. Prod.* 4 (2), 203–13.

(86) NEWSOM, I. E. (1952). *Sheep Diseases*, p. 94. Baltimore, Williams & Wilkins.

(87) NICHOLS, J. E. (1960). Wool improvement. *N.S.B.A. Yearbk.*, 1960, 35–6.

(88) O'HALLORAN, M. W. & SKERMAN, K. D. (1961). The effect of treating ewes during pregnancy with cobaltic-oxide pellets on the vitamin B_{12} concentration and the chemical composition of colostrum and milk and on lamb growth. *Brit. J. Nutr.* 15, 99.

(89) OLLERENSHAW, C. B. (1962). The control of fascioliasis—the need for a planned approach. *Outlook on Agric. III* (6), 278–81.

(90) OWEN, J. B. (1957). A study of the lactation and growth of hill sheep in their native environment and under lowland conditions. *J. agric. Sci.* 48, 387–412.

(91) OWEN, J. B. & SINCLAIR, K. B. (1961). The development of hypomagnesæmia in lactating ewes. *Vet. Rec.* 73 (50).

(92) PARNELL, I. W. (1962). Observations on the seasonal variations in the worm burdens of young sheep in S.W. Australia. *J. Helminth.* 36 (1 and 2), 161–88.

(93) PARRY, H. B. (1953). Disease and management of the in-lamb ewe. *Suffolk Sheep Soc. Yearbk.*, 1953, 26–34.

(94) PERRIN, D. R. (1958). The chemical composition of the colostrum and milk of the ewe. *J. Dairy Res.* 25 (1), 70–4.

(95) PHILLIPSON, A. T. (1961). The nutrition of sheep. *N.S.B.A. Yearbk.*, 1961, 34–40.

(96) PIERCE, A. E. (1962). Antigens and antibodies in the newly born. *In* Animal Health and Production, 189–206. *Proc. Colston Res. Soc.* London, Butterworth.

(97) PLOMMET, M. & RICORDEAU, G. (1960). Mammite staphylococcique de la brebis. Influence des modes de traite et de sevrage, du nombre d'agneaux, du strade de lactation et de la production laitière sur le déclenchement de l'infection. *Ann. zootech.* 9, 225–40.

(98) PORTER, J. W. G. (1961). Vitamin synthesis in the rumen. Ch. 19 in *Digestive Physiology and Nutrition of the Ruminant*, p. 226. Ed. Lewis D. London, Butterworth.

(99) PURSEL, V. G. & GRAHAM, E. F. (1962). Induced estrus in anestrous ewes by use of progestogens and follicle stimulating hormone. *J. Anim. Sci.* 21 (1), 132–6.

(100) Purser, A. F. & Young, G. B. (1959). Lamb survival in two hill flocks. *Anim. Prod. 1* (1), 85–91.
(101) Reid, R. L. (1960). Pregnancy toxaemia in ewes. *Proc. VIII int. Grassld Congr.* (Reading), 657.
(102) Reid, R. L. (1961). Energy requirements of ewes in late pregnancy. Ch. 17 in *Digestive Physiology and Nutrition of the Ruminant*, p. 198. Ed. Lewis, D. London, Butterworth.
(103) Reid, R. L. (1962). Studies on the carbohydrate metabolism of sheep. XV and XVI. *Aust. J. agric. Res. 13* (2), 296–306; 307–19.
(104) Reid, R. L. & Hinks, N. T. (1962). Studies on the carbohydrate metabolism of sheep. XVIII and XIX. *Aust. J. agric. Res. 13* (6), 1112–23; 1124–36.
(105) Reid, R. L. & Mills, S. C. (1962). Studies on the carbohydrate metabolism of sheep. XIV. *Aust. J. agric. Res. 13* (2), 282–95.
(106) Ricordeau, G. & Denamur, R. (1962). Production laitière des brebis Préalpes du sud pendant les phases d'allaitement, de sevrage et de traite. *Ann. zootech. 11* (1), 5–38.
(107) Ritchie, N. S., Hemingway, R. G., Inglis, J. S. S. & Peacock, R. M. (1962). Experimental production of hypomagnesæmia in ewes and its control by small magnesium supplements. *J. agric. Sci. 58*, 399.
(108) Robinson, J. F. (1959). Concentrates for hill ewes in pregnancy. *N.S.B.A. Yearbk.*, 1959, 15–19.
(109) Robinson, J. F., Currie, D. C. & Peart, J. N. (1961). Feeding hill ewes. *Trans. R. Highld agric. Soc. Scotl.*, 1961, *6*, 31–46.
(110) Rowlands, W. T. (1959). Disease factors limiting production by sheep during winter- and spring-grazing. *J. Brit. Grassld Soc. 14* (2), 131–6.
(111) Rowlands, W. T. (1960). Hypomagnesaemia. *N.S.B.A. Yearbk.*, 7–8.
(112) Ryle, M. (1962). Early reproductive failure of ewes in a hot environment. II. *J. agric. Sci. 58*, 137.
(113) Sharafeldin, M. A. & Mostageer, A. (1961). Suckling in Ossimi & Rahmani lambs. *J. Anim. Prod. U.A.R. 1*, 53–9.
(114) Smith, I. D. (1962). The effect of plane of nutrition upon the incidence of oestrus in the Merino ewe in Queensland. *Aust. vet. J.*, 338–40.
(115) Sönmez, R. (1961). A study of the Sakiz sheep and its comparison with other milk sheep. *Atatürk Üniv. (Erzurum)*, 49–77. (In *Anim. Breed. Abstr. 31*, 207.)
(116) Soulsby, E. J. L. (1962). Immunity to helminths and its effect on helminth infections. In *Animal Health and Production*, p. 165. Ed. Grunsell & Wright. London, Butterworth.
(117) Spedding, C. R. W. (1953). Variation in the nematode egg content of sheep faeces from day to day. *J. Helminth. 27*, 9–16.
(118) Taylor, E. L. (1961). Control of worms in ruminants by pasture management. *Outlook on Agric. 3*, 141.
(119) Terrill, C. E. (1962). The reproduction of sheep. Ch. 14 in *Reproduction in Farm Animals*. p. 240. Ed. E. F. H. Hafez. London, Baillière, Tindall & Cox.
(120) Tribe, D. E. & Seebeck, R. M. (1962). Effect of liveweight and liveweight change on the lambing performance of ewes. *J. agric. Sci. 59*, 105.
(121) Underwood, E. J. (1962). *Trace Elements in Human and Animal Nutrition.* 2nd ed. London and New York, Academic Press Inc.
(122) Velloso, G. (1963). Pôrto Alegre, Brazil. Personal communication.
(123) Wallace, L. R. (1948). The growth of lambs before and after birth in relation to the level of nutrition. *J. agric. Sci. 38*, 93, 243, 367.
(124) Wallace, L. R. (1961). Influence of liveweight and condition on ewe fertility. *Proc. Ruakura Fmrs' Conf. Week*, 14–23.
(125) Watson, W. A. (1962). Ovine abortion. *Brit. vet. Ass. 80th Ann. Congr.*, Scarborough.
(126) White, J. B. (1960). Clinical hypomagnesaemia. *B.V.A. Conf. on Hypomesaemia*, p. 39.

(127) WHITTLESTONE, W. G. (1957). Intramammary pressure changes in the lactating ewe. I. *J. Dairy Res. 24* (2), 165–70.

(128) WILLIAMS, B. M. (1962). Black disease of sheep: observations on the disease in mid-Wales. *Vet. Rec. 74* (52), 1536–43.

(129) WILSON, A. L. (1962). Trace elements in sheep nutrition. *Outlook on Agric. 3* (4), 160–6.

(130) WILSON, R. K., MAGUIRE, M. F. & POOLE, D. B. R. (1962). A field trial of a heavy magnesium pellet. *Vet. Rec. 74* (39), 1041–43.

(131) WOLTON, K. M. (1963). Fertilizers and hypomagnesaemia. *N.A.A.S. quart. Rev. 14* (59), 122–30.

(132) YEATES, N. T. M. (1958). Foetal dwarfism in sheep—an effect of high atmospheric temperature during gestation. *J. agric. Sci. 51* (1), 84–9.

(133) AMIR, D. & VOLCANI, R. (1965a). Seasonal fluctuation sin the sexual activity of Awassi, German Mutton Merino, Corriedale, Border-Leicester and Dorset Horn rams. I, II and IV. *J. agric. Sci. 64,* 115, 113, 121.

(134) AMIR, D. & VOLCANI, R. (1965b). The sexual season of the Awassi fat-tailed ewe. *J. agric. Sci. 64,* 83–85.

(135) A.R.C. (1965). Nutrient requirements of farm livestock. No. 2, *Ruminants.* Agricultural Research Council: H.M.S.O.

(136) ARDRAN, G. M. & BROWN, T. H. (1964). X-ray diagnosis of pregnancy in sheep with special reference to the determination of the number of foetuses. *J. agric. Sci. 63,* 205.

(137) ARMSTRONG, D. G., ALEXANDER, R. H. & MCGOWAN, M. (1964). *Proc. Nutr. Soc. Abstr. Communic.* 162nd Mtg., 26.

(138) BARNARD, A. (1962). *The Simple Fleece.* Ed. Barnard, A. A.N.U. and M.U.P.

(139) BARRETT, J. F., GEORGE, J. M. & LAMOND, D. R. (1965). Reproductive performance of merino ewes grazing red clover (*Trifolium pratense* L.), improved pasture, or native pasture. *Aust. J. agric. Res. 16,* 189–200.

(140) BENNETT, D., AXELSEN, A. & CHAPMAN, H. W. (1964). The effect of nutritional restriction of sheep during early pregnancy on numbers of lambs born. *Proc. Aust. Soc. Anim. Prod. 5,* 70.

(141) BENNETTS, H. W., UNDERWOOD, E. J. & SHIER, F. L. (1946). A specific breeding problem of sheep on subterranean clover pastures in Western Australia. *Aust. vet. J. 22,* 2–12.

(142) BENZIE, D. (1951). X-ray diagnosis of pregnancy in ewes. *Brit. vet. J. 107,* 3–6.

(143) BETTS, J. E., NEWTON, J. E. & DENEHY, H. L. (1969). Out of season breeding in sheep. *Vet. Rec. 84,* 358-9.

(144) BICKOFF, E. M. (1968). Oestrogenic constituents of forage plants. *C.A.B. Rev. Series No. 1.* Comm. Bur. Past. and Fld. Crops.

(145) BIGGERS, J. D. (1958). Plant phenols possessing oestrogenic activity. In *The Pharmacology of Plant Phenolics.* Ed. Fairbairn, J. W., Academic Press, London.

(146) CAMPBELL, F. R. (1962). Influence of age and fertility of Rambouillet ewes on lamb and wool production. Texas Agric. Exp. Sta., Texas A and M Univ.

(147) CONNAN, R. M. (1968). Studies on the worm populations in the alimentary tract of breeding ewes. *J. Helminth. 42* (1/2), 9–28.

(148) COOP, I. E. (1966a). The response of ewes to flushing. *World Rev. Anim. Prod. 4,* 69–75.

(149) COOP, I. E. (1966b). Effect of flushing on reproductive performance of ewes. *J. agric. Sci., Camb. 67,* 305–23.

(150) COPENHAVER, J. S. & CARTER, R. C. (1966). Earlier weaning and multiple lambing. *In* Livestock Research. *Prog. Rep. Va Agric. Exp. Stn* 1965–6, 54–7.

(151) CRESSWELL, E. (1963). A modified sheep stell. *Scot. Agric. 43,* 68–72.

(152) CUNNINGHAM, J. M. M., DEAS, D. W. & FITZSIMONS, J. (1967). Synchronisation of oestrus in ewes. *Vet. Rec. 80* (20), 590–1.

(153) DALY, R. A. & CARTER, H. B. (1955). Fleece growth of young Lincoln, Corriedale, Polwarth and fine wool Merino maiden ewes under housed conditions and unrestricted and progressively restricted feeding on a standard diet. *Aust. J. agric. Res. 6*, 476–513.
(154) DAWES, G. S. & PARRY, H. B. (1965). Premature delivery and survival in lambs. *Nature, Lond. 207*, 330.
(155) DOBROTVORSKAJA, Z. L. (1965). Making full use of the early maturity of Romanov ewes. *Ovtsevodstvo 11*, 18–19.
(156) DONALD, H. P. & READ, J. L. (1967). The performance of Finnish Landrace sheep in Britain. *Anim. Prod. 9*, 471–6
(157) DONEY, J. M. (1963). The effects of exposure in Blackface sheep with particular reference to the role of the fleece. *J. agric. Sci., Camb. 60*, 267–73.
(158) DONEY, J. M. (1964). The fleece of the Scottish Blackface sheep. IV. *J. agric. Sci. 62*, 59.
(159) DONEY, J. M. & SMITH, W. F. (1964). Modification of fleece development in Blackface sheep by variation in pre- and post-natal nutrition. *Anim. Prod. 6* (2), 155–67.
(160) DOWNEY, N. E. (1968). Worm egg counts of ewes after artificially-induced ovulation. *J. Helminth. 42* (1/2), 33–6.
(161) DUNCAN, D. L. & PHILLIPSON, A. T. (1951). The development of motor responses in the stomach of the foetal sheep. *J. Exp. Biol. 28*, 32.
(162) DUNSMORE, J. D. (1965). *Ostertagia* spp. in lambs and pregnant ewes. *J. Helminth. 39* (2/3), 159–84.
(163) EDEY, T. N. (1967). Early embryonic death and subsequent cycle length in the ewe. *J. Reprod. Fert. 13*, 437–43.
(164) EDGAR, D. G. (1963). The place of ram testing in the sheep industry. *N.Z. vet. J. 11* (5), 113–15.
(165) EDGAR, D. G. & BILKEY, D. A. (1963). The influence of rams on the onset of the breeding season in ewes. *Proc. N.Z. Soc. Anim. Prod. 23*, 79–87.
(166) EDWARDS, C. M. (1968). Liver fluke in sheep. *Vet. Rec. 82* (25), 718–28.
(167) EVERITT, G. C. (1964). Maternal undernutrition and retarded foetal development in Merino sheep. *Nature, Lond. 201*, 1341–2.
(168) EVERITT, G. C. (1966). Maternal food consumption and foetal growth in Merino sheep. *Proc. Aust. Soc. Anim. Prod. 6*, 91–101.
(169) FERGUSON, K. A., HEMSLEY, J. A. & REIS, P. J. (1967). Nutrition and wool growth. *Aust. J. Sci. 30* (6), 215–17.
(170) FORBES, J. M. (1966). Problems in feeding the pregnant ewe. *Beef and Sheep Farming 3* (7), 20.
(171) FORBES, J. M. (1967). Factors affecting the gestation length in sheep. *J. agric. Sci., Camb. 68*, 191–4.
(172) FORBES, J. M. (1968). The water intake of ewes. *Brit. J. Nutr. 22*, 33–43.
(173) FORBES, J. M., REES, J. K. S. & BOAZ, T. G. (1967). Silage as a feed for pregnant ewes. *Anim. Prod. 9* (3), 399–408.
(174) FORBES, T. J. & ROBINSON, J. J. (1967). The effect of source and level of dietary protein on the performance of in-lamb ewes. *Anim. Prod. 9* (4), 521–30.
(175) FORD, E. J. H., CLARK, J. W. & GALLUP, A. L. (1963). The detection of foetal numbers in sheep by means of X-rays. *Vet. Rec. 75*, 958.
(176) FRASER, A. F. & ROBERTSON, J. G. (1968). Pregnancy diagnosis and detection of foetal life in sheep and pigs by an ultrasonic method. *Brit. vet. J. 124*, 239.
(177) GIBSON, T. E. (1966). The ecology of the infective larvae of *Trichostrongylus colubriformis*. In *Biology of Parasites*. Ed. Soulsby, E. J. L., New York, Academic Press.
(178) GORDON, I. (1963). The induction of pregnancy in the anestrous ewe by hormonal therapy. I. Progesterone-pregnant mare's serum therapy during the breeding season. *J. agric. Sci., Camb. 60*, 31–41.

(179) GRAHAM, N. McC. (1964). Energy exchanges of pregnant and lactating ewes. *Aust. J. agric. Res.* *15* (1), 127–41.

(180) HAFEZ, E. S. E. (1952). Studies on the breeding season and reproduction in the ewe. *J. agric. Sci. Camb.* *42*, 189–231 and 232–65.

(181) HASHIMOTO, H. (1962). Diagnosis of pregnancy in the ewe. *Can. J. comp. Med.* *25*, 51–3.

(182) HULET, C. V., BLACKWELL, R. L., ERCANBRACK, S. K., PRICE, D. A. & WILSON, L. O. (1962). Mating behaviour of the ewe. *J. Anim. Sci.* *21*, 870.

(183) HUNTER, G. L. (1968). Increasing the frequency of pregnancy in sheep. 1. Some factors affecting rebreeding during the *post-partum* period. *Anim. Breeding Abstr.* *36* (3), 347–78.

(184) HUNTER, G. L. & LISHMAN, A. W. (1967). Effect of the ram early in the breeding season on the incidence of ovulation and oestrus in sheep. *Proc. S. Afr. Soc. Anim. Prod.* *6*. In Press.

(185) JONES, R. C. (1968). The occurrence of oestrus in ewes and the duration and detection of pregnancy in artificially inseminated ewes. *Aust. J. exp. Agric. Anim. Husb.* *8*, 9–12.

(186) NEWTON, J. E. & Betts J. E. (1968). Factors affecting litter size in the Scotch Half-bred ewe. II. *J. Reprod. Fert.* *17*, 485–93.

(187) KRALJ, J. (1952). Rentgenska diajagnostika gravidnosti kod ovce. *Veterinaria, Sara.* *1*, 240 & 255.

(188) LAING, J. A. (1968). Studies on fertility and infertility in cattle, sheep and pigs. *Vet. Rec.* *83*, 65–9.

(189) LAMOND, D. R. (1963). Diagnosis of early pregnancy in the ewe. *Aust. Vet. J.* *39* (5), 192.

(190) LARKS, S. D., HOLM, L. W. & PARKER, H. R. (1960). A new technique for the demonstration of the foetal electrocardiogram in the large domestic animal. *Cornell Vet.* *50*, 459.

(191) LEES, J. L. (1964). Inhibitory effect of lactation on the breeding activity of the ewe. *Nature, Lond.* *203*, 1089–90.

(192) LEES, J. L. (1965). Seasonal variation in breeding activity of rams. *Nature, Lond.* *207*, 221–2.

(193) LEES, J. L. (1966). Variations in the time of onset of the breeding season in Clun ewes. *J. agric. Sci., Camb.* *67*, 173–9.

(194) LEES, J. L. (1967). Effect of time of shearing on the onset of breeding activity in the ewe. *Nature, Lond.* *214*, 743–4.

(195) LERNER, I. M. & DONALD, H. P. (1966). Modern Developments in Animal Breeding. London and New York, Academic Press.

(196) LINDAHL, I. L. (1966). Detection of pregnancy in sheep by means of Ultrasound. *Nature, Lond.* *212*, 642–3.

(197) LINDSAY, D. R. & FLETCHER, I. C. (1968). Sensory involvement in the recognition of lambs by their dams. *Anim. Behav.* *16*, 415–17.

(198) MATTNER, P. E., BRADEN, A. W. H. & TURNBULL, K. E. (1967). Studies in flock mating of sheep. 1. Mating behaviour. *Aust. J. exp. Agric. Anim. Husb.* *7*, 103–9.

(199) MATTNER, P. E. & BRADEN, A. W. H. (1967). Studies in flock mating of sheep. 2. Fertilization and prenatal mortality. *Aust. J. exp. Agric. Anim. Husb.* *7*, 110–16.

(200) MAULEON, P. & DAUZIER, L. (1965). Variations in the duration of lactation anoestrus in ewes of the Ile-de-France breed. *Annls. Biol. anim. Biochim. Biophys.* *5*, 131–43.

(201) McDONALD, P., EDWARDS, R. A. & GREENHALGH, J. F. D. (1966). *Animal Nutrition.* Edinburgh and London, Oliver and Boyd.

(202) MICHEL, J. F. & OLLERENSHAW, C. B. (1963). Parasitic gastroenteritis. Ch. 18 in *Animal Health, Production and Pasture.* Ed. A. N. Worden, K. C. Sellers & D. E. Tribe. London, Longmans.

(203) MORLEY, F. H. W., AXELSEN, A. & BENNETT, D. (1963). Effects of oestrogen on fertility in ewes. *Nature, Lond.* *199*, 403–4.
(204) MORRIS, I. (1961). *Sheep recording and progeny testing.* London, Min. Agric. Fish. and Fd., H.M.S.O.
(205) MULLANEY, P. D. (1966). Pre-natal losses in sheep in Western Victoria. *Proc. Aust. Soc. Anim. Prod.* *6*, 56–9.
(206) MUNRO, J. (1962). The use of natural shelter by hill sheep. *Anim. Prod.* *4*, 343–9.
(207) National Livestock Breeding Conference (1962). British Livestock Breeding: The way ahead. Summary of Proceedings, H.M.S.O., London.
(208) NELSON, W. A. & SLEN, S. B. (1968). Weight gains and wool growth in sheep infested with the sheep ked *Melophagus ovinus*. *Expl. Parasit.* *22*, (2), 223–6.
(209) NEWTON, J. E. (1967). Effects of hormonal synchronisation and P.M.S. on breeding in ewes. *Vet. Rec.* 422–5.
(210) NEWTON, J. E. & BETTS, J. E. (1966). Factors affecting litter-size in the Scotch Half-bred ewe. I. *J. Reprod. Fert.* *12*, 167–75.
(211) NEWTON, J. E. & BETTS, J. E. (1967). Breeding performance of Dorset Horn ewes augmented by hormonal treatment. *Expl. Agric.* *3*, 307–13.
(212) NEWTON, J. E. & BETTS, J. E. (1968). Seasonal variation in oestrogenic activity of various legumes. *J. agric. Sci., Camb.* *70*, 77–82.
(213) NEWTON, J. E., BETTS, J. E. & GIBSON, J. H. (1969). Oestrogens in pasture. *Ann. Rep. Grassld Res. Inst., 1968.*
(214) OLLERENSHAW, C. B. & SMITH, L. P. (1966). An empirical approach to forecasting the incidence of nematodiriasis over England and Wales. *Vet. Rec.* *79* (19), 536–40.
(215) PEART, J. N. (1964). Winter coats for hill sheep. 3rd Rep. p. 90, Hill Farming Res. Organisation.
(216) PEART, J. N. (1968). Some effects of live weight and body conditions on the milk production of Blackface ewes. *J. agric. Sci., Camb.* *70*, 331–8.
(217) PEPELKO, W. E. & CLEGG, M. T. (1965). Influence of season of the year upon patterns of sexual behaviour in male sheep. *J. Anim. Sci.* *24* (3), 633–7.
(218) PRETORIUS, P. S. (1966). Influence of lactation on body weight, wool production and *post-partum* anoestrus of Merino ewes. *S. Afr. J. agric. Sci.* *9*, 823–33.
(219) PRUD'HON, M., DENOY, I. & DESVIGNES, A. (1968). Etude des résultats de six années d'élevage des brebis 'Merinos d'Arles' du domaine du Merle. *Annls Zootech.* *17* (2), 159–68.
(220) RADFORD, H. M. (1961). Photoperiodism and sexual activity in Merino ewes. II. *Aust. J. agric. Res.* *12*, 139–46.
(221) RAHMAN, SHEIKH SAIF-UR & KITTS, W. D. (1967). Hormonal control of reproduction in ewes during their normal breeding season. *Can. J. Anim. Sci.* *47*, 71–6.
(222) RAPIC, S. & ILIJAS, B. (1963). Roentgen diagnosis of pregnancy in domestic animals. *Vet. Arh.* *33*, 151.
(223) REID, R. L. (1963). The nutritional physiology of the pregnant ewe. *J. Aust. Inst. Agric. Sci.* *29* (4), 215–23.
(224) RICHTER, J. & GOTZE, R. (1960). Tiergeburtshilfe, Berlin-Hamburg, quoted by S. Rapić & B. Ilijaš, 1963. X-ray diagnosis of pregnancy in domestic animals. *Veterimarski Archiv*, *33* (5–6), 151–6.
(225) ROBINSON, T. J. (1951). The augmentation of fertility by gonadrotrophin treatment of the ewe in the normal breeding season. *J. agric. Sci. Camb.* *41*, 6–63.
(226) ROBINSON, T. J. (1965). Use of progestagen-impregnated sponges inserted intravaginally or subcutaneously for the control of the oestrous cycle in the sheep. *Nature, Lond.* *206*, 39.
(227) ROBINSON, T. J. (1966). Control of reproduction in sheep and cattle. *Proc. Aust. Soc. Anim. Prod.* *6*, 10–18.
(228) ROBINSON, T. J. (1967). *The Control of the Ovarian Cycle in the Sheep.* Sydney Univ. Press.

(229) ROBINSON, J. J. & FORBES, T. J. (1967). A study of the protein requirements of the mature breeding ewe, 2. *Brit. J. Nutr. 21*, 879–91.

(230) ROBINSON, J. J. & FORBES, T. J. (1968). The effect of protein intake during gestation on ewe and lamb performance. *Anim. Prod. 10* (3), 297–309.

(231) ROBINSON, T. J. & LAMOND, D. R. (1966). Control of reproduction in sheep and cattle. *Proc. Aust. Soc. Anim. Prod. 6*, 10.

(232) RUSSEL, A. J. F., GUNN, R. G. & DONEY, J. M. (1968). Components of weight loss in pregnant hill ewes during winter. *Anim. Prod. 9* (3), 399–408.

(233) RYDER, M. L. & STEPHENSON, S. K. (1968). *Wool Growth.* London and New York, Academic Press.

(234) SACKER, G. D. & TRAIL, J. C. M. (1966). Production characteristics in a flock of East African Blackheaded sheep. *E. Afr. agric. For. J. 31*, 392–8.

(235) SCALES, G. H. (1968). Lactation performances of Romney, Corriedale, and Merino ewes in a tussock grassland environment. *N.Z. Jl agric. Res. 11*, 155–70.

(236) SELJARNIN, G. I. (1960). Breeding Romanov sheep in the district of Urals and Trans Urals. *Ovtsevodstvo 6*, 24–7.

(237) SHELTON, M., MORROW, J. T. & BUTLER, O. D. (1966). *Reproductive Efficiency of Fine-wool Sheep.* Texas Agric. Exp. Sta., Texas A and M Univ.

(238) SILANGWA, S. M. & TODD, A. C. (1964). Vertical migration of Trichostrongylid larvae on grasses. *J. Parasit. 50* (2), 278–85.

(239) SMITH, I. D. (1964). Post-parturient anoestrus in the Peppin Merino in western Queensland. *Aust. vet. J. 40*, 199–201.

(240) SMITH, I. D. (1966). The onset of the breeding season in Southdown ewes in subtropical Australia. *J. agric. Sci. 66*, 295–6.

(241) SMITH, I. D. (1967a). The breeding season in British breeds of sheep in Australia. *Aust. vet. J. 43*, 59–62.

(242) SMITH, I. D. (1967b). Breed differences in the duration of gestation in sheep. *Aust. vet. J. 43*, 63–4.

(243) STAROVOITENKO, V. & ELIN, G. J. (1965). Ram lambs of multifoetal ewes should be reared. *Ovtsevodstvo 11*, 14–15.

(244) SYMINGTON, R. V. & OLIVER, J. (1966). Observations on the reproductive activity of tropical sheep in relation to the photoperiod. *J. agric. Sci., Camb. 67*, 7–12.

(245) TASSELL, R. (1967). The effects of diet on reproduction in pigs, sheep and cattle. Pts. III and IV. *Brit. vet. J. 123*, 257–64, 364–71.

(246) THOMSON, W. & AITKEN, F. C. (1959). Diet in relation to reproduction and the viability of the young. Pt. II. Sheep. *Comm. Bur. Anim. Nutr., Tech. Comm. No. 20.*

(247) THWAITES, C. J. (1965). Photoperiod control of breeding activity in the Southdown ewe with particular reference to the effects of an equatorial light regime. *J. agric. Sci. 65*, 57–64.

(248) THWAITES, C. J. (1967). Embryo mortality in the heat stressed ewe. *J. Reprod. Fert. 14*, 5–14.

(249) TURNER, H. N., HAYMAN, R. H., TRIFFITT, E. K. & PRUNSTER, R. W. (1962). Response to selection for multiple births in the Australian Merino: a progress report. *Anim. Prod. 4* (1), 165–76.

(250) WALLACE, L. R. (1948). The growth of lambs before and after birth in relation to the level of nutrition. *J. agric. Sci. 38*, 93, 243, 367.

(251) WIGGAN, L. S. & CLARK, J. B. K. (1968). The effect of synchronisation of oestrus on the fertility of ewes inseminated artificially. *Brit. vet. J. 124*, 460.

(252) WILLIAMS, H. Ll., HILL, R. & ALDERMAN, G. (1965). The effects of feeding kale to breeding ewes. *Brit. vet. J. 121*, 2–17.

(253) WILSON, I. A. N. & NEWTON, J. E. (1969). Pregnancy diagnosis in the ewe: a method for use on the farm. *Vet. Rec., 84*, 356–8.

(254) WINFIELD, C. J., BROWN, W. & LUCAS, I. A. M. (1968). Some effects of compulsory exposure over winter onin-lamb Welsh Mountain ewes. *Anim. Prod.* *10* (4), 451–63.
(255) Wintzer, H. J. (1964). On the diagnosis of pregnancy in pigs and sheep with the aid of X-rays. *Dt. tierärztl. Wschr.* *71*, 153.
(256) WISHART, D. F. (1966). The induction of earlier breeding activity in sheep. *Vet. Rec.* *79* (13), 356–8.
(257) WISHART, D. F. (1967). Synchronisation of oestrus in sheep: the use of pessaries. *Vet. Rec.* *80*, 276–87.
(258) WODZICKA-TOMASZEWSKA, M., HUTCHINSON, J. C. D. & BENNETT, J. W. (1967). Control of the annual rhythm of breeding in ewes: effect of an equatorial daylength with reversed thermal seasons. *J. agric. Sci., Camb.* *68*, 61–7.
(259) YEATES, N. T. M. (1949). The breeding season of the sheep with particular reference to its modification by artificial means using light. *J. agric. Sci., Camb.* *39*, 1–43.
(260) ABRAMS, J. T. (1956). Mineral supplements for farm animals. *B.V.A. Conf. on Supplements and Additives in Animal Feeding-stuffs.*
(261) ARMSTRONG, R. H. & CAMERON, A. E. (1959). Hexoestrol implantation of Blackface wether lambs. *Anim. Prod.* *1* (1), 37–40.
(262) ARMSTRONG, R. H. & CAMERON, A. E. (1961). Further studies of hexoestrol implantation of Blackface wether lambs. *Anim. Prod.* *3* (3), 295–300.
(263) BURGESS, T. D. & LAMMING, G. E. (1960). The effect of diethylstilboestrol, hexoestrol and testosterone on the growth rate and carcase quality of fattening beef steers. *Anim. Prod.* *2* (1), 83–92.
(264) COOPER, M. M. (1958). The expansion of animal production. In *The Biological Productivity of Britain.* London, Inst. of Biol., 53–9.
(265) CUTHBERTSON, W. F. J. (1956). Vitamins in animal feeding-stuffs. *B.V.A. Conf. on Supplements and Additives in Animal Feeding-stuffs.*
(266) DICKINSON, A. G., HANCOCK, J. L., HOVELL, G. J. R., TAYLOR, ST. C. S. & WIENER, G. (1962). The size of lambs at birth—a study involving egg transfer. *Anim. Prod.* *4* (1), 64–79.
(267) HAMMOND, J. (1957). Hormones in meat production. *Outlook on Agric.* *1* (6), 230–4.
(268) LAMMING, G. E. (1961). Endocrinology and ruminant nutrition. In *Digestive Physiology and Nutrition in the Ruminant.* Ed. Lewis. London, Butterworth.
(269) LAMMING, G. E. (1961). The use of hormones to induce extra-seasonal breeding in sheep. *Univ. of Nottingham, Sch. of Agric. Rep.*
(270) LAMMING, G. E., STOKES, R. M. & HORSPOOL, D. (1960). The influence on growth rate and slaughter characteristics of implanting fattening male hoggets with various levels of hexoestrol and diethylstilboestrol. *Anim. Prod.* *2* (2), 197–208.
(271) LUXTON, H. W. B. & LOADMAN, M. (1959). Fat sheep production in Devon, 1957/58. *Rep. No. 115, Univ. of Bristol.*
(272) PORTER, J. W. G. (1956). Antibiotics as dietary supplements in ruminants and other herbivores. *B.V.A. Conf. on Supplements and Additives in Animal Feeding-stuffs.*
(273) PRESTON, T. R. (1962). Antibiotics for the young ruminant. In *Antibiotics in Agriculture.* London, Butterworth.
(274) PRESTON, T. R. & GEE, I. (1957). Oestrogens in lamb and mutton production. *Proc. Brit. Soc. Anim. Prod.*, 41–8.
(275) PRESTON, T. R., GREENHALGH, I. & MACLEOD, N. A. (1959). The effect of hexoestrol implants on sucking and weaned lambs fattened on farms in the north-east of Scotland. *Anim. Prod.* *1* (2), 135–8.
(276) PRESTON, T. R., GREENHALGH, I. & MACLEOD, N. A. (1960). The effect of hexoestrol on growth, carcase quality, endocrines and reproductive organs of ram, wether and female lambs. *Anim. Prod.* *2* (1), 11–26.

(277) Scarisbrick, R. (1960). Problems of feeding additives and implantation on the growth and fattening of farm livestock. *J. Fmrs' Club.* Pt. I, 119–30.
(278) Shepherd, R. W. (1962). Hexoestrol implantation of very young lambs. *Exp. Husb.* 7, 72–6.
(279) Shorthose, W. R. & Lamming, G. E. (1961). The effect of implanting testosterone alone or with oestradiol on the growth rate and carcase quality of fattening female hoggets. *Anim. Prod.* 3 (1), 97–102.
(280) Taylor, J. H. (1956). The mode of action of antibiotics in promoting animal growth. *B.V.A. Conf. on Supplements and Additives in Animal Feeding-stuffs.*

V

THE LAMB—ITS GROWTH AND DEVELOPMENT

The development of the lamb is continuous from conception until maturity or slaughter, but birth is a major landmark in this process. It represents the beginning of an independent life and is accompanied by enormous changes in the environment to which the lamb is subjected. The adjustments required of the newly born animal are often considerable and the adaptability of the lamb is impressive. From the protected situation of the foetus, it is suddenly exposed to the rigours of life in the open. Furthermore, the coat is soaked in foetal fluid when the lamb is born. The temperature regulation of the new-born lamb and the effect of environmental factors on heat-loss and metabolic rate, have been studied and described in detail by Alexander and others (2, 3, 4, 5, 6).

The changes in structure of the foetus during its development within the ewe are very great (152) and it is worth remembering that, as lambs increasingly tend to be slaughtered for meat at earlier ages, even before they are 3 months old, the time spent in this period (gestation is normally about 147 days) may greatly exceed that spent in independent life.

The growth and development of the lamb have been intensively studied by Hammond (72) and his co-workers. Naturally, much attention has been given to the development of those parts of the lamb that ultimately become products of agricultural value. The lamb consists of a nervous system, a skeleton, internal organs, muscles, fat and wool. Nervous tissues develop early and are relatively untouched by starvation. The skeleton is developed before the muscles and fat is the last tissue to be laid down. The rate at which any tissue increases thus depends on the physiological age of the animal as well as on nutrition. Fat, muscle and bone differ in value to the butcher but tissues cannot be completely evaluated in this way.

At birth, the lamb has a large head, long legs and well-developed sensory organs. These parts have little value for meat but are of the utmost importance to a new-born lamb that has to see, hear, smell and follow an active ewe which constitutes its supply of milk, warmth and, in some cases, protection. Just as the foregoing attributes can have a high survival value, so the skeleton may be an important factor in muscle development and the size of the digestive tract may have an influence on the rate at which more valuable tissues, like muscle and fat, can be grown. Some idea of the considerable changes in the relative proportions of the component parts of the body that occur between birth and slaughter (39) may be gained from Table XXII.

TABLE XXII

BODY COMPOSITION OF THE LAMB

(as a percentage of empty body weight)

	Carcase	Head, skin, feet	Total empty gut	Viscera*
Birth . .	47·4	31·4	3·6	11·2
16 weeks of age	56·7	14·8	9·4	10·1

(Source : R. V. Large, from an experiment at the Grassland Research Institute)

* Viscera includes: Heart, lungs, liver, kidneys and blood.

Although not all parts of the live sheep are of equal value to the consumer, it must also be remembered that in order to produce the more valuable parts efficiently most parts of the body are required. Ruminants possess a rumen, in which roughage can be fermented (14, 29, 38, 104, 165, 166). This extension of the digestive system is capable of much earlier development than that normally associated with the suckled lamb (168, 169, 170) and is an example where the greater development of an organ of low ultimate value can influence the growth of the whole animal or the nature of the food on which growth can take place.

The rates at which the various parts of the alimentary tract develop (shown in Fig. 5.1) are of importance chiefly because the capacity of the tract, or parts of it, markedly influences the quantity of food that can be dealt with. It will be seen that, while the whole tract increases to some 1900 per cent of its weight at birth, the reticulo-rumen

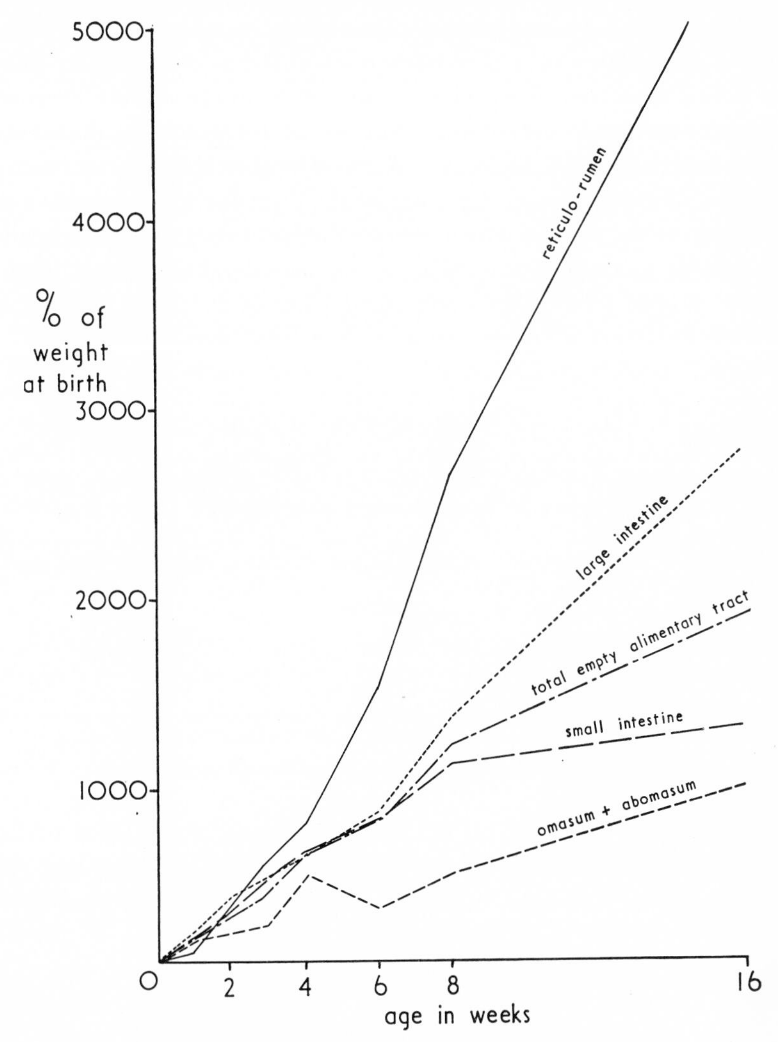

FIG. 5.1—*Development of the alimentary tract of the lamb*

(Source : data from R. V. Large (Ref. 174))

increases to nearly three times this extent. At slaughter only 45–50 per cent of the unfasted liveweight actually remains in the carcase: this is called the dressing-out or killing-out percentage, terms which are frequently abbreviated to D.O. per cent and K.O. per cent. This percentage is an important figure but the value of the carcase is chiefly affected by the proportions of the different joints in it. These proportions may be greatly affected by the rate at which the animal is grown (27, 28) and by the shape of the growth curve.

Boccard & Duplan (27) studied the carcase composition of 22 lambs in two groups, with average daily growth rates of 345 g and 256 g respectively, when the animals reached a carcase weight of 17·5 kg. The average proportions of the different regions of the body did not differ significantly but there were differences in the amount and distribution of the various tissues. The carcases of the more rapidly grown lambs had, on average, less muscular tissue and more fatty tissue than those of the lambs with the slower growth rate.

In general, the composition of the fat-free carcase varies very little with rate of growth; faster-growing animals simply add more fat. As the plane of nutrition increases, therefore, the rate of growth of the fat-free body also increases: when a point is reached at which the energy intake is greater than that which such growth can utilize, fat is deposited (thus altering the *percentage* of muscle and bone in the entire body). The subject of growth and development in mammals has recently been reviewed (see Lodge & Lamming (199)).

Growth rate

One of the difficulties of discussing growth is to describe it in relevant terms (33, 84, 96, 151, 171, 172). Kunkel (84) has pointed out that, although growth is more than an increase in size, it is most easily visualized, for statistical purposes, as the gain in total body mass. In agricultural science it is customary to describe the growth of sheep in terms of liveweight gain and this is probably the most useful way to do it. It is as well, however, to bear in mind that other ways of expressing or describing growth may be more useful for some purposes. There is no doubt that both the rate and the pattern of growth of the lamb are of crucial importance in meat production. Frequently, however, the *average* growth rate of a group of lambs may be quite different from the actual performance of any one of them. Equally, the mean daily growth rate for one lamb over several months may only be really useful if its growth can be represented by a straight line. If it cannot, there are innumerable curves that would each give the same mean but would not necessarily lead to the same conformation in the live animal or to the same proportions of different tissues in the carcase.

It is at present extremely difficult to obtain much information on the growth rate of the carcase, since this involves the slaughter of

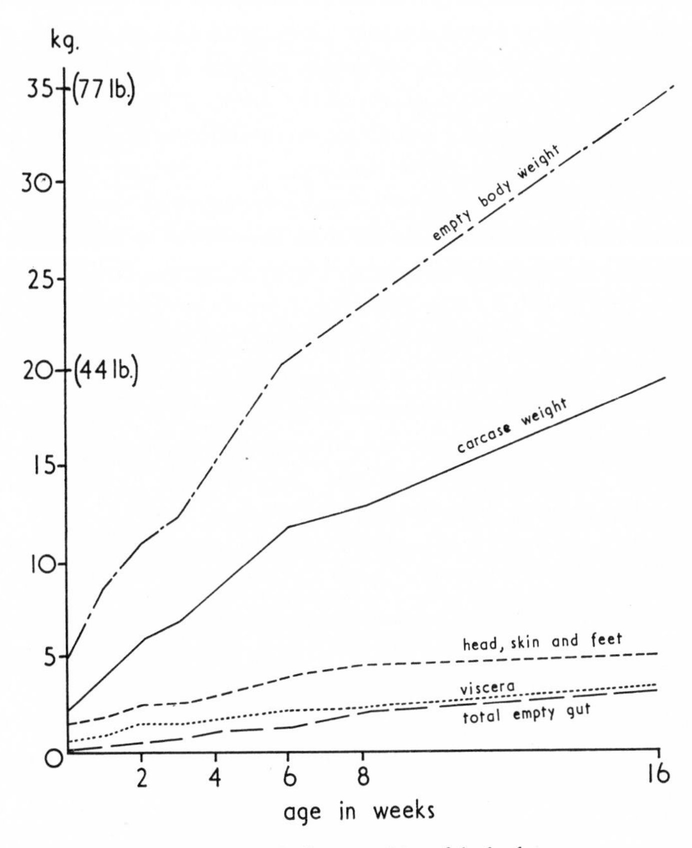

FIG. 5.2—*Body composition of the lamb*

(Source : data from R. V. Large (174))

frequent samples from birth onwards. In considering liveweight gain it has to be remembered that the composition of the weight gained may change greatly during the life of a lamb (see Figs. 5.2 and 5.3). These two figures also illustrate the different impressions obtained according to whether changes in body composition are expressed in percentages or absolute terms. The increase in weight of the whole alimentary tract, for example, appears quite small in absolute terms (Fig. 5.2) but enormous as a percentage of its weight at birth (Fig. 5.3).

The growth curves, expressed as liveweight, characteristic of single,

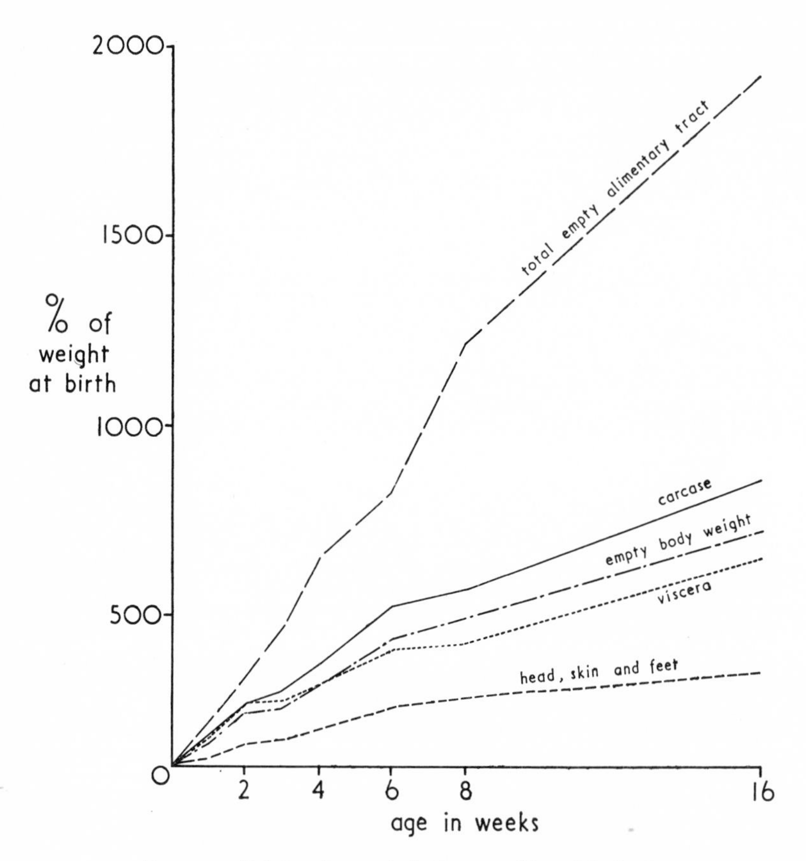

FIG. 5.3—*Relative changes in body composition of the lamb*

(Source : data from R. V. Large (174))

twin and triplet lambs sucking their dams under grazing conditions, are shown in Fig. 5.4. They have been chosen to illustrate three points. First, singles grow faster than twins (35, 86, 90) and twins faster than triplets (72). Secondly, lambs generally grow most rapidly in the first few weeks after birth (88, 89). Thirdly, growth is normally linear for a period of about 10 weeks, after which the rate declines, producing the curves shown. The curves plotted are for Suffolk × Half-bred lambs; comparable data for Clun Forest lambs are shown in Fig. 5.5. The actual rates of growth could be quite different with different breeds or with different feeding but the features noted would probably survive large variations in both.

Significance of growth rate

Before considering the major factors which influence the growth of lambs the fundamental significance of growth rate should be made

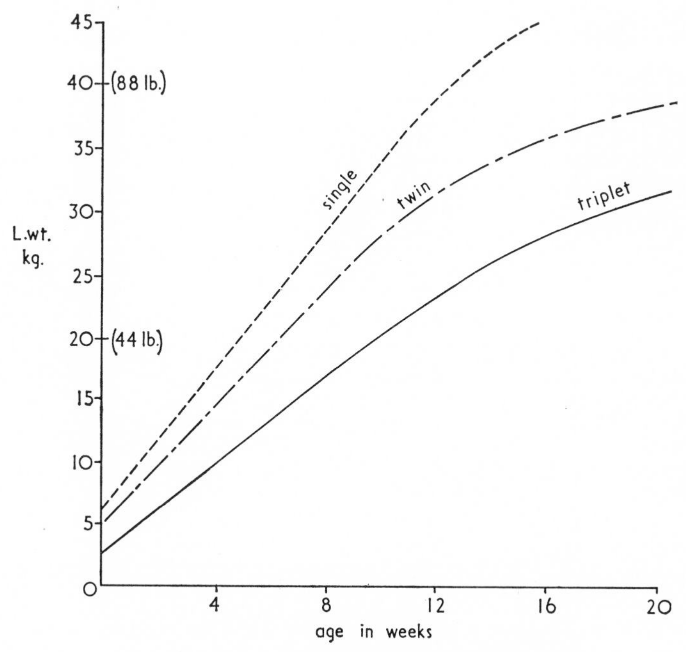

FIG. 5.4—*Growth curves of Suffolk × Half-bred lambs*

clear. Any growth rate might conceivably be the best for a particular purpose but two things are generally true. First, the more rapidly a lamb grows the sooner it will reach whatever stage may be regarded as a product. This may have great economic significance in the production of fat lambs for slaughter or stores for sale; it may have great ecological significance because the period for which lambs are kept on the farm has implications for disease control and the provision of food. Secondly, a high rate of growth generally represents a more efficient conversion of food. This is chiefly because food is required not only to produce the fat lamb but also to maintain it whilst it is

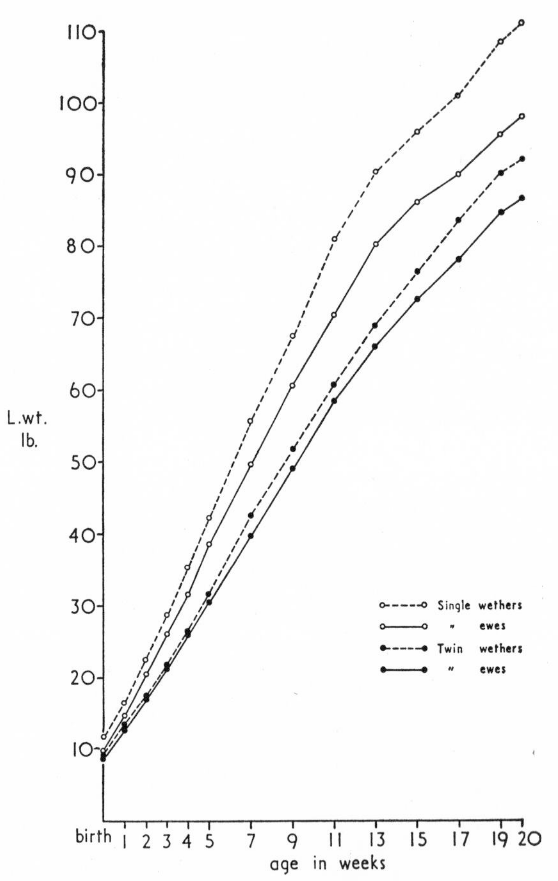

FIG. 5.5—*Growth curves of Clun Forest lambs*

(Source : Large, R. V. & Tayler, J. C. (90))

growing. This latter, maintenance requirement, is simply the theoretical amount of food on which the lamb would gain no weight. In total, it is governed by the size of the animal, or some function of its size, and the length of time for which the latter has to be maintained. In general, then, the longer a lamb takes to reach a given weight, the more food it will require. This is illustrated diagrammatically in Fig. 5.6. It does not follow that the maximum rate of

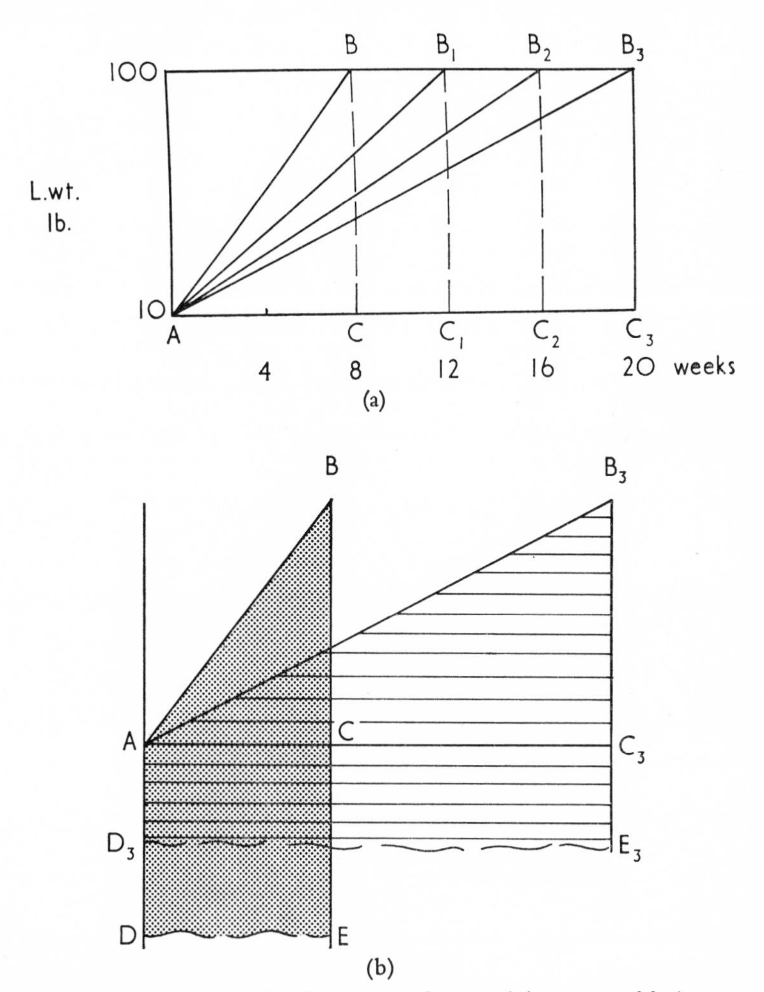

FIG. 5.6 *a & b—The relationship between growth rate and the amount of food required to gain a given weight*

gain is necessarily the most efficient or that a higher rate of gain is necessarily worth having if it requires a different, and perhaps more expensive, diet to produce it. If a high rate of gain does not produce the desired kind of carcase, however, it may simply indicate that it is not appropriate to that particular strain or breed of sheep.

In Fig. 5.6 (*a*), AB, AB_1, AB_2 and AB_3 represent the simplified growth curves of lambs, from birth at A to 100 lb liveweight, growing at four different rates and therefore reaching 100 lb at different ages (B, B_1, B_2, B_3). Since the 'maintenance' requirement at any one time is related to the liveweight at that time, it may be represented by the vertical dropped from this point on the growth curve to the base line. The possible errors involved in this assumption should not affect the general validity of the illustration.

The total food required for 'maintenance' purposes is then represented by the areas of the triangles (A B C, A B_1 C_1, etc.).

Food required for production may be thought of as the additional amount needed to build a 100 lb animal (from birth): it would therefore be the same (approximately) for all four animals.

The diagram can be redrawn, as shown in Fig. 5.6 (*b*), for the two extreme rates of growth, where these additional quantities are represented by A C D E and A C_3 D_3 E_3 (which are equal in area).

The total requirements are thus represented by the areas A B E D and A B_3 E_3 D_3. The areas shown are purely for illustration and it is important to appreciate the influence that the size of the additional area (for production) has on the importance of the difference between the 'maintenance' values.

Finally, the distinction should always be made between an individual growing nearly as fast as it can, which is likely to be good, and one kind of animal growing faster than another of a different breed. In the second case, the faster-growing animal might be eating twice as much for a relatively small increase in growth rate.

A recent illustration of the effect of growth rate on the efficiency of food conversion is shown in Table XXIII. Lambs growing faster, because they were on a higher plane of nutrition, consumed less food to produce the same carcase weight. Food conversion efficiency, calculated from birth to slaughter (although the differences in nutrition and growth rate only applied from 6 weeks of age), was much higher, therefore, for the faster-growing lambs. Table XXIII also shows, however, how much smaller is the effect when diluted by the food required by the ewe. The significance of growth rate for food conversion efficiency thus depends on the size of the ewe and the number of lambs in the litter, both of which will alter the proportion of the total food that is consumed by the lambs.

TABLE XXIII

EFFECT OF LAMB GROWTH RATE ON E (kg carcase/per 100 kg D.O.M.)

Lamb growth rate		Lamb only	Ewe+lamb
	L.W.G. g/day	162	162
	kg Carcase gained	14·8	
Low	kg Carcase produced		17·0
	kg D.O.M. consumed	87·6	387·6
	E	16·9	4·4
	L.W.G. g/day	256	256
	kg Carcase gained	14·9	
High	kg Carcase produced		17·3
	kg D.O.M. consumed	69·8	369·8
	E	21·3	4·7
Improvement in E due to increased growth rate		26%	6·8%

(Source : Spedding, C. R. W. (1969) Agric. Prog. In press)

FACTORS AFFECTING GROWTH RATE

These may be conveniently grouped as nutrition and disease.

Nutrition

The nutrients available to the lamb depend primarily on the quantity eaten (intake) and the digestibility of the food. In addition, the value of the nutrients absorbed from the alimentary tract may vary according to the nature of the diet. A complete food is one which contains all that the lamb requires, but these requirements are not static (110, 173). Bacterial activity in the rumen (107) not only breaks down complex plant structures into relatively simple substances but also synthesizes complex substances that were not present in the original diet (127, 128), notably vitamins. Thus the functional ruminant does not require vitamins of the B complex in its food (24). On the other hand, since the bacteria need the raw materials for vitamin synthesis, a shortage of some substances can result in vitamin deficiencies which may be very serious, for instance lack of cobalt will result in 'pining' due to shortage of vitamin B_{12} (24, 55, 77, 129). The rumen also contains ciliate protozoa but it does not appear likely that they have a major effect on lamb performance (179). These relationships are characteristic of the ruminant, but the lamb in its first few weeks depends primarily on milk digested in the abomasum (75) and uses its rumen very little. As it grows, rumination becomes increasingly important and the needs of the lamb that have to be supplied by its food change.

A relatively recent disease of sheep in Britain, cerebrocortical necrosis (CCN), appears to interfere with the synthesis of absorption of thiamine in the rumen (184, 200). It is possible that this may be caused by the presence of thiaminase in grasses (188, 189).

Dietary requirements

At all times the food must supply the necessary vitamins (47, 48, 53, 71), minerals (7, 108, 118), sufficient protein and abundant energy (24). Lambs deprived of any of the essential minerals would fail to thrive but, in fact, mineral deficiencies are not common in the lamb (7). The most likely exceptions are deficiencies of cobalt, sodium, iodine and phosphorus (97, 105, 108, 129): responses to supplementary selenium have also been recorded (65, 132, 133, 164). This does not mean that, under changed husbandry conditions or in different environments, mineral deficiencies in growing stock might not occur.

Under good grazing conditions protein is not normally in short supply. Where mature herbage is conserved *in situ*, however, as 'foggage' for grazing during periods when pasture growth is negligible, a shortage of protein can occur. This is well known under Australian winter conditions (8) where Milford (100) has found that natural pasture may contain as little as 2 to 3 per cent crude protein. Energy is normally the limiting dietary component for lamb growth and much remains to be learned as to the optimum form in which it should be present in the food.

Digestibility of food

Raymond (113) and his colleagues have stressed the importance of the digestibility of the food (see also Chapter II) and clearly the proportion of the intake that is absorbed by the animal must have a considerable effect on its nutritive value. Some animals can digest a given food more efficiently than others but the differences, even between sheep and cattle (Blaxter & Wainman (25)), are normally very small. Digestibility (64, 102, 103, 104) can therefore be regarded as an attribute of the food (115, 116, 117). A highly digestible food can be used to produce a high-growth rate or, by feeding less of it, a lower rate. If it is costly to produce it may be uneconomic to use it except for high-growth rates. A food of low digestibility cannot be used for high-growth rates because the intake of a lamb is limited and insufficient nutrients are absorbed from such a food. Superficially it might seem that a lamb could eat more of the less digestible food, but this is not necessarily so. In general, voluntary intake is greater on foods of high digestibility (24, 26), because intake is governed to a considerable extent by the rate at which food passes along the alimentary tract and this, in turn, depends on the digestibility of the food. Of course, highly digestible food may be disliked by the lamb because of some internal or external attribute. Intake may be reduced by disease or by behavioural reactions to the weather, flies or other stock, as well as by low herbage availability (9) (see Chapter VII). At this point it must suffice to emphasize that intake can be of major importance (15, 44), particularly where a high-growth rate is required. This discussion will be pursued further in Chapters VI and VII: the effect of hormone implantation (13, 56, 130) on growth rate is discussed in Chapter XV.

The foregoing considerations concerning the factors governing intake apply to all the foods offered to the lamb, although none of them may normally be available entirely without restriction. Thus the supply of milk, digestibility 92 to 95 per cent (126), is frequently limited, especially to lambs in litters of two or more. Grass may be more or less 'available' (see Chapter VII) to the grazing animal according to its distribution on the ground, or the supply to the individual may be governed by the stocking rate. Other foods, such as concentrate mixtures, are normally rationed when fed as a supplement to grazing. The relative values of milk, herbage and concentrates depend, of course, on the composition of each and on the physiological development of the lamb.

Weaning and growth rate

In its first week the lamb must have milk and will consume little else, but it is not easy to say for how long milk is essential. Although, on a pasture stocked with ewes and lambs, it is generally true that the more milk a lamb receives the faster it will grow, there is increasing

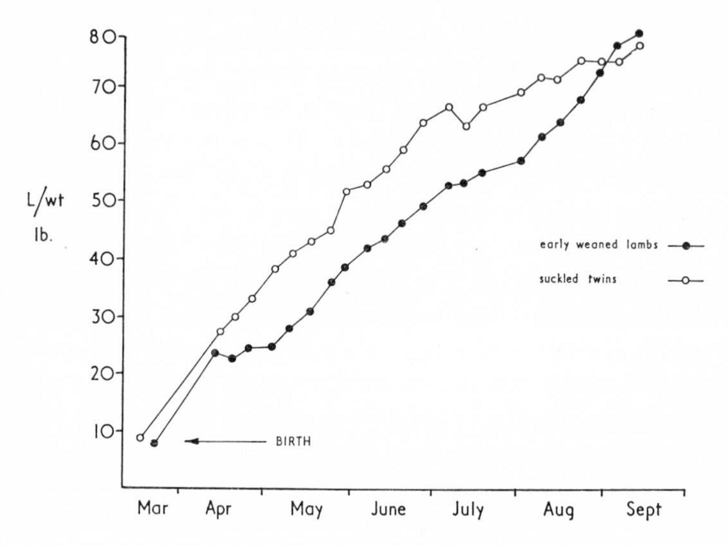

FIG. 5.7—*The growth of early weaned lambs compared with that of suckled twins*

(Source: from T. H. Brown (176))

evidence that, under many conditions, early weaning, at ages between 8 and 12 weeks, results in little, if any, reduction in growth rate. This has been shown in the work of Bonsma and others (31, 32) in South Africa and by that of Barnicoat *et al.* (16) and Clarke (41) in New Zealand. Recent contributions to the subject have come from Cameron & Hamilton (35) in Canada, and Wardrop, Tribe & Coombe (170) in Australia. Southcott & Corbett (212) have reported that weaning Merino lambs at 4 or 6 weeks helped to control roundworm infection, compared with weaning at 11 weeks of age. Benefits have also been obtained by separation of ewes and lambs (178). Experimentally, Suffolk × Half-bred lambs have been weaned abruptly from their ewes as early as 15 to 21 days of age (145) and thereafter fattened on herbage alone (149). It took from 169 to 228 days for them to reach 75–85 lb liveweight, however, and one of the reasons for this was that they failed to make any appreciable weight gain for about 3 weeks after weaning (Fig. 5.7). Their rate of gain after this period was similar to that of the normally suckled twins shown, but it must be emphasized that the latter were set-stocked at 6 ewes per acre.

The growth of such early weaned lambs does not normally equal that of well-managed twins. The early weaned lambs referred to in Fig. 5.7 grazed an entirely separate area and it is unlikely that their performance would have been as good if they had been grazed with older sheep. Artificially reared Welsh Mountain lambs have been weaned even earlier, at 10 to 12 days old, but not on to grass alone. Studies of such early weaning, including some hundreds of Dorset Horn lambs, have demonstrated that lambs can live without milk from a remarkably early age (30) but that they cannot eat enough herbage, at least of the kinds so far investigated, for reasonable growth until they are about 5 to 6 weeks old. It has been found that they can sometimes consume enough of a concentrate pellet to grow normally at earlier ages, whether weaned from the ewe or from a bucket rearing system: this has not always proved to be the case, however (21). The practical implications of these findings will be discussed later (see Chapters VI, IX and XV). Here it is sufficient to note that the age, or perhaps better, the weight, at which a lamb can do without milk altogether, depends on what are regarded as satisfactory criteria for 'doing without'.

In general, the more milk a lamb receives the faster it will grow and this may remain true for several months. This basically is why singles grow faster than lambs suckled in larger litters (23). Single lambs are normally somewhat heavier at birth (Table XXIV) and may differ in other respects from lambs born as multiples, but it is the milk supply that dominates their growth. This was clearly shown by Wallace (167) when he transferred twin lambs to be reared singly on ewes that had borne single lambs.

Similarly, the early weaning of one member of a twin pair at 2, 3 or 4 weeks, leaving the remaining lamb with an increased supply of milk, has demonstrated the same point. Lambs born as twins but reared as singles from an early age grow like singles (51), both in respect to rate and pattern of growth. This is clearly shown in Fig. 5.8, which is based on data from an experiment carried out by T. H. Brown at Hurley. The removal of one lamb from each pair of twins, at 3 weeks of age, resulted in growth rates of the remaining lambs which did not differ significantly from those of singles. Indeed, in the example illustrated, the two groups were very similar in weight and condition at slaughter: the fact that one group of ewes produced and

TABLE XXIV

MEAN BIRTH WEIGHTS OF SUFFOLK × HALF-BRED LAMBS (LB)

	SINGLE	TWIN	TRIPLET
Birth weight . . .	12·5	10·0	8·5
Total weight of litter .	12·5	20·0	25·5

(Source : data from the Grassland Research Institute)

reared, to 3 weeks of age, an equal number of lambs in addition, has considerable implications.

It seems likely that lamb growth would be greatly affected by the composition of the milk (see Table XIX), but little is known of these relationships or of the optimal values for milk constituents when fed in conjunction with various kinds of herbage.

Relative intakes of different foods

The quantitative relationships between the intake of various foods by the lamb are beginning to emerge from recent experimental work (148). It is clear that young lambs receiving as much milk as they will voluntarily consume, eat very little solid food. When the quantity of milk fed is reduced the lamb increases its intake of other foods as soon as it is old enough to do so (34, 87): this is also true of calves (106). This happens even when, on a constant daily milk supply, the lamb's relative milk intake is reduced simply because it has grown bigger. It has been possible to assess this reaction to a reduced milk supply accurately, under conditions of artificial rearing, by feeding different quantities of a liquid milk substitute and measuring the associated intake of grass or other solid food. Fig. 5.9 shows the negative relationship between the intakes of milk and of cold-stored grass* for lambs between 4 and 6 weeks of age: the experiment was carried out in indoor pens at Hurley.

In experiments where the quantity of milk was sharply varied, the intake of grass rose when the milk was reduced but did not decrease when the milk ration was increased. This is partly because, as the lamb grows, its total absolute intake increases anyway: if it receives more milk, the intake of other foods increases more slowly. It is because total voluntary intake is related to body weight that the

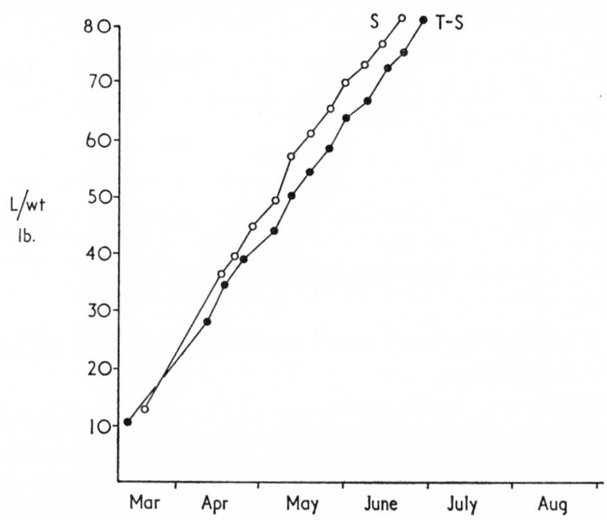

FIG. 5.8—*The growth of lambs born and reared as singles (S) and of lambs born as twins but reared as singles from 3 weeks of age (T-S)* (Source: from T. H. Brown (176))

intakes shown in Fig. 5.9 are expressed per unit of liveweight. Presumably, if the milk ration was increased sufficiently, the intake of grass would decline, unless the development of the grass-eating habit in the young lamb had been firmly established by treatment. This suggests the intriguing possibility of inducing an early intake of herbage by reducing the milk supply for a short period only, on the assumption that the herbage intake will be at least maintained even when the milk intake is fully restored. This possibility could readily be explored with artificially reared lambs.

Since the amount of milk being received can markedly influence the intakes of both herbage and concentrates it follows that deficiencies

* Cold-stored or frozen grass has many experimental advantages (109, 114); for example, the entire crop can be cut at the same time.

in dietary solids may only be expressed when, and if, the milk supply drops below a given level. Thus, if herbage of poor nutritive value is grazed by lambs the effect on their growth may well depend on how much milk they are receiving.

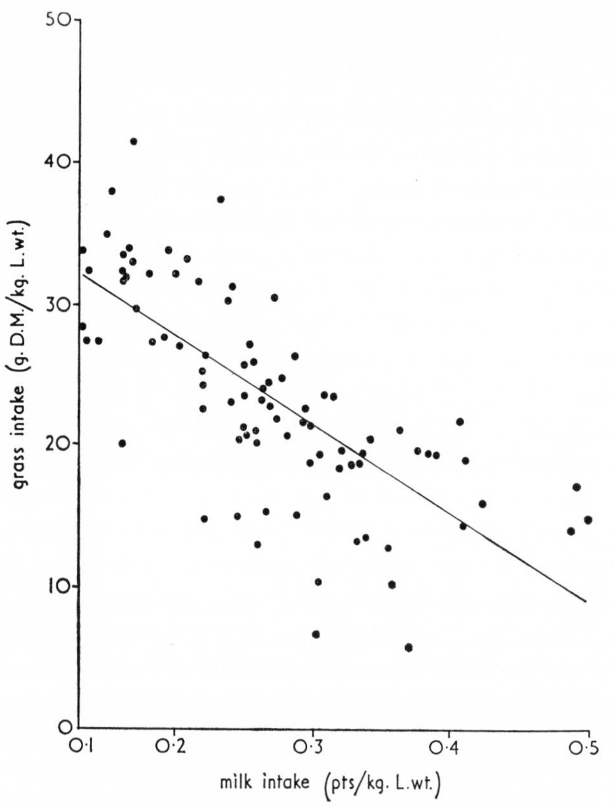

FIG. 5.9—*The relationship between milk intake and grass intake by the lamb*

Regression equation for the line shown:

$$y=a+bx$$

where $a=39{\cdot}9$

and $b=-61{\cdot}0$ (S.E. $=\pm 6{\cdot}73$)

(Source: data from C. R. W. Spedding, T. H. Brown and R. V. Large (1963), *Proc. Nutr. Soc.* 22 (1), 32)

In a similar way, supplementary feeding frequently reduces herbage intake (10, 76), although, on good pasture, the reverse can also happen (123). Thus supplementary feeding is difficult to evaluate under grazing conditions because an unknown proportion of it may simply be substituting for herbage intake. The implications of these interactions are considerable and nowhere more so than in the sphere of animal health and disease.

Early Weaning

Early weaning of some lambs can reduce the litter to whatever number the ewe and its environment can sustain at a satisfactory level of performance. It does not follow, of course, that lambs born as part of a large litter could necessarily be weaned as early (at 2 to 3 weeks) as those born as singles or twins. A progressive weaning procedure can be visualized, however, involving the removal of lambs at, perhaps, weekly intervals until the desired number is left suckling. Since it seems unlikely that lambs could be weaned on to solid food much before 2 weeks of age, this would be the earliest weaning point and a ewe would have to be capable of suckling all her lambs at least up to this age. Those that she could not would be better removed at birth and reared artificially. Early weaning could also be used to shorten the lactation period by removal of all lambs at some time between 3 and 8 weeks. The probable object of this would be related to more frequent lambing, though whether it would be necessary to terminate lactation in order to mate the ewe again is by no means certain (see Chapter IV). It seems likely that the major use for early weaning would be, not so much to release the ewe from stress or for further breeding, but to ensure that the growth rates of the suckled lambs remained satisfactory. This can also be attempted for the lambs removed, by offering more food than was previously available; it is doubtful, however, if early weaning could have the effect of increasing nutrient intake. Furthermore, if a sufficiently nutritive solid food was available that would result in lamb growth rates as rapid as those obtained on ewe's milk, it could just as easily be offered as a supplement to the suckling lamb. Indeed, it is as well not to ignore the possibility of even liquid supplementation of large litters. The main effect of early weaning some lambs would be to restore or maintain the growth rate of those remaining on the ewe. Those removed then represent a separate problem which can be approached from the point of view that these are 'extras'. If a ewe can suckle two lambs in the normal fashion, with normal production, and, in addition, give birth to 1, 2 or 3 extras, they might repay a good deal of additional rearing cost, because of their considerable effect on the productive efficiency of the ewe/lamb unit.

Artificial Rearing

There are two main reasons why it may be desirable to rear lambs artificially; (*a*) because the milk supply to the lamb is inadequate and (*b*) because it is necessary to remove the progeny in order to make full use of the ewe (in relation to breeding frequency, for example).

(*a*) There are many different reasons, however, as to why the milk supply may be inadequate. The most important are related to there being too little milk or too many lambs.

A poor milk yield may be associated with breed, plane of nutrition in pregnancy or lactation (see Chapter IV), disease or lack of suckling stimulus, possibly due to small or not very vigorous lambs at birth. Very high stocking rates can therefore result in lowered milk yields and may represent a category in which removal of some lambs at some earlier-than-normal stage will be sensible. Only if this led to lambs being removed at birth, or within a very few days of birth, would this involve artificial rearing, however.

The extreme case of inadequate milk supply occurs where the ewe dies: the resulting orphan lambs have often been artificially reared by shepherds and farmer's wives, using a variety of methods. In the case of weak orphan lambs, there is considerable advantage in introducing a milk feed by a plastic stomach tube, as used successfully by Dr J. Yuill-Walker of Oxfordshire.

Too many lambs in relation to a normal milk yield will generally be due to an increase in litter size.

(*b*) Removal of some lambs at birth may have little effect on the ewe, although a reduction to one lamb per ewe may decrease the amount of milk removed and thus reduce the strain of lactation.

Removal of all lambs, provided milk is not withdrawn in some other fashion, is associated with a complete cessation of lactation and this has some effect on the physiological state of the ewe. As mentioned in Chapter IV, lactation may cause a state of anoestrus and prevent more frequent breeding. Where this is so and where the aim is to mate the ewe as soon as possible after parturition, all lambs may be removed simply to make the latter possible.

Lamb performance

It is not possible to generalize sensibly about whether lamb performance is improved or not under artificial compared with natural rearing, since each is capable of giving both good and bad results. The chief justification for artificial rearing, however, apart from (*b*) above, is that it will produce better lamb performance in the lambs removed or in those left on the ewe, or in both. This will rarely be so for single lambs on ewes with a high potential milk yield and is most likely where large litters are combined with high stocking rates.

The possible performance in artificial systems is now very high. Developments in diet, method of feeding and disease control, have been rapid over the last few years and, in general, it is now possible to rear lambs artificially at growth rates at least as high as, and frequently higher than those obtained on the ewe, even with singles.

The highest growth rates so far obtained at Hurley are given in Table XXV.

TABLE XXV

GROWTH RATE OF LAMBS FED *AD LIB.* ON COLD MILK SUBSTITUTE WITH SOME HAY AND CONCENTRATES

	Lamb no.	Age (A) when removed from the ewe	Sex	Breed*	Wt. at A (kg)	Wt. at slaughter (kg)	Age at slaughter (days)	L/wt gain g/day
Group I	71	2 days	♀	D.H. × S.HB	5·6	31·5	60	432
	73	1 day	♂	D.H.	3·9	35·5	59	536
	74	1 day	♂	D.H.	4·2	24·5	56	362
	75	1 day	♂	D.H. × S.HB	5·4	34·0	56	511
	Mean				4·8	31·4	58	460
Group II	76	1 day	♀	D.H. × S.HB	4·9	37·0	69	465
	77	1 day	♀	D.H. × S.HB	4·1	30·0	69	375
	78	2 days	♂	D.H.	5·3	35·5	67	451
	79	4 hr	♀	D.H.	3·8	32·0	65	434
	80	4 hr	♂	D.H.	3·8	32·5	65	442
	Mean				4·4	33·4	67	433

* D.H. = Dorset Horn S.HB = Scottish Half-bred

Since such results can be obtained using the diets and methods described in the legend, the reasons for rearing in other ways are primarily economic. It is necessary to know whether current prices of dietary components will change, whether cheaper diets can be used with similar results or, if not, with what reduction in growth rate or food conversion efficiency.

Effect of diet

Colostrum is at least as vital for lambs being reared artificially as for those suckled (177) and possibly more so. It has two main functions; as an initial concentrated source of nutrients and as a carrier of protective antibodies from the ewe. Substitutes are available, if required, for the dietary component (177) but there is as yet no completely satisfactory substitute for the protective part. Immunity is required specifically for those infections that the lamb is most likely to encounter early in its life: prominent amongst these is *E. coli*, but this occurs in many different strains.

Natural colostrum can be supplied in several different ways, the simplest is to allow the lamb to suck it naturally. This involves leaving the lamb with the ewe for about 24 hours. Probably there is a critical period within which the lamb is able to absorb antibodies from colostrum. Lambs are reported to absorb large amounts of γ-globulin from colostrum during the 29 to 48 hours following birth (201, 202) and the levels of γ-globulin in blood samples of dead lambs on hill farms were found by Halliday (190) to be lower than in live lambs. Halliday (191) has also found that the concentration of γ-globulins in the sera of Finnish Landrace lambs was higher than for Scottish Blackface, Merino or Merino × Cheviot lambs: in addition, the concentrations in Finnish twin and triplets were as high as in single lambs. If the critical period for absorption was more precisely defined and if the quantity of colostrum required by each lamb was known, it might be possible to organize the provision of colostrum with greater efficiency. As it is, leaving the lamb with the ewe for the first day works well with small litters, where artificial rearing is less likely to be required, and much less well with the larger litters where artificial rearing is most needed.

A single lamb can safely be left with the ewe and, at the end of its first day, it can be assumed that it has had sufficient colostrum. If the same thing is done with a litter of six lambs, however, there are much greater risks that some will be crushed or injured in some way and it cannot be assumed that every lamb will have received enough colostrum.

It may be better, therefore, although more laborious, to remove the lambs at birth, milk out the colostrum and feed it in a controlled fashion to each lamb. Perhaps, if timing could be improved, the colostrum could be diluted or given in very small doses. A bank of stored colostrum might make the whole process easier: a completely satisfactory synthetic substitute would be ideal, provided the cost was not prohibitive.

There are good reasons for wishing to bring this first 24-hour period under the same degree of control as can now be applied to the remainder of the artificial rearing process.

It is well known that the longer the delay, after birth, in transferring a lamb to its final rearing system, the more difficult is the transition and the more laborious the training required. It is also evident that this training is likely to prove the most labour-demanding part of the whole procedure and may represent the main bottleneck in terms of the number of lambs that one man can handle.

Yet the newly born lamb generally needs no assistance in learning to suck its ewe, which moves about, often hinders as much as it helps and may be engaged in further lambing at the same time.

In short, the newly born lamb is naturally adapted to teat-sucking behaviour and, once it has successfully fed itself, it becomes progressively fixed in its habits and recalcitrant to change. In such circumstances, it appears to be unnecessarily complicated to allow one pattern of behaviour to become fixed, only to spend time and effort in changing it.

There seems to be no good reason why a feeding device should not be designed so that a new-born lamb would train itself and use it right from birth onwards. It might have to have some features that would prove unnecessary after a few days but this would simply imply a 'training' pen for new entrants to the rearing system.

The main factor inhibiting such a development at the present time, is the problem of providing the colostrum. If it has to be milked out by hand, it might as well be given directly to the lamb. A special feeding device only makes sense for an independent supply of colostrum or substitute, even if this is derived from the storage of diluted natural colostrum.

Milk substitute

A variety of liquid feeds have been used in rearing the lamb.

Bottle feeding of ewe's milk (203) is no great departure from the natural situation: the feeding of cow's milk, on the other hand, is a different proposition. Ewe's milk (see Chapter IV) differs in composition from cow's milk, with a dry matter content of 18–19 per cent and a mean fat content of about 7·0 per cent. Cow's milk is thus more dilute and, when fed infrequently, proves an inadequate source of nutrients, since the lamb becomes full of liquid before it has taken in enough nutrients. If it is fed *ad lib.*, however, this problem is overcome and cow's milk has been used successfully in this way. Another way of overcoming the same difficulty is to use dried cow's milk, so that the D.M. concentration of the liquid fed can be adjusted. Large (195, 196) obtained satisfactory results using full-cream dried milk reconstituted with warm water and fed four times a day, and showed that there was little effect of concentration of the milks over a range of D.M. content, from 10–25 per cent, provided that the lambs were fed to appetite on each occasion. Pinot & Teissier (207) compared the growth of lambs on ewe's milk, full-cream dried cow's milk and various milk replacers which contained some tallow and potato starch and which were reconstituted to the same D.M. content as ewe's milk. The key factor appeared to be feeding to appetite at each feed. It is clear that the quantity of nutrients obtained by the lamb is the major factor, within quite a wide range of satisfactory diets, and this is greatly influenced by the method of feeding.

Spray-dried full-cream cow's milk has proved perfectly satisfactory, fed warm or cold (196, 197, 206), as have many proprietary substitutes. The substitute used on a large scale with success at Hurley was similar to that used by Owen & Davies (204); it was manufactured by L. E. Pritchitt & Co. Ltd, and contained 30 per cent fat in the dry matter. The added fat consisted of a blend of animal and vegetable fats homogenized into low temperature preheated spray-dried skim milk with the following additives per kg of dry powder: 7,000 i.u. vitamin A, 1,750 i.u. vitamin D_3, 20 mg vitamin E; 0·01 per cent butylated hydroxyanisole included as an antioxidant. The milk substitute has generally been reconstituted with warm water to 20 per cent dry matter and fed either warm or cold (for the first use of the latter, see Walker (217)). The results have generally been satisfactory either way but rather better with cold milk (see Table XXVI), depending on the method of feeding.

It is important to emphasize that other products have been used with success and a good deal of experience has now been gained with them in several parts of the world (208, 220). Further developments are extremely likely in the near future, since most existing products are

TABLE XXVI

GROWTH OF LAMBS ON WARM AND COLD MILK SUBSTITUTE

Type of reconstituted 'milk'	Group	No. of lambs	Live wt. gain (kg over 21 days) S.E.*	Milk D.M. consumed (kg)	F.C.E.†
Full cream, cold . .	1	6	7·50±0·22	7·74	0·97
(fed *ad lib.*) . . .	2	5	7.87±0·44	9·36	0·84
	3	6	7·61±0·40	7·66	0·99
Full cream, warm . (to appetite 4 times/day)	4	12	6·98±0·36	7·75	0·90
Milk substitute, cold .	5	6	7·65±0·64	8·32	0·92
(fed *ad lib.*) . . .	6	5	5·54±0·27	6·52	0·85
Milk substitute, warm . (to appetite 4 times/day)	7	12	5·45±0·41	7·13	0·76

* From within group variation only

† Food conversion efficiency $= \frac{\text{Live weight gain}}{\text{Milk D.M. consumed}}$

(Source: Large & Penning (197))

expensive. The procedures are now capable of straightforward application giving excellent and predictable results; a lower-cost substitute would therefore transform the practical situation. This would be extremely likely to remain based on skim milk, since young lambs are unable to utilize carbohydrates other than lactose and glucose (218, 219).

It is primarily the cost of the presently available substitutes that provides the biggest incentive to early weaning within artificial rearing systems, in order to get the lambs on to a much cheaper diet at an early stage.

Solid foods

The relationships between the intake of solid foods and the intake of milk have already been referred to: whilst both are fed, such interactions will operate.

Quite a lot of research has now been carried out on the effect of solid supplements to a liquid diet and the effect of different solid diets, without any liquid, on the growth rate and food conversion efficiency of lambs. The feeding of the weaned lamb is dealt with later in this chapter, however, and this discussion is solely concerned with supplements to a liquid diet within artificial rearing systems.

Davies & Owen (183) fed lambs from birth to 11·4 kg liveweight on two levels of liquid feeding, restricted (to 1·14 l. per day) and *ad lib.* (as much as each could consume at each of 4 daily feeds) and measured growth rate and consumption of proprietary creep pellets (crude fibre 7 per cent, crude protein 18 per cent, ether extract 3·5 per cent, a starch equivalent of 65, digestible crude protein 14·2 per cent). They found that growth rates were somewhat lower on the restricted treatment but more pellets were consumed and the total food costs were lower; subsequent performance was also affected, however.

Cunningham, Edwards & Simpson (182) found better growth in lambs on a synthetic milk than on cow's milk but also found some compensation by greater consumption of a meal mixture (35 per cent flaked maize, 20 per cent bruised oats, 15 per cent dried skim milk, 2·5 per cent white fish meal, 17·5 per cent decorticated groundnut meal, 10 per cent molassine meal, with mineral and vitamin supplements).

Effect of environment

The main components of the environment for artificial rearing are (*a*) temperature, (*b*) humidity, (*c*) light.

Little is known as to the optimum values of these components for lamb rearing.

Wide variations are obviously tolerable but the consequences to growth rate or food conversion efficiency are not known.

The optimum temperature probably lies between 20°–30° C but it would be unwise to consider temperature without regard to both humidity and ventilation.

Alexander & Williams (175) give the zone of thermal neutrality for new-born Merino lambs as 25°–30° C. Observations at Hurley have shown that lambs being reared artificially seek a heat source even when ambient temperature is about 16° C.

The requirements for housed fattening lambs are stated (198) to be between 13° and 15° C (certainly less than 26° C) with a relative humidity of 85 per cent, and Hurst (193) reported that ewes with lambs up to 5 weeks of age were kept at a temperature of 14° C in France.

Possible ventilation requirements have been estimated at 0·3 to 0·6 cu. ft/min/lb liveweight (maximum) and 0·1 cu. ft/min/lb (minimum), by Charles (180), and Hjulstad (192) has quoted 17–24 cu. ft/min per animal under Norwegian conditions.

Effect of method of feeding

The two main components of feeding method are (*a*) the size of each feed and (*b*) the frequency of feeding.

(*a*) *Size of feed.* This can be determined by the amount offered or, where the supply is *ad lib.*, by the voluntary intake of the lamb.

Where the object is to economize by rationing, no problems from overfeeding should arise (provided large individual feeds are avoided) but growth rate may be reduced. Lowered liquid intake, however, will lead to increased intake of solids and it is under these conditions that abomasal bloat has frequently caused trouble (215).

Where the object is to encourage high intakes in order to obtain high growth rates, the problems are largely concerned with digestive upsets following overfeeding, leading to scouring.

Voluntary intake of liquid at any one feed will depend upon the size of the lamb and how hungry it is: it will therefore depend upon the previous feeding pattern.

It is not clear whether the *rate* of liquid intake is important, independent of the size of a single feed, but it is possible that it could influence such things as the effectiveness of the closure of the oesophageal groove.

In general, the size of an individual feed is only likely to be important where the frequency of feeding is low. Thus, in the feeding of cold milk substitute, for example, if the feeding frequency is high the individual feed size is likely to be small and the temperature unimportant. On the other hand, with a low frequency (e.g. once or twice per day) it is easy to imagine circumstances where the intake of a large volume of cold 'milk' at one time could be disadvantageous.

(*b*) *Frequency of feeding.* Many different feeding regimes have been successfully used. For young lambs, the minimum frequency may be regarded as about 4 feeds per 24 hours and more frequent feeding would often lead to higher growth rates. With liquids that are fed at very low temperatures (e.g. near 0° C) or at very low D.M. concentrations, it is probable that better performance would be associated with increased frequency of feeding.

There are complex interactions involved in these feeding patterns, however. This is well illustrated by an experience at Hurley with a prototype automatic feeding device.

This device was first set to supply 'milk' *ad lib.* and the total daily intake measured. Using the same lambs, the device was then set to make the intervals between feeds progressively longer and the feeding periods progressively shorter. It was not until a combination of 2-hourly intervals and 30-second periods had been passed, that daily intake began to be reduced: at this combination intake was still equal to that obtained under *ad lib.* conditions. The importance of frequency can be appreciated when it is realized that this combination only gave a total of 6 minutes feeding time per 24 hours.

Ad lib. feeding may be regarded as the extreme case for frequency but, in fact, it can only result in the maximum frequency that a lamb will voluntarily adopt. It is not certain that this frequency will necessarily be as great as could be achieved by some imposed systems: indeed, it is very unlikely in the case of systems involving severe rationing, where very small individual feeds could be combined with very high feeding frequencies. This might not be any more difficult to achieve within automated systems and might have some advantages.

So far most experience suggests that *ad lib.* feeding has given maximum intakes and maximum performance.

Rearing methods and equipment

The physical methods of liquid feeding and the associated equipment vary from simple bowls and bottles to costly automated pipeline systems with electronic controls.

Lambs can be trained to drink from bowls or troughs (204) and these represent the simplest possible feeding devices: they are only appropriate to certain limited situations, however.

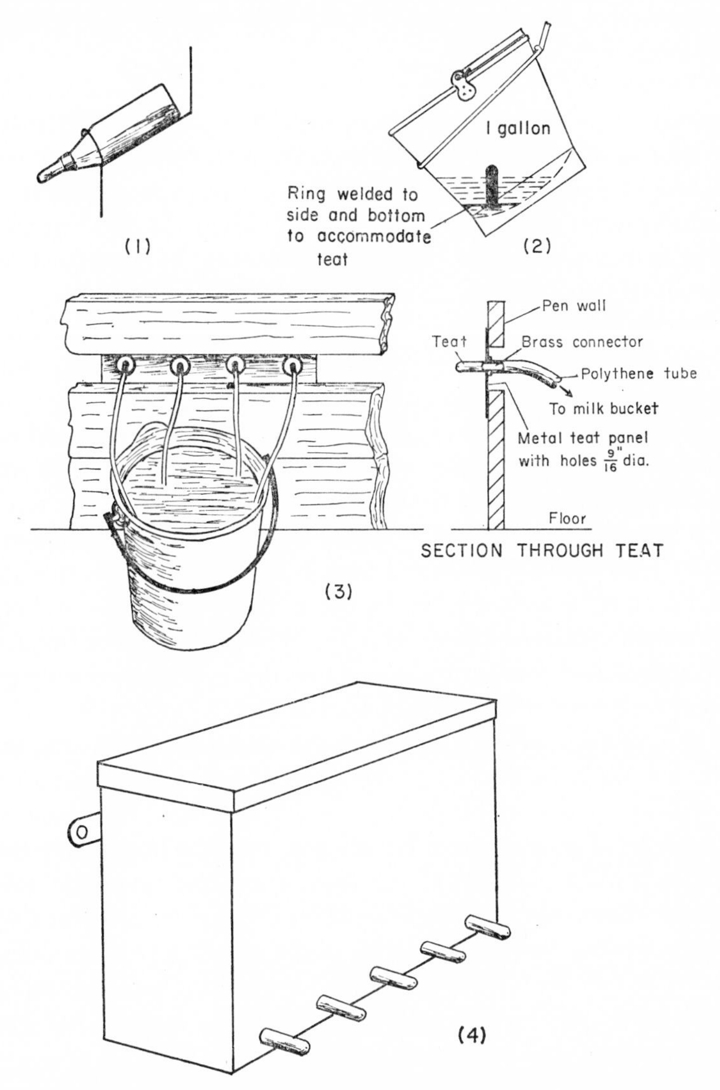

FIG. 5.10—*Simple feeding devices for lamb rearing* (for details see caption to Fig. 5.11 on p. 162)

Bottles fitted with teats are the simplest of the suckling devices and have been much used by shepherds, farmers and their wives for rearing

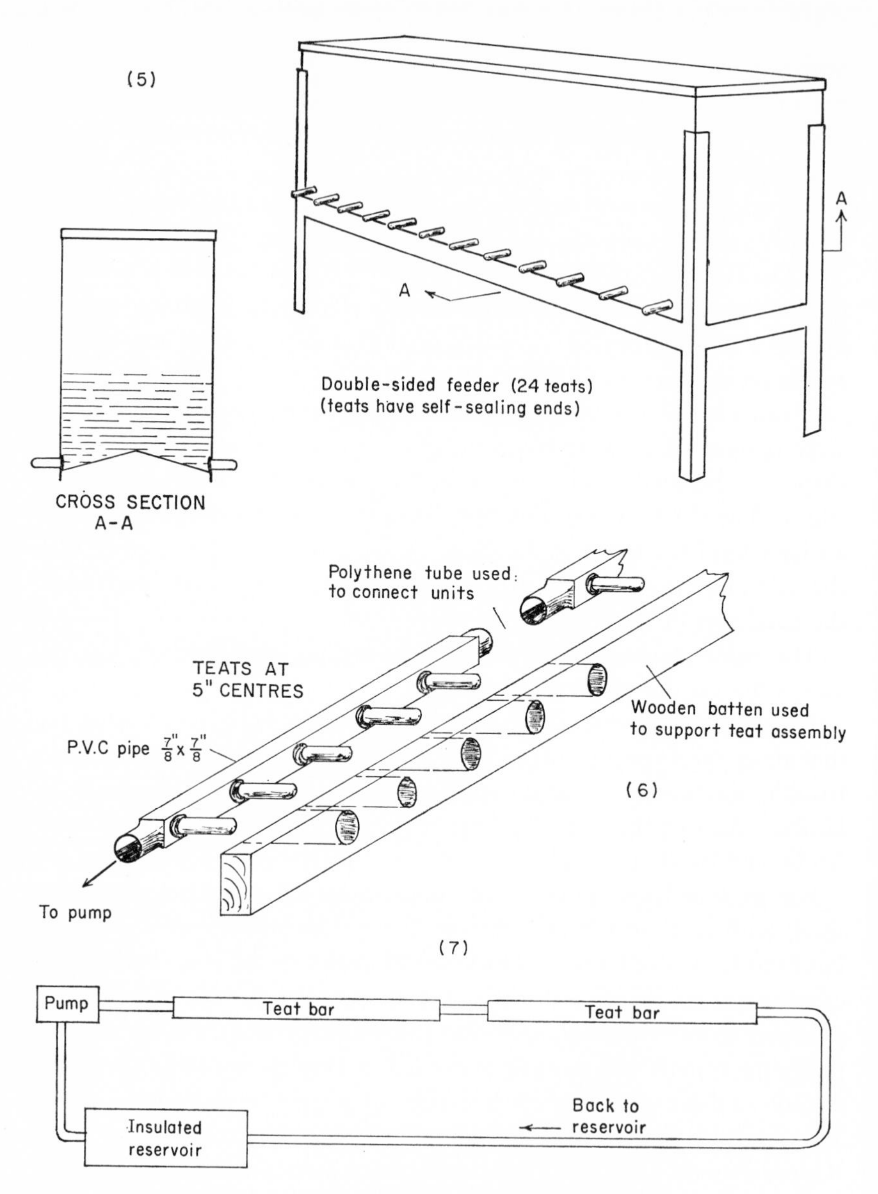

FIG. 5.11—*Simple feeding devices for lamb rearing*

1. Bottle
2. Bucket
3. Bucket with plastic tubes
4. Metal box with teats
5. Box for 24 lambs (as currently used at Hurley)
6. Section of a pipe-line
7. Plan of a pipe-line

orphan lambs. Their simplicity is deceptive, however, for bottles are not simple to clean or to use for large numbers of lambs. Supporting stands have been designed but they represent a quite unnecessary degree of complexity.

Historically, the next step was to use devices which combined several (usually 12) teats and tubes connected to one container (usually circular). The disadvantage of the principle involved is that no milk is received until after an appreciable sucking interval. This is the last thing wanted during a training period: a quick reward is required for any correct action on the part of the lamb.

In order to achieve this and to avoid the complexity of tubes, devices were designed that were based on gravity-feeding: containers of any shape can be used, with teats projecting from the lower surfaces or edges. The risk here is of leakage, of a minor kind from the teats or of a major kind if a teat was damaged or knocked out. The former was the more common and non-return valves were often used, inserted in the basal part of the teat.

The most recent version is the simplest yet and depends not on valves but on the structure of the teat.

A selection of these devices is illustrated in Figs. 5.10 and 5.11, including the type in current use at Hurley. This has given little trouble and has been satisfactorily used to rear many hundreds of lambs. Any of these devices can be used for *ad lib.* feeding of cold 'milk', replaced once daily.

For feeding large numbers of lambs, quite long containers can be used, with rows of teats on the two long sides, allowing 20 or more lambs to feed at one time. This leads naturally to the idea of a square-cross-sectioned pipeline through which cold 'milk' can flow or be pumped from a central reservoir (insulated or not). A prototype pumping system was devised and tried at Hurley, by P. D. Penning in 1966, and demonstrated considerable advantages for large-scale work. This has now been taken considerably further, in conjunction with Ripper Robots Ltd, and a fully-automated system is currently in use, capable of feeding 480 lambs at any one time. For experimental purposes, the feeding frequency can be controlled from *ad lib.* to once-daily; the interval between feeds can be varied from 20 minutes to 24 hours; and the length of the feeding period can be varied from 30 seconds to 15 minutes.

An impression of the system is given in Fig. 5.12.

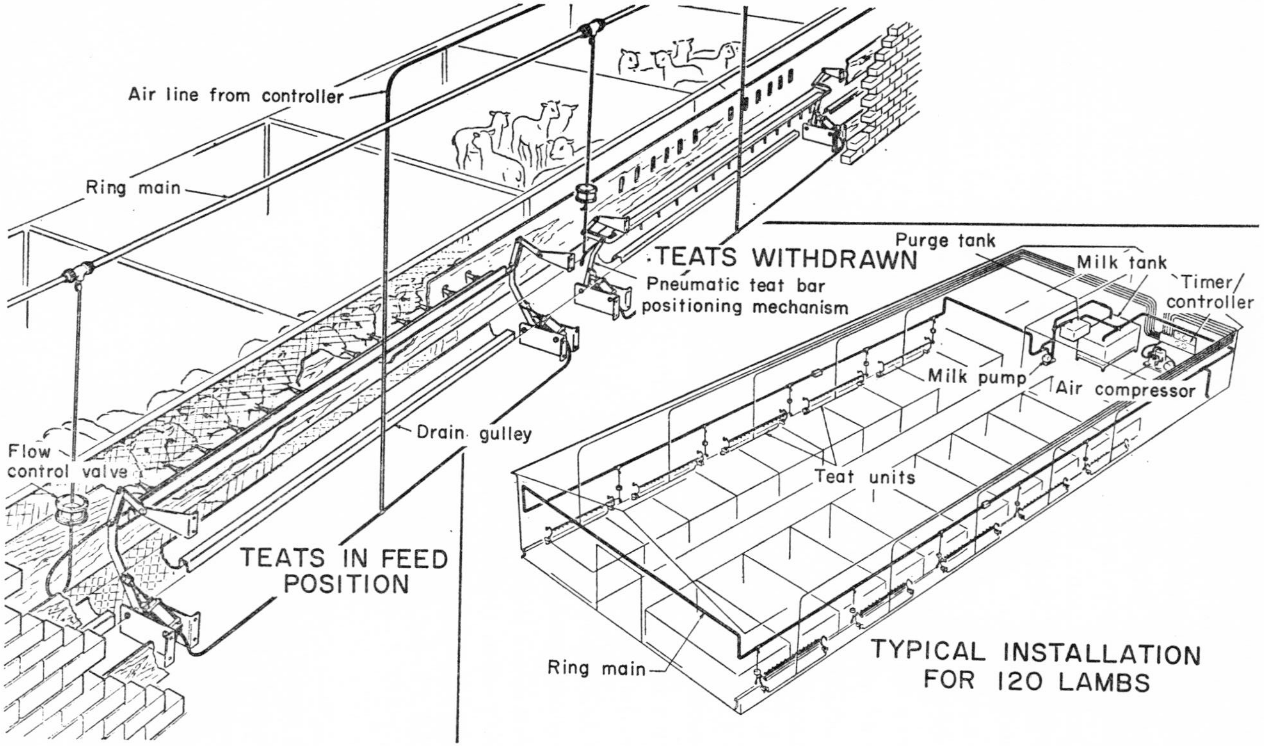

Air line from controller
Ring main
TEATS WITHDRAWN
Pneumatic teat bar positioning mechanism
Flow control valve
Drain gulley
TEATS IN FEED POSITION
Purge tank
Milk tank
Timer/controller
Milk pump
Air compressor
Teat units
Ring main
TYPICAL INSTALLATION FOR 120 LAMBS

FIG. 5.12

The economics of lamb rearing

It is, of course, too early to say what version of this equipment, and at what cost, can be justified in precisely what lamb-rearing system. Factors of major importance in this calculation would be the cost of milk substitute, the cost of the lamb at birth, the cost of any necessary housing and the price of the product.

There is no point in discussing the particular situation at the time of writing but there are some underlying principles that are worth mentioning.

It now seems reasonable to suppose that automated rearing systems will be available that will allow lamb rearing, or, indeed, lamb production to the point of slaughter, to be carried out in a fairly predictable manner, on a large scale, with good disease control, using little labour and obtaining high performance and efficient food conversion. The questions of importance will then be economic.

It is likely that the cost of a new-born lamb will depend chiefly on litter size for a given ewe, but also on the cost of the ewe and the costs of keeping it. This could be extremely variable, therefore, but should become lower. Total food costs could certainly be reduced but it is too early to put figures to this. The other costs can only be estimated very approximately and are very sensitive to the scale of operation.

It is not possible to generalize about the economics of lamb rearing because of the large variations in most parts of the calculation, not only at the present time but presumably also in the future. In order to illustrate the kind of calculation that has to be made, however, *and solely for this purpose*, P. D. Penning has prepared the following account, stating first the main assumptions on which it is based.

Costs of rearing new-born triplet lambs

Assumptions

Intensively managed grassland flock of 100 ewes (e.g. Scottish Half-bred)
No housing
Flock life of 4 years for ewes
Two rams per 100 ewes, with life of 5 years
Ewes stocked at 20 per acre
All lambs artificially reared
300 per cent lambing

Variable costs	£	s.	d.
Foods:			
Concentrates 40 cwt at £28 per ton (approx. 1 lb/head for 6 weeks)	56	0	0
Hay 8 tons at £12 per ton (approx. 3 lb/head for 90 days)	144	0	0
Grazing 5 acres at £6 per acre	30	0	0
Other costs:			
Labour (shearing and lambing bonus only)	20	0	0
Miscellaneous expenses at 15s. per ewe	75	0	0
Depreciation cost of ewes (assuming initial purchase price of £12 and cull price of £4. 15s. Costs over 4 years)	180	0	0
Ram costs per annum (assuming initial cost of £45 each and cull value of £6 each over 5 years)	16	0	0
Cost of P.M.S. and sponge	90	0	0
Seeds and fertilizer at sowing	14	0	0
Returns:			
97 fleeces at $6\frac{1}{2}$ lb and 4s. 6d. per lb	142	0	0
Total variable costs for 300 lambs	£483	0	0
Fixed costs			
Rental for 5 acres at £10 per acre	50	0	0
Tractor running costs for production (3 hours per acre at 5s. 10d. per hour)	4	10	0
Fencing, watering and feeding equipment (depreciated over 5 years)	30	0	0
Share of farm overheads (water, bank charges etc.)	7	0	0
Labour 6 hours per ewe per annum at 6s. per hour	180	0	0
Labour costs for establishment of grassland and seeds and application of fertilizer	4	10	0
Total fixed costs	276	0	0
Plus variable cost of	483	0	0
Total cost for 300 lambs	£759	0	0
∴ cost per lamb at birth (*a*)	£2	10	9

Variable costs for the artificial rearing of one lamb

	£	s.	d.
Milk substitute per lamb 21 lb at 1s. 6d./lb	1	11	6
Early weaning pellets 15 lb at 4d./lb	0	5	0
Barley beef mix 190 lb at 3d./lb (based on Owen & Davies (1965), 3·5: 1 F.C.R. and 54 lb l/wt gain required to give 40 lb e.d.c.w.)	2	7	6
Share of lamb mortality (6% mortality)	0	3	0
Medicines etc.	0	2	0
Total (*b*)	£4	9	0

Fixed costs for artificial rearing of lambs

	£	s.	d.
An *estimate* of cost per lamb (*c*) (assuming 1,500 lambs per year reared in a building capable of holding 500 lambs, one person employed full time)	£1	10	0

N.B.—Time required to fatten a lamb using this sytem would be 102 days. Thus, if lambs are housed only 3 times, the capacity of the house is equal to the total number of lambs that can be reared per annum.

Total cost per lamb ($a+b+c$)= £8. 9s. 9d.

It can be seen that, in this example, the total cost is high and would require a high price for the product in order to make the enterprise worth while.

Larger litters or more frequent lambing could greatly improve the outcome by raising output; cheaper, smaller ewes and cheaper milk substitute could greatly reduce costs.

Diseases Affecting Growth Rate

Disease frequently affects the growth of the lamb either by reducing intake or by decreasing the efficiency with which food is converted. Indeed, any impairment of the health of the lamb, from physical injury to infectious disease, which affects growth, must do so in one or both of these ways. In the interests of efficient animal production, disease would be better eliminated, wherever this is possible.

The elimination of disease, however, does not necessarily involve

TABLE XXVII
ROUNDWORMS PARASITIC IN SHEEP

Species	Common names	Length in mm (total range ♂ and ♀)	Prepatent[1] period (days)
A. *Adults living in the Abomasum*			
HAEMONCHUS contortus[2]	Barber's pole worm; Large stomach worm; Twisted stomach worm; Common stomach worm; Wireworm	10·0–30·0	15–21
OSTERTAGIA circumcincta, trifurcata, ostertagi[2]	Medium stomach worm; Brown stomach worm; Small stomach worm; Brown hairworm	6·5–12·2	15–17
TRICHOSTRONGYLUS axei[2]	Stomach hairworm; Bankrupt worm; Small stomach worm	4·5–7·0	17–21
B. *Adults living in the Small Intestine*			
TRICHOSTRONGYLUS vitrinus, colubriformis[2]*, probolurus, rugatus, falculatus, capricola*	Bankrupt worm; Black scour worm	4·0–7·0	17–21
COOPERIA curticei, oncophora[2]	Small intestinal worm	4·5–9·0	14
NEMATODIRUS filicollis[2]*, spathiger*[2]*, battus*[2]	Thread-necked strongyles; Thin-necked intestinal worm; Thin-necked bowel worm	10–23	21–29
BUNOSTOMUM trigonocephalum[3]	Hookworm	12–26	21–70
STRONGYLOIDES papillosus[2, 3, 4]	Threadworm	3·5–6·0	7
C. *Adults living in the Colon and Caecum*			
OESOPHAGOSTOMUM columbianum[5]	Common nodular worm	12–21·5	30–56
OESOPHAGOSTOMUM venulosum	Lesser nodular worm	11–24	
CHABERTIA ovina[2]	Large-mouthed bowel worm	13–20	48–54
TRICHURIS ovis[2, 4]	Whipworm	35–80	30
D. *Adults living in the Respiratory Tract*			
DICTYOCAULUS filaria	Thread lungworm; Sheep lungworm	30–100	28–42
PROTOSTRONGYLUS rufescens[6]	Red lungworm; Smaller sheep lungworm	16–35	28–42
MUELLERIUS capillaris[6]	Hair lungworm; Nodular lungworm	11–30	21–42

TABLE XXVIII

FLUKES AND TAPEWORMS OF SHEEP

Species	Common name	Length in mm	Prepatent period (weeks)	Part of sheep inhabited by adult
A. *Flukes*				
Fasciola hepatica [2]	Common liver fluke	30	10–14	Liver and bile ducts
Fasciola gigantica [2, 5]	Common liver fluke	250–750	10–14	Liver and bile ducts
Fascioloides magna [2, 5]	Large liver fluke	100	20	Liver and bile ducts
Dicrocoelium dendriticum [2]	Lancet fluke Lesser liver fluke	6–10	11	Liver and bile ducts
Paramphistomum cervi [2]	Conical fluke Rumen fluke	5–12	8–16	Reticulo-rumen
B. *Tapeworms*				
Moniezia expansa [2] *Moniezia benedeni* [2]	Broad tapeworm	Up to 6,000	5–6	Small intestine
Thysanosoma actinoides [2, 5]	Fringed tapeworm	150–300	5–6	Small intestine (sometimes in gall ducts)

C. *Bladderworms* (Larval tapeworms)

	Common name	Size of cyst in mm.	Period from ingestion to fully developed cyst (months)	Site of cyst
Coenurus cerebralis[3] (larvae of *Toenia multiceps*, a tapeworm of dog, fox and jackal)	Gid parasite	50 diameter	7–8	Brain and spinal cord
Cysticercus tenuicollis [2] (larvae of *Toenia hydatigena*, a dog tapeworm)	Thin-necked bladder-worm	50 diameter	1	Liver and peritoneal cavity
Echinococcus granulosus [2] (larvae of a dog tapeworm of the same name)	Hydatid worm	50–100 diameter	5–6	Liver and lungs
Cysticercus ovis (larvae of *Toenia ovis*, a dog tapeworm)	Sheep measles worm	10 × 20 oval	3	Heart, diaphragm and muscles (esp. of jaws and tongue)

All the species in this Table XXVIII require a secondary host in order to complete their life cycles.

[1] Period between ingestion of infective larva and sexual maturity of adult.
[2] Also occur in cattle.
[3] These species can infect the host by penetrating the skin.
[4] Not particularly pathogenic but extremely common in the United Kingdom.
[5] Not normally found in the United Kingdom.
[6] These species require a secondary host to complete their life cycle.

the elimination of all organisms capable of producing it. The encouragement of immunity may be the best insurance against the risk of serious disease even where such resistance is obtained at some slight cost in terms of lowered efficiency. Apart from this consideration, which is not in fact so very different in its long-term aim, the object must be to eradicate disease from the production process. The elimination of all dietary faults, too much of one thing (36, 49, 160) or too little of another (7, 22), must await more knowledge of animal requirements and how to satisfy them. Many diseases, such as braxy, pulpy kidney, lamb dysentery, blackleg and tetanus, can be effectively eliminated by vaccination (150). Certain infectious diseases, such as foot-rot, could and should be eliminated by management, including foot-bathing (52, 91, 92, 101).

The most important and widespread diseases of the lamb that cannot, as yet, be completely eliminated under practical conditions are those caused by internal parasites, primarily those of the lungs, the stomach and the intestines (139, 143, 146). Tables XXVII and XXVIII list those that are most common or most important and indicate their size and the part of the sheep where the adults are found: Plate III shows what some of them look like. Most of them have been shown to be capable of reducing the liveweight gain of a lamb.

Lungworms

The lungworms (95, 98) do their damage (122) partly in the lungs where the adults live and partly in other parts of the body during the migrations of the immature stages (121). If present in sufficient numbers they can cause parasitic bronchitis or husk: this is comparable to the same disease in cattle but is usually caused by a different species of worm. The disease can be mild, with occasional coughing, particularly when sheep are driven; or it can be devastatingly severe, resulting in rapid weight loss, loss of appetite, and, frequently, death (66). Plate IV (a) shows the appearance of an affected lung.

One of the most serious results of such lung infections may be secondary bacterial invasion of the damaged tissues. *Pasteurella haemolytica* is probably the commonest of the secondary organisms (68, 69, 70) and can give rise to appalling losses. This superimposition of secondary troubles may frequently increase the severity of the total consequences of an initial infection. Since lungworms often require

PLATE V

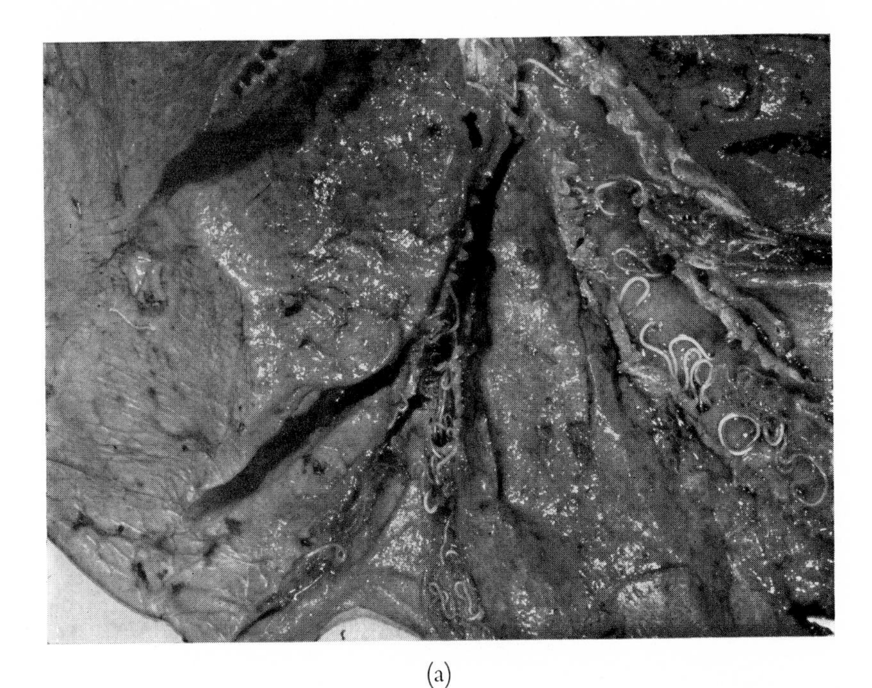

(a)

A lamb's lung dissected to show the presence of adult lungworms (this lamb had a total of 159).

(b)

Two Suffolk × Half-bred lambs of similar age: the lamb on the right carried a heavier though moderate worm-burden.

PLATE VI

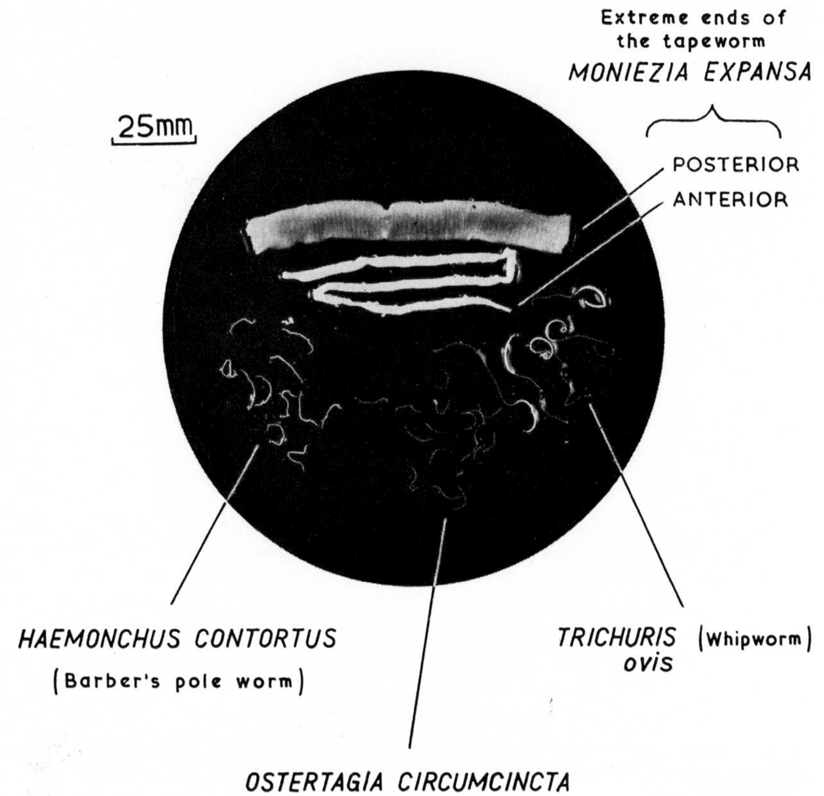

Some common worms of the alimentary tract of sheep.

secondary hosts, such as slugs, in which to complete their life cycles (119, 120), the effects of grazing management are somewhat complex (see Chapter IX). The occurrence and spread of secondary infections may be much affected by the conditions, such as humidity, housing (1) and degree of overcrowding, under which the sheep are kept (85) (see Chapter X). Such management considerations are of great importance in the control or prevention of coccidiosis in sheep (93, 105).

Coccidiosis

The coccidial populations of normal sheep have been described by Pout *et al.* (210) and by Joyner *et al.* (194).

Pout (209) showed that experimental infection with *Eimeria crandallis* decreased weight gain in young lambs and caused abdominal pain and lassitude.

Treatment with amprolium and ethopabate has been successful in natural infections (211), resulting in greater weight gains even where the level of infection (with *E. arloingi*) was judged to be 'normal'. However, these lambs were weaned at 4 weeks of age and it seems probable that this would exacerbate the effects of coccidia.

As a problem, coccidiosis appears more likely in circumstances where milk supplies are, for one reason or another, reduced. Thus intensive stocking of early weaned lambs represents a situation in which the risk may be substantial.

Parasites of the liver, stomach and intestines

Parasites of the liver, stomach and intestines (67) include the tapeworms (111) and flukes, or flatworms (112), and the nematodes, or roundworms (11, 46, 60, 79, 125); the last are generally by far the more important. The common tapeworms and flukes of lambs all require intermediate hosts, which are often oribatid mites and snails, respectively, whereas the common nematodes all have a direct life cycle. When present in large numbers tapeworms can cause a reduction in growth rate; sometimes they cause physical blockage of the gut. Hansen and others (73, 74) found that lambs infected with *Moniezia expansa* were retarded in growth. This began with maturity of the tapeworms and was accompanied by some depression in the haemoglobin content. The average duration of a single infection of tapeworms was found to be about 60 days but one lamb remained infected for 228 days. Although infected lambs passed softened faeces, no diarrhoea developed. Stampa (216) also found that *Moniezia expansa* caused a reduction in growth rate of lambs, especially under arid and generally less nematode-infested conditions.

The liver fluke, *Fasciola hepatica*, like the tapeworms, requires an intermediate host, without which it cannot develop (205). In this case it is a small snail, and, in Britain, it is limited to one species (*Limnaea truncatula*). The implications for management inherent in this type of life cycle will be considered in Chapter IX. Liver flukes may cause 'acute' disease due to massive infestation of the liver by immature stages and death may occur suddenly. Otherwise, since flukes take 10–12 weeks to mature (105), the problem is not only restricted to certain areas and mainly in certain years but is also chiefly a problem in older sheep (see Chapter IV).

Nematodes are most serious when present in large numbers, and almost astronomical numbers are commonly associated with such parasites (45, 156). Nevertheless remarkably few may cause significant damage, particularly in a susceptible lamb. Gibson (57) has pointed out that between 30,000 and 40,000 *Trichostrongylus axei* would just about fill a teaspoon, yet numbers of this order can make a difference of 30 lb in the liveweight gain of lambs during a period of one year. A great deal of research in recent years has been devoted to the study of levels of infestation which may cause no obvious symptoms of disease but yet reduce performance (138, 155, 157). Gordon (62), writing not only of Australian conditions, concluded that 'the significance of subclinical infections is now generally recognized'.

The way in which this damage is done is incompletely understood. Nematodes of many species have been shown to reduce liveweight gain (12, 131, 161, 162, 163) and wool growth (37, 63, 83, 124, 144), even at subclinical levels (19, 142), and have been shown to be capable of affecting carcase composition by altering the rate of growth. Southcott, Heath & Langlands (213) found that in dry, mature Merino sheep, elimination of worms by drenching with thiabendazole resulted in greater wool production. The untreated sheep not only grew less wool but also appeared to consume more food: the gross efficiency of wool production was therefore depressed, from 14·1 to 10·8 of wool per kg of organic matter ingested. Many species are blood-suckers and cause anaemia, others feed on the tissue of the gut wall. A recent estimate of the blood loss resulting from *Haemonchus contortus* infection in sheep was made by Clark, Kiesel & Goby (40). They used radioactive chromium and iron to label the red blood cells and found that labelled blood first appeared in the faeces 6 to 12 days after the sheep were given the *H. contortus* larvae. The average blood loss per parasite per day was estimated at 0·049 ml.

Many nematode species cause a reduction in appetite. This has been shown for *Trichostrongylus axei* by Gibson (57, 58) and for *Nematodirus spathiger* by Kates & Turner (80). Gordon (61) also reported effects on appetite and later (63) showed that a single dose of 50,000 *Trichostrongylus colubriformis* larvae could result in a 50 per cent reduction in food intake within 6 weeks of infection. Gordon also reported (63) a decrease in food intake from over 2·5 lb daily to 1 lb within a month of the administration of 4,900 *Oesophagostomum columbianum* larvae. The pre-infestation level of food intake was not regained for almost a year. Some species have also been found to reduce the 'apparent digestibility' of the food eaten (54, 131, 140, 153). This latter term expresses the difference between the amount of food ingested and the amount of residue, voided in the faeces, as a proportion of the amount eaten. A reduction in apparent digestibility, therefore, does not necessarily mean that the animal has not *digested* its food as efficiently: it may mean, and probably does in this case, that less has been absorbed. Dobson (185), working with *Oesophagostomum columbianum*, found that anorexia and depressed 'apparent digestibility' were associated with low infestation doses (500 larvae); at higher levels (2,000–5,000 larvae), depressed water intake and diarrhoea also contributed to the loss of weight that was suffered by the sheep.

Since worms can cause considerable inflammation of the wall of the abomasum and enteritis of the intestine (81, 82), it seems reasonable to postulate interference with absorption as one of the consequences. Damage of this kind might conceivably exercise an adverse influence on the growth of the lamb, even after the worms causing it have been eliminated. Little is known of such long-term consequences. Normally, loss of productivity, through reduced liveweight gain, for example, is permanent in the sense that the lamb may remain that much lighter than comparable, uninfected sheep. In other words, it will take longer to reach a given weight. Occasionally, by what is known as 'compensatory growth' (see Bassett (17) and Chapter VI), lambs may actually catch up (50). Such growth is difficult to assess because an animal which gains weight rapidly after a temporary set-back, whether caused by disease or not, may simply be regaining its normal content of ingesta (154). The contents of the alimentary

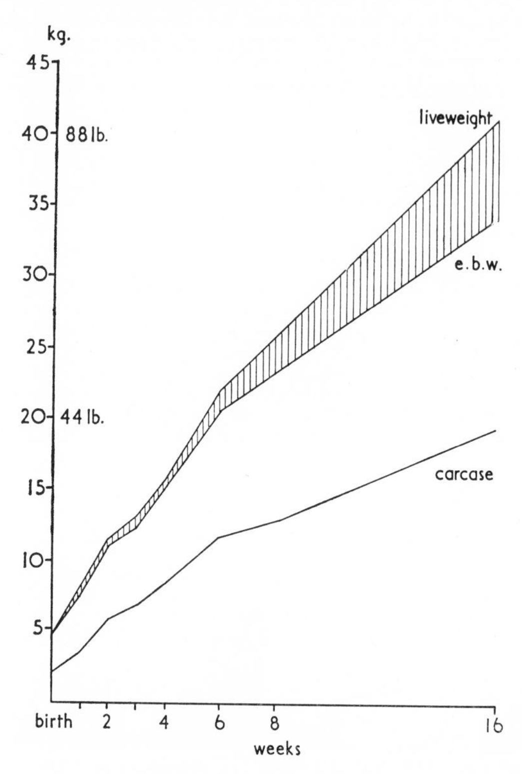

FIG. 5.13—*Relative proportions of carcase and empty body weight (E.B.W.) in the liveweight of a lamb. The shaded area represents the contents of the alimentary tract*

(Source : data from R. V. Large (174))

tract may amount to 15 per cent. of the lamb's liveweight (see Fig. 5.13), the absolute quantity of ingesta increasing as the lamb grows bigger. Liveweight can also be rapidly regained by restoring water to a dehydrated body. Dehydration is a common result of excessive scouring, or diarrhoea, such as that caused by heavy infestations of *Nematodirus* spp. (18, 20).

The effect of moderate worm-infestation on the growth and appearance of lambs is illustrated in Fig. 5.14 and Plate IV (b). Fig. 5.14 shows the growth of early weaned, artificially reared lambs separated, at random, into two groups (A and B). From 8 May, Group A grazed within a 'worm-free' area; Group B grazed in a normal field, which had been only lightly grazed by sheep. Both groups had a plentiful supply of pure S. 23 ryegrass available at all times. The marked (and statistically significant ($P<0{\cdot}001$)) depression of the

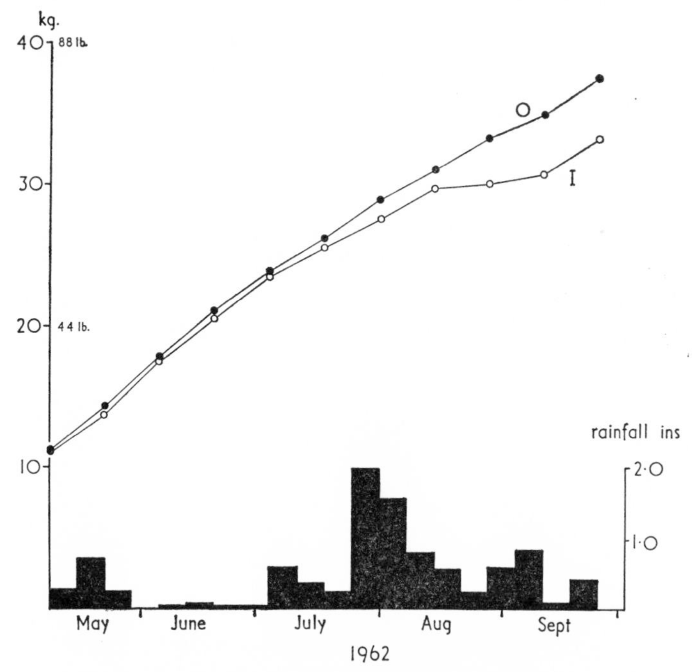

FIG. 5.14—*The growth of weaned 'worm-free' (O) and infected (I) lambs. Rainfall is shown as the total for each week*

(Source : data from C. R. W. Spedding, T. H. Brown and R. V. Large, *J. Agric. Sci.* In Press)

growth of Group B lambs after the drought ended could only be attributed to worm-infestation by reference to the growth of Group A. This provides a good illustration of the need for worm-free controls, not because the growth of lambs is *necessarily* worse in a normal environment but because it is almost impossible to be sure when worms are significantly affecting growth and when they are not.

The Interaction of Nutrition and Disease

There is little doubt that, nutritionally, milk dominates the performance of the suckling lamb and that, from a health point of view, the most widespread dominant factor is infestation with nematode parasites. There are times and places where other factors are of greater importance, but, in general, milk and worms are the major ones. It is of considerable interest, therefore, that these two factors interact with each other to the extent that they do.

It has been pointed out that single lambs grow faster because they receive more milk, and, in the field, it is noticeable that this superior growth rate is usually maintained for a remarkably long time. Indeed, at four months of age, when milk is only a part of the diet, the difference in favour of single lambs may be greater than it was earlier. Thus the ratio of single to twin growth rates frequently increases as the lambs get older (147). The reasons for this are probably the same as those responsible for the decline in growth rate, from about 10 weeks of age, that is observed in most lambs. As lambs grow older they normally become fatter; and fat, having a higher calorific value than other tissues, requires more food per unit of weight gained; but, at least in several fat-lamb producing breeds, the decline in growth rate is not primarily due to changes in the kind of tissue being laid down, though these occur, of course. In the absence of parasites, and of other diseases, and with plentiful feeding, lambs of the several breeds studied in this way can grow at a constant rate, or very nearly so, up to about 100 lb liveweight. The two common reasons for the marked curvature in cumulative liveweight growth are inadequate nutrition and too many worms.

Inadequate nutrition

Under grazing conditions inadequate nutrition is most frequently due to a decline in the milk yield of the ewe, a decline in the quality of the herbage on offer, or a shortage of pasture due to drought, inadequate manuring or over-stocking. When these deficiencies occur growth rates normally decline; when they are corrected growth rates may still prove unsatisfactory if worms are too numerous.

Worm burden

Infective larvae tend to be more numerous on the herbage as the season progresses (59) and the spring-born lamb is often exposed to an increasing number as it grows older. This gradual exposure may be beneficial in the sense that it may lead to the development of immunity, acquired resistance (94, 134, 135, 136, 159), before the lamb is exposed to very large doses. The gradual reduction in milk intake and its replacement by herbage have the effect of a graduated exposure even where larval populations are constantly high. This will be further discussed in Chapter IX. Here it is sufficient to note

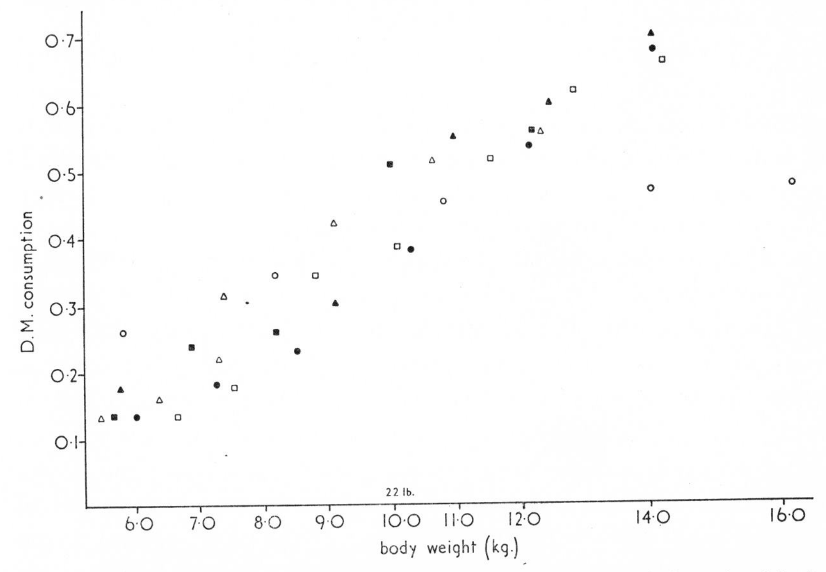

FIG. 5.15—*The relationship between daily D.M. consumption (kg.) and bodyweight of lambs (kg.). The different symbols refer to different feeding régimes, all involving both milk substitutes and solid pellets*

(Source: C. R. W. Spedding, T. H. Brown and R. V. Large (1963), *Proc. Nutr. Soc.* *22* (1), 32)

that, since the lamb is born free from worms, the worm-burden tends to increase for the first few months of its life.

The roundworms that establish themselves in the lamb's alimentary tract, however, have to be ingested as infective larvae on the feed. With few exceptions, this is their only means of entry and, without exception, there is no multiplication within the host. Worms reproduce within the host animal and lay great numbers of eggs, but these must pass through stages outside the sheep before they can complete their life cycle. Not all the larvae ingested become adult and successfully establish themselves. The proportion that do so may vary very widely (from 1 to 100 per cent (158)), depending on the number of larvae being ingested, sometimes on the nature of the diet and considerably on the health status of the sheep. The plane of nutrition can therefore affect the process. The total number established depends also on the number ingested and this is determined by the number on each unit weight of herbage and by the amount eaten. The total amount of dry matter consumed daily varies with the weight or size of the animal (see Fig. 5.15), but, as described earlier in this chapter, at any given body weight, the intake of herbage tends to be negatively related to the milk intake.

Importance of the milk supply

The milk supply to the lamb has important effects on: (*a*) growth rate (directly and particularly when milk forms the major part of the diet); (*b*) herbage intake (the importance of the extent to which a lamb is dependent on this for nutrient supply varies with the quantity and quality of the herbage available); (*c*) intake of infective larvae (which can result in disease or immunity according to the magnitude and distribution of larval intake).

The interactions of these factors are difficult to disentangle in a context as complex as the grazing situation. It has been noted that lambs reared as singles, whether born as twins or as singles, have fewer worms at slaughter than do lambs reared as twins, even when grazed on the same pasture (148). The lambs receiving less milk take longer to fatten, however, and may have more worms because they have been exposed for a longer period. This is not always the explanation; twins have been found in one recent experiment at Hurley to carry nearly twice as many adult worms, predominantly *Ostertagia* spp., as recovered *post mortem* from singles, when all lambs were slaughtered at the same age (about three months). Much work remains to be done on these problems but the following conclusions appear justified:

Lambs receiving a very high proportion of milk, or other high-quality, non-infected feed, in their diet, under grazing conditions (*a*) are less likely to suffer from serious worm-infestation; (*b*) grow rapidly and may thus need to remain in a given environment for a relatively short time (this is particularly relevant to lambs intended for sale); (*c*) are less dependent on the quality and quantity of the herbage available.

It needs emphasizing that, at present, it appears to be the proportion of the daily and total food supply contributed by milk that is important. Clearly this need not have anything to do with whether a lamb is a single or a twin, although it is likely to do so within a breed. In general, the proportion of the total diet contributed by milk will be affected by: (*a*) the milk yield of the ewe; (*b*) the number of lambs sharing the milk supply; (*c*) the ultimate size of the lamb (or, more strictly, the weight to be gained from birth). This is greatly influenced, in fat lamb production, by the desired slaughter weight.

Much practical observation and many experimental results have failed to distinguish between the effects, for example, of the nutritional value of a feed and the quantity being consumed, the effects of disease and of the consumption of other feeds than those under investigation (62, 137). This is not stated in any derogatory sense: the difficulties are still considerable. In fact, the central difficulty in such ecological studies is to disentangle without destroying the situation being studied. A recent contribution to this topic (181) reported an attempt to demonstrate passive resistance in lambs to three species of gastro-intestinal nematodes as a result of sucking the ewe. No evidence was found, however, that passive resistance was transferred to the lambs by the colostrum of the ewe.

The influence of plane of nutrition

There is one further general concept which must be considered. This is the idea, which, as Hunter (78) has pointed out, has been widely held in relation to helminths, that a high level of nutrition has a beneficial influence in disease control. Clearly, in the case of milk, exposure to parasitism may be reduced by what normally represents a high plane of nutrition, but it may not be entirely attributable to this aspect of the diet. It is possible that ewe's milk has properties of direct value in disease prevention or control. Certainly it has effects on such things as the acidity (pH) of the abomasum which could conceivably influence, for example, larval establishment.

It is also possible that a high milk supply could be associated with less bloat, though this is not a major problem in sheep in Britain, less enterotoxaemia or a reduced incidence of any other disorder which may be associated with herbage intake.

Concentrate feeding may also represent a high plane of nutrition, as judged by lamb growth rate, and may also result in reduced incidence or severity of disease for indirect reasons. Inadequate feeding, on the other hand, is bound to place the lamb at a disadvantage in meeting any stress whether due to disease or not. Where a short-term stress would have only mild consequences in a well-fed sheep with plentiful reserves of energy, it might result in serious disturbance to the poorly fed animal, extremely sensitive to any further reduction in its nutrient intake. Obviously, nutrition can never be regarded as unimportant, but it would be equally erroneous to think that it is all that matters. For example, a high plane of nutrition will not necessarily prevent serious worm-infestation. It may, however, make the consequences less serious because a given adverse effect will represent a smaller proportion of the gain, weight, wool yield, appetite or other affected attribute of the well-fed lamb. Fig. 5.16 attempts to illustrate this diagrammatically. However small the effect on performance that is caused by disease, at very low growth rates, i.e. at low planes of nutrition, this effect could prevent any growth taking place or result in weight loss. Thus, if an animal's diet is only adequate for maintenance, any reduction in performance will ultimately result in death if the effect is sufficiently prolonged. It is likely that a high plane of nutrition should be the aim in lamb-rearing, with the possible exception, in some circumstances, of ewe lambs being reared for the replacement of breeding stock.

An illustration of the complexity of the interactions between nutrition and parasitism may be found in the work of Downey (186, 187), in which infected lambs sometimes grew faster if they were also on a cobalt-deficient diet.

The Weaned Lamb

Much of the foregoing is specially important in relation to the suckled lamb. There are some additional features, however, which are particularly characteristic of the weaned lamb.

The main feature of the weaned lamb is, of course, that it has no ewe with it and therefore no milk supply. This results in an all-herbage diet, unless concentrates are used to replace the milk component of the ration. The implications are clear. First, the lamb may be dependent on herbage for all its nutrients; the quality of the herbage may therefore be more critical. Secondly, since the entire intake may be derived from pasture, it becomes even more important

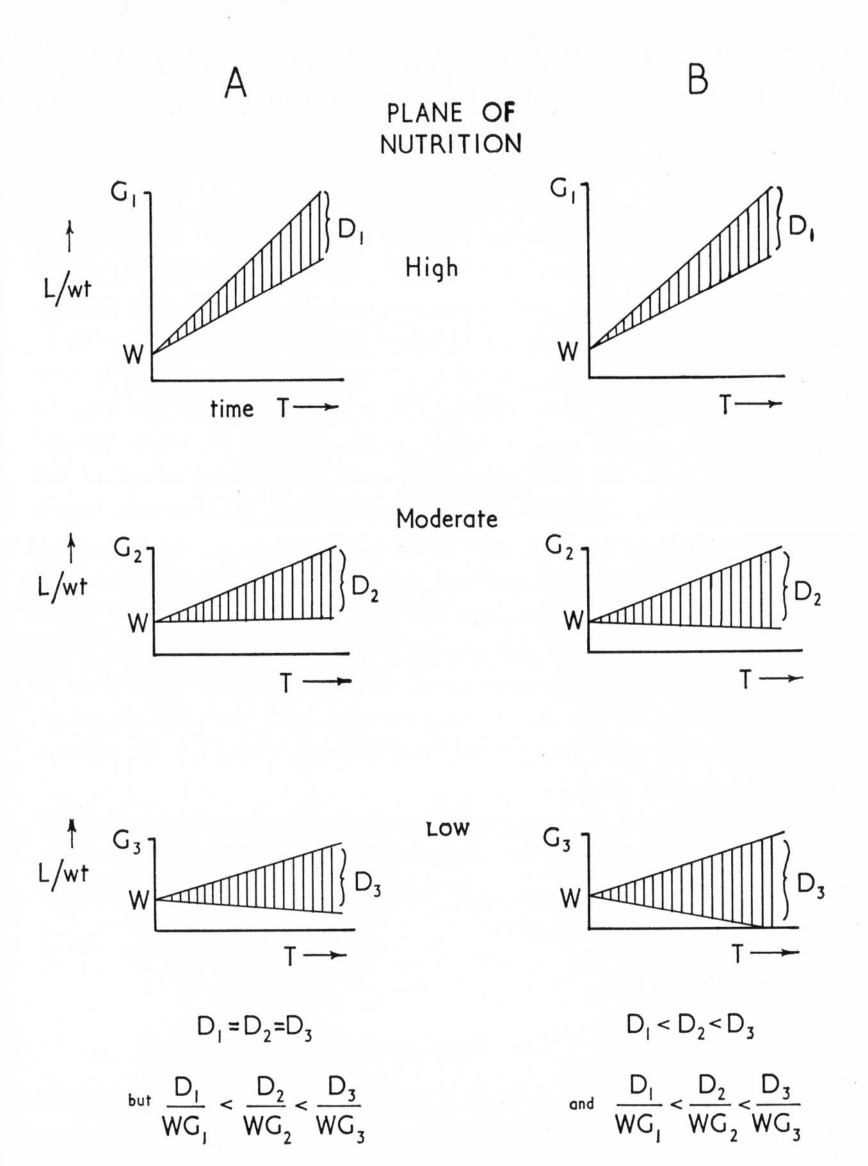

FIG. 5.16—*The influence of plane of nutrition on the effect of disease on animal growth rate, (A) where the effect of disease is assumed to be absolutely constant and (B) where it is assumed to increase as the plane of nutrition decreases*

D = Absolute reduction in growth rate due to disease.
WG = Weight gained by disease-free animals.

The upper line in each diagram represents the growth of disease-free animals; the lower line that of diseased animals.

At lower planes of nutrition, both diseased and disease-free animals gain less weight, but the effect of disease (D) increases as a percentage of the weight gained (WG) by the disease-free group. Thus the importance of the effect is greater at lower planes of nutrition, whether D is assumed to increase or remain constant.

that the number of infective nematode larvae on the herbage should not be excessive. Thirdly, the ewe is not present and cannot therefore affect the parasitological status of the pasture (see Chapter IV). All these features vary in importance according to the age and development of the lamb when weaned.

There are, however, conditions, such as the hogget ill-thrift of New Zealand (42), where parasites are regarded as playing a secondary role (43) and where the pasture is apparently of high nutritive value, in which lamb growth may be markedly reduced. As Clarke & Filmer (43) have shown, marked differences in the growth rate of weaned lambs may be associated with differences in the botanical composition of the pasture and have little to do with worm-infestation. These workers found that the growth of hoggets on white clover pastures was much more rapid than that of similar animals on ryegrass. Recent work at Hurley has suggested that this may be so for 'worm-free', early weaned lambs also.

Although there may be marked differences in the nutritive value of different foods to the early weaned lambs at 2 to 6 weeks of age, they may chiefly reflect differential intake. After about 6 weeks of age, growth may be quite as rapid on herbage alone as that often obtained with suckling lambs under normal grazing conditions. Furthermore, in a 'worm-free' environment (see Chapter XIV), grazing lambs tend to grow at a relatively constant rate over a considerable period. In short, the value of milk appears to be quite different in a context which includes parasites and in one which does not.

It is just beginning to be possible to distinguish between the food value of different kinds of pasture in terms of lamb growth. It would be unwise, in the present rapidly changing state of knowledge of this subject, to regard the results quoted here as more than illustrations of the kind of variation to be expected, even within the rather narrow spectrum of herbage species currently employed in sown pastures. It has recently been shown that S. 24 ryegrass has a higher nutritive value to the lamb than S. 37 cocksfoot, even when fed at the same level of digestible organic matter intake (99). This kind of controlled feeding experiment is essential in order to provide a precise evaluation of herbage species. It is also necessary, however, to evaluate them under grazing conditions and in order to do this it has been found desirable to carry out experiments free from interference from worms. This is, of course, only a special case of the general proposition that, if animal performance is used as an index of the nutritive value of a food, the animal should be healthy. The use of 'worm-free' lambs (141, 147) is further discussed in Chapter XIV. Fig. 5.17 shows the growth curves obtained with a standard type of lamb, over a standard part of its life, grazing at different times of year, or on different herbage species. Bearing in mind the growth rate of which a well-fed single lamb is capable, it is clear that the rates illustrated are below the

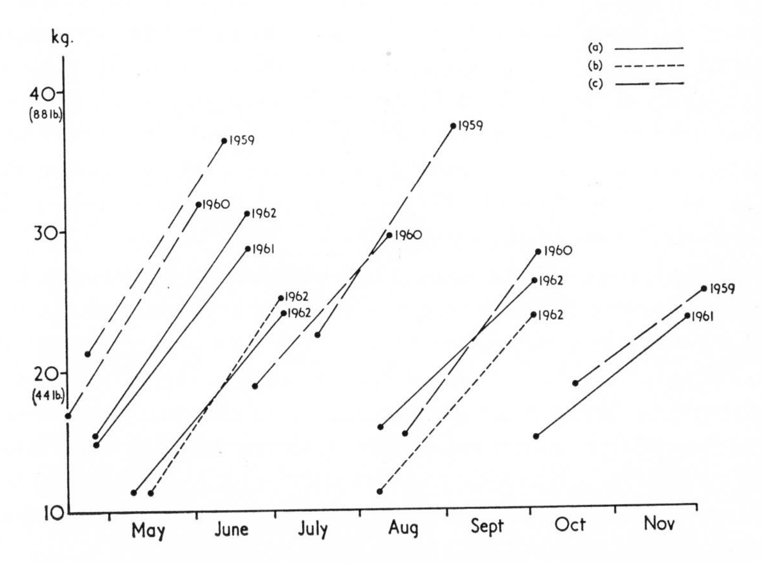

FIG. 5.17—*The growth of 'worm-free' lambs at pasture. The data have been derived from grazing experiments on (a) S. 23 ryegrass; (b) predominantly white clover swards; and (c) S. 23 ryegrass/white clover*

(Source : Grassland Research Institute, Hurley, 1959–62)

genetic potential. The elimination of any possibility of worm-infestation having influenced the results, however, does enable comparisons to be made between species and seasons. The growth curves shown in Fig. 5.17 were all derived from groups of lambs that always had a surplus of herbage available to them. It is uncertain, therefore, to what extent selective grazing may have influenced the result.

Later experiments at Hurley on a wider range of herbage species resulted in growth rates that varied with time of starting and finishing the experiment, with slaughter weight and whether the immediate post-weaning check was included. The main results may be summarized as follows, however:

Crops compared		*Lamb growth rates g/day*	
Lucerne (Eynsford)	Cocksfoot (S.37)	200	159
Perennial ryegrass (S.23)	Cocksfoot (S.37)	195	172
White clover (S.100)	Perennial ryegrass (S.23)	272	241

It cannot be assumed that similar results would occur if the herbage was fed in conjunction with a normal milk supply. In one of the few comparisons made (Treacher, 1966, unpublished data), in which lambs received a ration of 600 g (D.M.) of liquid milk substitute per day, growth rates were best on sainfoin (276 g/day), with only small differences between S.23 ryegrass (162 g/day), S.37 cocksfoot (156 g/day), *Agrostis gigantea* (145 g/day) and S.100 white clover (162 g/day): timothy (*Phleum pratense*) gave intermediate growth rates (207 g/day).

It is difficult to ensure in such experiments that the quantity of herbage is not limiting, especially on very different herbage species. However, a similar experiment in the previous year also resulted in the best lamb growth occurring on sainfoin, with only slightly lower rates on lucerne. In this year, growth rates were much higher on these species than on perennial ryegrass.

REFERENCES

(1) Aikman, C. D. (1962). In-wintering ewe hogs. *Farming Rev. 21* (17).

(2) Alexander, G. (1961). Shelter for new-born lambs. p. 119. *VIII int. Congr. Anim. Prod.*, Hamburg.

(3) Alexander, G. (1961). Temperature regulation in the new-born lamb. III. Effect of environmental temperature on metabolic rate, body temperatures and respiratory quotient. *Aust. J. agric. Res. 12* (6), 1152–1174.

(4) Alexander, G. (1962). Temperature regulation in the new-born lamb. IV. The effect of wind and evaporation of water from the coat on metabolic rate and body temperature. *Aust. J. agric. Res. 13* (1), 82–99.

(5) Alexander, G. (1962). Temperature regulation in the new-born lamb. V. Summit metabolism. *Aust. J. agric. Res. 13* (1), 100–121.

(6) Alexander, G. & Brook A. H. (1960). Loss of heat by evaporation in young lambs. *Nature (Lond.) 185* (4715), 770–771.

(7) Allcroft, Ruth (1961). The use and misuse of mineral supplements. *Proc. B.V.A. 79th annual Congr.* (Oxford).

(8) Allden, W. G. (1959). The summer nutrition of weaner sheep. The relative

roles of available energy and protein when fed as supplements to sheep grazing mature pasture herbage. *Aust. J. agric. Res. 10* (2), 219–236.

(9) Allden, W. G. (1962). Rate of herbage intake and grazing time in relation to herbage availability. *Proc. Aust. Soc. Anim. Prod. IV*, 163–166.

(10) Allden, W. G. & Jennings, A. C. (1962). Dietary supplements to sheep grazing mature herbage in relation to herbage intake. *Proc. Aust. Soc. Anim. Prod. IV*, 145–153.

(11) Allen, Rex W. (1956). Nodular worms of sheep and goats. *U.S. Yearbk. of Agriculture*, 399–400.

(12) Andrews, J. S., Kauffman, A. B. & Davis, R. E. (1944). Effects of the intestinal nematode, *Trichostrongylus colubriformis*, on the nutrition of lambs. *Am. J. vet. Res. 5* (14), 22–29.

(13) Armstrong, R. H. & Cameron, A. E. (1961). Further studies of hexoestrol implantation of Blackface wether lambs. *Anim. Prod. 3*, Part 3, 295–300.

(14) Balch, C. C. (1958). Digestion in the Rumen. *Outlook on Agric. II* (1), 33–43.

(15) Balch, C. C. & Campling, R. C. (1962). Regulation of voluntary food intake in ruminants. N.I.R.D. Paper No. 2563. *Nutr. Rev. 32*, 669–686.

(16) Barnicoat, C. R., Murray, P. F., Roberts, E. M. & Wilson, G. S. (1956). Milk secretion studies with New Zealand Romney ewes. Part VIII. Experiments on the early weaning of lambs. *J. agric. Sci. 48*, Part I, 22–24.

(17) Bassett, J. M. (1960). The effect of interruptions in growth on subsequent growth in liveweight and on food consumption of Suffolk × Half-bred lambs. *Anim. Prod. 2* (2), 214.

(18) Baxter, J. T. (1957). The effects on lambs of infection by the intestinal parasite *Nematodirus*. *Res. and Exp. Rec., Min. of Agr., N. Ireland, VI*, 45–51.

(19) Baxter, J. T. (1957). Some aspects of *Nematodirus* disease in Northern Ireland. *Vet. Rec. 69* (43), 1007–1009.

(20) Baxter, J. T. (1958). On the pattern of *Nematodirus* infection on pasture and in lambs. *Res. and Exp. Rec., Min. of Agr., N. Ireland, VII*, 147–155.

(21) Bindloss, A. A. & Cuthbert, N. H. (1962). Some field experiments on techniques of early weaning of lambs to dry feeding systems. *Anim. Prod. 4* (2), 299.

(22) Blaxter, K. L. (1961). Minerals in relation to disease. *Vet. Ann.*, 230–238. Ed. Pool, W. A. Wright, Bristol.

(23) Blaxter, K. L. (1961) Lactation and the growth of the young. In *Milk, Vol. II*, p. 334. Ed. Kon & Cowie. New York, Academic Press Inc.

(24) Blaxter, K. L. (1962). The energy metabolism of ruminants. London, Hutchinson.

(25) Blaxter, K. L. & Wainman, F. W. (1961). The utilization of food by sheep and cattle. *J. agric. Sci. 57*, 419–425.

(26) Blaxter, K. L., Wainman, F. W. & Wilson, R. S. (1961). The regulation of food intake by sheep. *Anim. Prod. 3*, Part I. 51–61.

(27) Boccard, R. & Duplan, J. M. (1961). Étude de la production de la viande chez les ovins. III. Note sur l'influence de la vitesse de croissance sur la composition corporelle des agneaux. *Ann. zootech. 10* (1), 31–38.

(28) Boccard, R., Dumont, B. L., Le Guelte, P. & Arnoux, J. (1961). Étude de la production de la viande chez les ovins. IV. Relation entre la forme et la composition du membre posterieur. *Ann. zootech. 10* (3), 155–160.

(29) Boda, J. M., Riley, P. & Wegner, T. (1962). Tissue glycogen levels in relation to age and some parameters of rumen development in lambs. *J. Anim. Sci. 21* (2), 252–257.

(30) Bonelli, P. (1961). Early weaning and artificial feeding of lambs. First experimental contribution. *Riv. zootech. 34*, 412–417.

(31) Bonsma, H. C. & Engela, D. J. (1941). Weaning lambs at various ages. *Farming in S. Afr. 16*, 321–332.

(32) BONSMA, S. W. & BOSMAN, H. C. (1944). Early weaning of lambs for saving winter cereal pasture. *Farming in S. Afr. 19*, 573–580.
(33) BRODY, S. (1945). *Bioenergetics and Growth.* New York, Reinhold Pub. Co.
(34) BROWN, T. H. (1959). Parasitism in the ewe and the lamb. *J. Brit. Grassld Soc. 14* (3), 216–220.
(35) CAMERON, C. D. T. & HAMILTON, L. S. (1961). Effect of age at weaning of Shropshire lambs on weight gains and carcase score. *Can. J. Anim. Sci. 41*, 180–186.
(36) CAMPBELL, J. B., DAVIS, A. N. & MYHR, P. J. (1954). Methæmoglobinæmia of livestock caused by high nitrate contents of well water. *Can. J. comp. Med. XVIII* (3), 93–101.
(37) CARTER, H. B , FRANKLIN, M. C. & GORDON, H. McL. (1946). The effect on wool production of a mild infestation with *Trichostrongylus colubriformis* in sheep. *J. Counc. sci. indust. Res. Aust. 19* (1), 61.
(38) CHRISTIAN, K. R. & WILLIAMS, V. J. (1957). Rumen studies in sheep. IV. The rumen metabolism of fresh and dried grass. *N.Z. J. Sci. Tech., Sec. A. 38* (10), 1003–1010.
(39) CHURCH, D. C. (1962). Stomach development in the suckling lamb. *Am. J. vet. Res. 23* (93), 220–225.
(40) CLARK, C. H., KIESEL, G. K. & GOBY, C. H. (1962). Measurements of blood loss caused by *Hæmonchus contortus* infection in sheep. *Am. J. vet. Res. 23* (96), 977–980.
(41) CLARKE, E. A. (1954). Early weaning of lambs on Hill country. *J. Dep. Agric. N.Z. 89*, 471–476.
(42) CLARKE, E. A. & FILMER, DAISY B. (1958). Studies in hogget rearing. I. General characteristics of hogget ill-thrift. *N.Z. J. agric. Res. 1* (2), 249–264.
(43) CLARKE, E. A. & FILMER, DAISY B. (1958). Studies in hogget rearing. II. The role of parasites in hogget ill-thrift. *N.Z. J. agric. Res. 1* (3), 382–417.
(44) CRAMPTON, E. W., DONEFER, E. & LLOYD, L. E. (1960). A nutritive value index for forages. *Proc. VIII int. Grassl Congr.* (Reading), 462–466.
(45) CROFTON, H. D. (1955). Nematode parasite populations in sheep on lowland farms. II. Worm egg counts in lambs. *Parasitology 45*, 99–115.
(46) CROFTON, H. D. & THOMAS, R. J. (1951). A new species of *Nematodirus* in sheep. *Nature (Lond.) 168*, 559.
(47) CURRAN, S. & CROWLEY, J. P. (1961). Supplementary vitamin D_3 for lambs. *Ir. J. agric. Res. 1*, 43–48.
(48) DALGARNO, A. C., HILL, R. & MCDONALD, I. (1962). Vitamin D in the blood of sheep. *Brit. J. Nutr. 16*, 91–97.
(49) DIVEN, R. H., REED, R. E., TRAUTMAN, R. J., PISTOR, W. J. & WATTS, R. E. (1962). Experimentally induced nitrite poisoning in sheep. *Am. J. vet. Res. 23* (94), 494–496.
(50) DONALD, C. M. & ALLDEN, W. G. (1959). The summer nutrition of weaner sheep. The deficiencies of the mature herbage of sown pasture as a feed for young sheep. *Aust. J. agric. Res. 10* (2), 199–218.
(51) DONEY, J. M. & MUNRO, JOAN (1962). The effect of suckling management and season on sheep milk production as estimated by lamb growth. *Anim. Prod. 4*, Part 2, 215–220.
(52) FORSYTH, B. A. (1958). Footrot in Sheep. *Outlook on Agric. II* (2), 86–91.
(53) FRANKLIN, M. C. (1953). Vitamin D requirements of sheep with special reference to Australian conditions. *Aust. vet. J. 29* (11), 302–309.
(54) FRANKLIN, M. C., GORDON, H. McL. & MCGREGOR, C. H. (1946). A study of nutritional and biochemical effects in sheep of infestation with *Trichostrongylus colubriformis*. *J. Counc. sci. indust. Res. Aust. 19* (1), 46–60.
(55) FULLER, G. R. & MCALPINE, V. W. (1961). *Cobalt in Animal Feeding.* Booklet edited by Centre d'information du Cobalt, Brussels.

(56) GALLOWAY, J. H., PULVIRENTI, G. & SILBERMAN, G. (1962). Effect of hormone implants on the growth rate and wool growth of lambs on pasture. *Vet. Rec.* *74* (20), 574–577.
(57) GIBSON, T. E. (1951). The pathogenesis of *Trichostrongylus axei* in sheep. Personal communication.
(58) GIBSON, T. E. (1955). Studies on *Trichostrongylus axei*. IV. Factors in the causation of pathogenic effects by *T. axei*. *J. comp. Path.* *65*, 317–324.
(59) GIBSON, T. E. (1956). The hazards of parasitic gastro-enteritis in sheep running under conditions of intensive stocking. *Emp. J. exp. Agric.* *24* (96), 278–294.
(60) GOLDBERG, A. (1956). Lungworms in sheep and goats. *U.S. Yearbk. of Agric.*, 401–403.
(61) GORDON, H. McL. (1950). Some aspects of parasitic gastro-enteritis of sheep. *Aust. vet. J.* *26* (2), 14, 46.
(62) GORDON, H. McL. (1957). Helminthic Diseases. *Adv. vet. Sci.* *3*, 288–351.
(63) GORDON, H. McL. (1958). The effect of worm parasites on the productivity of sheep. *Proc. Aust. Soc. Anim. Prod.* *2*, 59–68.
(64) GREENHALGH, J. F. D. & CORBETT, J. L. (1960). The indirect estimation of the digestibility of pasture herbage. *J. agric. Sci.* *55*. I. Nitrogen and chromogen as fæcal index substances, 371–376. II. Regressions of digestibility on faecal nitrogen concentration; their determination in continuous digestibility trials and the effect of various factors on their accuracy, 377–386.
(65) GREIG, W. A. (1961). The vitamins. *Vet. Ann.*, 238–244. Ed. Pool. Bristol, Wright.
(66) GROVES, T. W. (1957). Developments in the field of parasitic bronchitis. *Outlook on Agric.* *1* (6), 252–258.
(67) HALL, M. C., DIKMANS, G. & WRIGHT, W. H. (1946). Parasites and parasitic diseases of sheep. *U.S. Dep. of Agric. Farmers' Bull.*, 1330.
(68) HAMDY, A. H. & POUNDEN, W. D. (1959). Experimental production of pneumonia in lambs. *Am. J. vet. Res. XX* (74), 78–83.
(69) HAMDY, A. H. & SANGER, V. L. (1959). Characteristics of a virus associated with lamb pneumonia. *Am. J. vet. Res. XX* (74), 84–86.
(70) HAMDY, A. H., POUNDEN, W. D. & FERGUSON, L. C. (1959). Microbial agents associated with pneumonia in slaughtered lambs. *Am. J. vet. Res. XX* (74), 87–90.
(71) HAMILTON, P., PLUMMER, B. E. JR. & HOWARD, C. D. (1955). Measures of digestion in the ruminant. *Maine Agricultural Experiment Station Bull.*, 543.
(72) HAMMOND, J. (1932). *Growth and Development of Mutton Quality in Sheep.* London, Oliver & Boyd.
(73) HANSEN, M. F., KELLEY, G. W. & TODD, A. C. (1950). Observations on the effects of a pure infection of *Monezia expansa* on lambs. *Trans. Am. micrsc. Soc.* *69* (2), 148–155.
(74) HANSEN, M. F., TODD, A. C., KELLEY, G. W. & CAWEIN, M .(1950). Effects of a pure infection of the tapeworm *Monezia expansa* on lambs. *Bull. Ky. agric. Exp. Sta.*, 556.
(75) HILL, K. J. (1961). The Abomasum. *Vet. Rev. Annot.* *7*, Part 2, 83–106.
(76) HOLDER, J. M. (1962). Supplementary feeding of grazing sheep—its effect on pasture intake. *Proc. Aust. Soc. anim. Prod. IV*, 154–159.
(77) HOPKIRK, C. S. M. & PATTERSON, J. B. E. *The Story of Cobalt Deficiency in Animal Health.* London, The Mond Nickel Co. Ltd.
(78) HUNTER, G. C. (1952–3). Nutrition and host-helminth relationships. *Nutr. Rev.* *23*, 705–714.
(79) KATES, K. C., REX, W. A. & TURNER, J. H. (1956). Roundworms of the digestive tract. *U.S. Yearbk. of Agric.*, 389–399.
(80) KATES, K. C. & TURNER, J. H. (1953). Experimental studies on the pathogenicity of *Nematodirus spathiger*, a trichostrongylid parasite of sheep. *Am. J. vet. Res.* *14* (50), 72–81.

(81) KATES, K. C. & TURNER, J. H. (1953). A comparison of the pathogenicity and course of infection of two nematodes of sheep, *Nematodirus spathiger* and *Trichostrongylus colubriformis*, in pure and mixed infections. *Proc. helminth. Soc., Wash.* *20* (2), 117–124.

(82) KATES, K. C. & TURNER, J. H. (1955). Observations on the life cycle of *Nematodirus spathiger*, a nematode parasitic in the intestine of sheep and other ruminants. *Am. J. vet. Res.* *16* (58), 105–115.

(83) KAUZAL, G. (1936). Further studies on the pathogenic importance of *Chabertia ovina*. *Aust. vet. J.* *12*, 107–110.

(84) KUNKEL, H. O. (1961). Biochemical and fundamental physiological bases of genetically variable growth of animals. *Texas agric. Exp. Sta., Bull.*

(85) LAGACÉ, A., POUNDEN, W. D., BELL, D. S. & WEIDE, K. D. (1961). Factors influencing the incidence of chronic pneumonia in lambs. *Am. J. vet. Res.* *22* (19), 1015–1019.

(86) LARGE, R. V. (1959). Nutrition of the lamb. *J. Brit. Grassld Soc.* *14* (3), 212–215.

(87) LARGE, R. V. (1959). The artificial rearing of lambs. *Exp. Grassld Res. Inst. Hurley* *12*, 103–107.

(88) LARGE, R. V. (1960). The growth of Suffolk × Half-bred lambs. *Exp. Grassld Res. Inst. Hurley* *13*, 45.

(89) LARGE, R. V. & SPEDDING, C. R. W. (1957). The growth of lambs at pasture. I. A comparison of growth on long and short ryegrass swards. *J. Brit. Grassld Soc.* *12*, 235–240.

(90) LARGE, R. V. & TAYLER, J. C. (1954). Studies on the growth of Clun lambs. *Emp. J. exp. Agric.* *22* (86), 141–147.

(91) LITTLEJOHN, A. I. (1955). The use of formalin in the control of footrot in sheep. *Vet. Rec.* *67* (32), 599–602.

(92) LITTLEJOHN, A. I. (1961). Field trials of a method for the eradication of footrot. *Vet. Rec.* *73* (32), 773–780.

(93) LOTZE, J. C. (1956). Coccidiosis of sheep and goats. *U.S. Yearbk. Agric.* 1956, 387–389.

(94) MANTON, V. J. A., PEACOCK, R., POYNTER, D., SILVERMAN, P. H. & TERRY, R. J. (1962). The influence of age on naturally acquired resistance to *Haemonchus contortus* in lambs. *Res. vet. Sci.* *3* (3), 308–314.

(95) MAPES, C. R. & BAKER, D. W. (1950). Studies on the protostrongyline lungworms of sheep. *J. Am. Vet. med. Ass.* *CXVI* (879), 433–435.

(96) MCCANCE, R. A. & WIDDOWSON, ELSIE M. (1962). Nutrition and growth. *Proc. R. Soc. B.* *156*, 326–337.

(97) MCCLYMONT, G. L., WYNNE, K. N., BRIGGS, P. K. & FRANKLIN, M. C. (1957). Sodium chloride supplementation of high grain diets for fattening Merino sheep. *Aust. J. agric. Res.* *8* (1), 83–90.

(98) MICHEL, J. F. (1956). Studies on host resistance to *Dictyocaulus* infection. II. Reinfection experiments with *D. filaria* in sheep. *J. comp. Path.* *66* (3), 241–248.

(99) MILFORD, R. (1960). Measurement of the net retention of energy from herbage. *Exp. Grassld Res. Inst. Hurley* *13*, 77.

(100) MILFORD, R. (1960). Nutritional value of subtropical pasture species under Australian conditions. *Proc. VIII int. Grassld Congr.* (Reading), 474.

(101) MINISTRY OF AGRICULTURE, GOVERNMENT OF N. IRELAND (1958). *Sheep Husbandry.* Leaflet No. 47.

(102) MINSON, D. J. & KEMP, C. D. (1961). Studies in the digestibility of herbage. IX. Herbage and fæcal nitrogen as indicators of herbage organic matter digestibility. *J. Brit. Grassld Soc.* *16* (1), 76–79.

(103) MINSON, D. J., RAYMOND, W. F. & HARRIS, C. E. (1960). Studies in the digestibility of herbage. VIII. The digestibility of S. 37 cocksfoot, S. 23 ryegrass and S. 24 ryegrass. *J. Brit. Grassld Soc.* *15* (2), 174–180.

(104) MINSON, D. J., RAYMOND, W. F. & HARRIS, C. E. (1960). The digestibility of grass species and varieties. *Proc. VIII int. Grassld Congr.* (Reading), 470.

(105) NEWSOM, I. E. (1952). *Sheep Diseases.* Baltimore, Williams & Wilkins.

(106) NOLLER, C. H., DICKSON, I. A. & HILL, D. L. (1962). Value of hay and rumen inoculation in an early weaning system for dairy calves. *J. Dairy Sci.* XLV (2), 197–201.

(107) PHILLIPSON, A. T. (1948). Digestion in the ruminant. *N.A.A.S. quart. Rev.* (2), 55–62.

(108) PHILLIPSON, A. T. (1959). Sheep. *Scientific Principles of Feeding Farm Live Stock.* London, Fmr. & Stockbreeder Pub. Ltd.

(109) PIGDEN, W. J., PRITCHARD, G. T., WINTER, K. A. & LOGAN, V. S. (1961). Freezing—a technique for forage investigations. *J. Anim. Sci. 20* (4), 796–801.

(110) POPE, A. L., COOK, C. W., GARRIGUS, U. S., DINUSSON, W. E. & WEIR, W. C. (1957). *Nutrient requirements of domestic animals.* V. Nutrient requirements of sheep. Washington, Nat. Academy of Sciences. Nat. Res. Counc.

(111) PORTER, D. A. & KATES, K. C. (1956). Tapeworms and bladderworms. *U.S. Yearbk. Agric.*, 153–156.

(112) PRICE, E. W. (1956). Liver flukes of cattle and sheep. *U.S. Yearbk. Agric.*, 148–153.

(113) RAYMOND, W. F. (1951). The problem of measuring the nutritive value of herbage. *J. Brit. Grassld Soc. 6* (3), 139–146.

(114) RAYMOND, W. F., HARRIS, C. E. & HARKER, V. G. (1953). Studies in the digestibility of herbage. II. Effect of freezing and cold storage of herbage on its digestibility by sheep. *J. Brit. Grassld Soc. 8* (4), 315–320.

(115) RAYMOND, W. F., HARRIS, C. E. & KEMP, C. D. (1955). Studies in the digestibility of herbage. VI. The effect of level of herbage intake on the digestibility of herbage by sheep. *J. Brit. Grassld Soc. 10* (1), 19–26.

(116) RAYMOND, W. F., MINSON, D. J. & HARRIS, C. E. (1959). Studies in the digestibility of herbage. VII. Further evidence on the effect of level of intake on the digestive efficiency of sheep. *J. Brit. Grassld Soc. 14* (2), 75–77.

(117) REID, J. T., KENNEDY, W. K., TURK, K. L., SLACK, S. T., TRIMBERGER, G. W. & MURPHY, R. P. (1959). Symposium on forage evaluation. I. What is forage quality from the animal standpoint ? *Agron. J. 51*, 213–216.

(118) RIŠ, M. A. & MAHMUDOV, M. (1962). The trace element copper and its importance in sheep nutrition. *Ovtsevodstvo* 4, 17–19.

(119) ROSE, J. H. (1957). Observations on the larval stages of *Muellerius capillaris* within the intermediate hosts *Agriolimax agrostis* and *A. reticulatus. J. Helminth. 31* (1/2), 1–16.

(120) ROSE, J. H. (1957). Observations on the bionomics of the free living first stage larvæ of the sheep lungworm, *Muellerius capillaris. J. Helminth. 31* (1/2), 17–28.

(121) ROSE, J. H. (1958). Site of development of the lungworms *Muellerius capillaris* in experimentally infected lambs *J. comp. Path. 68* (3), 359–362.

(122) ROSE, J. H. (1959). Experimental infection of lambs with *Muellerius capillaris. J. comp. Path. 69* (4), 414–422.

(123) ROSS, C. V., KARR, M. L. & PAVEY, R. L. (1961). Creep-feeding studies with lambs. *University Missouri Agric. exp. Sta. Res. Bull.*, 772.

(124) SARLES, M. P. (1944). Effects of experimental nodular worm (*Oesophagostomum columbianum*) infection in sheep. *Tech. Bull. U.S. Dep. Agric.* No. 875.

(125) SARLES, M. P. & FOSTER, A. O. (1943). Nodular worm disease of sheep. *U.S. Dep. Agric. Leaflet*, 228.

(126) SCHNEIDER, B. H. (1947). Feeds of the world. *West Va. Univ. agric. exp. Sta. Bull.*, p. 251.

(127) SHAW, J. C. (1959). Symposium on forage evaluation. VIII. Relation of digestion end-products to the energy economy of animals. *Agron. J. 51*, 242–245.

(128) SHAW, J. C. (1961). Nutritional physiology of the rumen. *VIII int. Congr. Anim. Prod.* (Hamburg), General reports 29–51.

(129) SHEEHY, E. J. (1955). *Animal Nutrition.* London, Macmillan.

(130) Shepherd, R. W. (1962). Hexœstrol implantation of very young lambs. *Exp. Husb.* (7), 72–76.

(131) Shumard, R. F., Bolin, D. W. & Eveleth, D. F. (1957). Physiological and nutritional changes in lambs infected with the nematodes *Hoemonchus contortus, Trichostrongylus colubriformis* and *Nematodirus spathiger. Am. J. vet. Res. 18* (67), 330–337.

(132) Skerman, K. D. (1962). Observations on selenium deficiency of lambs in Victoria. *Proc. Aust. Soc. Anim. Prod.* IV, 22–27.

(133) Slen, S. B., Demiruren, A. S. & Smith, A. D. (1961). Note on the effects of selenium on wool growth and body gains in sheep. *Can. J. Anim. Sci. 41*, 263–265.

(134) Soulsby, E. J. L. (1956). Studies on the serological response in sheep to naturally acquired gastro-intestinal nematodes. I. Preparation of antigens and evaluation of serological techniques. *J. Helminth. 30* (2/3), 129–142.

(135) Soulsby, E. J. L. (1961). Symposium. Recent advances in the treatment and control of intestinal parasites. Immune mechanisms in helminth infections. *79th Annual Congr. Brit. vet. Ass.*, Oxford.

(136) Soulsby, E. J. L. (1962). Immunity to helminths and its effect on helminth infections. In *Animal Health and Production, Colston Papers* (13), 165. Ed. Grunsell, C. S. & Wright, A. I. London, Butterworth.

(137) Southcott, W. H. (1955). Observations on the removal of *Oesophagostomum columbianum* Curtice from sheep grazing on green oats and on pastures. *Aust. J. agric. Res. 6* (3), 456–465.

(138) Spedding, C. R. W. (1953). The effect of a subclinical worm-burden on the liveweight gain of lambs. *Emp. J. exp. Agric. 21* (84), 255–261.

(139) Spedding, C. R. W. (1954). Pasture management to control worms in sheep. *Agriculture (Lond.) 61*, 51–54.

(140) Spedding, C. R. W. (1954). Effect of a subclinical worm-burden on the digestive efficiency of sheep. *J. comp. Path. 64* (1), 5–14.

(141) Spedding, C. R. W. (1954). The production of worm-free lambs at pasture. *Nature (Lond.)*, 174, 611.

(142) Spedding, C. R. W. (1955). The effect of a subclinical worm-burden on the productivity of sheep. *J. Brit. Grassld Soc. 10* (1), 35–43.

(143) Spedding, C. R. W. (1956). The control of worm-infestation in sheep by grazing management. *J. Helminth. 29* (4), 179–186.

(144) Spedding, C. R. W. & Brown, T. H. (1957). A study of subclinical worm infestation in sheep. I. The effect of level of infestation on the growth of the lamb. *J. agric. Sci. 48* (3), 286–293. II. The 'tolerance' level of infestation. *J. agric. Sci. 49* (2), 223–228. III. The effect on wool production. *J. agric. Sci. 49* (2), 229–233.

(145) Spedding, C. R. W. & Brown, T. H. (1961). The effect of early weaning on the growth rate of lambs. *VIII int. Congr. Anim Prod.* (Hamburg). Points for discussion. p.180.

(146) Spedding, C. R. W., Brown, T. H. & Wilson, I. A. N. (1958). Growth and reproduction in worm-free sheep at pasture. *Nature (Lond.) 181*, 168–170.

(147) Spedding, C. R. W., Brown, T. H. & Large, R. V. (1960). Some factors affecting the significance of internal parasites in the utilization of grass by sheep. *Proc. VIII int. Grassld Congr.* (Reading), 718–722.

(148) Spedding, C. R. W., Brown, T. H. & Large, R. V. (1963). The effect of milk intake on nematode infestation of the lamb. *Proc. Nutr. Soc. 22* (1), 32–41.

(149) Spedding, C. R. W., Large, R. V. & Brown, T. H. (1961). Symposium on some modern feeding methods. I. The early weaning of lambs. *Vet. Rec. 73* (51), 1428–1432.

(150) Stamp, J. T. (1959). Clostridial diseases of sheep. *Outlook on Agric.* II (4), 185–191.

(151) STEPHENSON, S. K. (1962). Growth measurements and the biological interpretation of mammalian growth. *Nature (Lond.)* *196* (4859), 1070–1074.
(152) STEPHENSON, S. K. & ROBERTS, J. (1962). Specific gravity changes during the development of the sheep fœtus. *Nature (Lond.)* *196* (4856), 788.
(153) STEWART, J. S. (1933). The effects of nematode infestations on the metabolism of the host. *Part I.* Metabolism experiments. *Rep. Dir. Inst. Anim. Path. (Camb.)*, 58–76.
(154) TAYLER, J. C. (1959). A relationship between weight of internal fat, 'fill', and the herbage intake of grazing cattle. *Nature (Lond.)* *184*, 2021.
(155) TAYLOR, E. L. (1942). Subclinical helminthiasis of farm animals. *Vet. Rec.* *54* (38), 377–380.
(156) TAYLOR, E. L. (1947). The ecology of nematodes parasitic in farm animals. *Vet. Rec.* *59* (45), 624–625.
(157) TAYLOR, E. L. (1953). Diseases of animals. Part II. Parasitic diseases. *Jl R. agric. Soc.* *114*, 152–160.
(158) TETLEY, J. H. (1959). Development of *Hæmonchus contortus* in weaned and unweaned lambs. *J. Helminth.* *33* (4), 301–304.
(159) TODD, A. C. & HANSEN, M. F. (1951). The economic import of host resistance to helminth infection. *Am. J. vet. Res.* *12* (42), 58–64.
(160) TODD, J. R. (1962). Chronic copper poisoning in farm animals. *Commonwealth Bureau of Animal Health—Vet. Bull.* *32* (9), 573–580.
(161) TURNER, J. H. (1955). Preliminary report of experimental strongyloidiasis in lambs. *Proc. helminth. Soc. (Wash.)* *22* (2), 132–133.
(162) TURNER, J. H. (1959). Experimental strongyloidiasis in sheep and goats. I. Single infections. *Am. J. vet. Res.* *20* (74), 102–110.
(163) TURNER, J. H. & WILSON, G. I. (1958). Strongyloidiasis in lambs. *Vet. Med.* *53*, 242.
(164) WALKER, D. J., HARRIS, A. N. A., FARLEIGH, E. A., SETCHELL, B. P. & LITTLEJOHNS, I. R. (1961). Muscular dystrophy in lambs in N.S.W. *Aust. vet. J.* *37* (5), 172–175.
(165) WALKER, D. M. & LEE, BARBARA A. (1961). The rumen fermentation of lactose in the adult sheep. *J. agric. Sci.* *57*, 267–270.
(166) WALKER, D. M. & WALKER, GWEN J. (1961). The development of the digestive system of the young animal. V. The development of rumen function in the young lamb. *J. agric. Sci.* *57*, 271–278.
(167) WALLACE, L. R. (1948). The growth of the lambs before and after birth in relation to the level of nutrition. *J. agric. Sci.* *38*, 93–153, 243–302, 367–401.
(168) WARDROP, I. D. (1961). Some preliminary observations on the histological development of the fore-stomachs of the lamb. I & II. *J. agric. Sci.* *57*, 335–341; *57*, 343–346.
(169) WARDROP, I. D. & COOMBE, J. B. (1961). The development of rumen function in the lamb. *Aust. J. agric. Res.* *12* (4), 661–680.
(170) WARDROP, I. D., TRIBE, D. E. & COOMBE, J. B. (1960). An experimental study of the early weaning of lambs. *J. agric. Sci.* *55* (1), 133–136.
(171) WHITLOCK, J. H. (1951). The evaluation of animal growth. *Cornell Vet.* XLI (3), 254–266.
(172) WIDDOWSON, ELSIE M. & KENNEDY, G. C. (1962). Rate of growth, mature weight and life-span. *Proc. roy. Soc., B*, 156, 96–108.
(173) WRIGHT, P. L. (1961). *Some Aspects of Ewe and Lamb Nutrition.* Ph.D. Thesis, (University of Wisconsin).
(174) LARGE, R. V. (1964). The development of the lamb with particular reference to the alimentary tract. *Anim. Prod.* *6* (2), 169–178.
(175) ALEXANDER, G. & WILLIAMS, D. (1962). Temperature regulation in the new-born lamb. VI. *Aust. J. agric. Res.* *13*, 122–143.
(176) BROWN, T. H. (1964). The early weaning of lambs. *J. agric. Sci., Camb.* *36*, 191–204.

(177) Caldwell, D. W. (1957). Rearing orphan lambs. *N.Z. Jl Agric. 94*, 610.

(178) Cameron, C. D. T. & Gibbs, H. C. (1966). Effects of stocking rate and flock management on internal parasitism in lambs. *Can. J. Anim. Sci. 46*, 121–124.

(179) Chalmers, Margaret I., Davidson, J., Eadie, J. Margaret & Gill, J. C. (1968). Some comparisons of performance of lambs with and without rumen ciliate protozoa. *Proc. Nutr. Soc. 27* (2), 29A–30A.

(180) Charles, D. R. (1967). Intensive Livestock Housing. *The Electricity (EDA Division) Rural Electrification Conf.*

(181) Connan, R. M. (1968). An attempt to demonstrate passive resistance in lambs to three species of gastro-intestinal nematodes as a result of sucking the ewe. *Res. vet. Sci. 9*, 591–593.

(182) Cunningham, J. M. M., Edwards, R. A. & Simpson, M. E. (1961). Rearing lambs on a synthetic diet. *Anim. Prod. 3*, 105–109.

(183) Davies, D. A. R. & Owen, J. B. (1967). The intensive rearing of lambs. I. Some factors affecting performance in the liquid feeding period. *Anim. Prod. 9* (4), 501–508.

(184) Davies, E. T., Pill, A. H., Collings, D. F., Venn, J. A. J. & Bridges, G. D. (1965). Cerebro cortical necrosis in calves. *Vet. Rec. 77*, 290.

(185) Dobson, C. (1967). The effects of different doses of *Oesophagostomum columbianum* larvae on the body weight intake and digestibility of feed and water intake of sheep. *Aust. vet. J. 43*, 291–296.

(186) Downey, N. E. (1965). Some relationship between trichostrongylid infestation and cobalt status in lambs. *Haemonchus contortus* infestation. *Brit. vet. J. 121*, 362–370.

(187) Downey, N. E. (1966). Some relationship between trichostrongylid infestation and cobalt status in lambs: II and III. *Brit. vet. J. 122*, 201 and 316.

(188) Edwin, E. E., Lewis, G. & Allcroft, R. (1968). Cerebrocortical necrosis: a hypothesis for the possible role of thiaminases in its pathogenesis. *Vet. Rec. 83* (7), 176–178.

(189) Edwin, E. E., Spence, J. B. & Woods, A. J. (1968). Thiaminases and cerebrocortical necrosis. *Vet. Rec. 83* (16), 417.

(190) Halliday, R. (1968a). Serum γ-globulin levels in dead lambs from Hill flocks. *Anim. Prod. 10* (2), 177–182.

(191) Halliday, R. (1968b). Serum protein concentrations in 2-day-old Finnish Landrace, Scottish Blackface, Merino and Merino × Cheviot lambs. *J. agric. Sci., Camb. 71*, 41–46.

(192) Hjulstad, O. A. (1966). A Norwegian view of environmental control. *Farm Buildings, No. 2.*

(193) Hurst, D. (1963). Sheep housing in France. *Fmrs' Weekly*, May 24th, 95–97.

(194) Joyner, L. P., Norton, C. C., Davies, S. F. M. & Watkins, C. V. (1966). The species of coccidia occurring in cattle and sheep in the South-west of England. *Parasitology 56*, 531–541.

(195) Large, R. V. (1965a) The effect of concentration of milk substitute on the performance of artificially reared lambs. *Anim. Prod. 7*, 325–332.

(196) Large, R. V. (1965b). The artificial rearing of lambs. *J. agric. Sci., Camb. 65*, 101–108.

(197) Large, R. V. & Penning, P. D. (1967). The artificial rearing of lambs on cold reconstituted whole milk and on milk substitute. *J. agric. Sci., Camb. 69*, 405–409.

(198) Lee, A. (1967). *A Study of Sheep Housing.* M.Sc. Thesis, University of Reading.

(199) Lodge, G. A. & Lamming, G. E. (1968). *Growth and Development of Mammals.* London, Butterworths.

(200) Markson, L. M. & Terlecki, S. (1968). The aetiology of cerebrocortical necrosis. *Br. vet. J. 124*, 309–315.

(201) Mason, J. H., Dalling, T. & Gordon, W. S. (1930). Transmission of maternal immunity. *J. Path. Bact. 33*, 783–797.

(202) McCARTHY, E. F. & McDOUGALL, E. I. (1953). Absorption of immune globulin by the young lamb after ingestion of colostrum. *Biochem. J.* 55, 177–182.

(203) MUMFORD, F. B. (1901). Breeding experiments with sheep. *Bull. Mo. Agric. Exp. Sta. No.* 53.

(204) OWEN, J. B. & DAVIES, D. A. R. (1965). Artificial rearing of lambs. *Agriculture, Lond.* 72, 54–57.

(205) PANTELOURIS, E. M. (1965). *The Common Liver Fluke*. London, Pergamon Press.

(206) PENNING, P. D. (1966). Artificial rearing of lambs. *Ann. Rep. Grassld Res. Inst.*, 86–92.

(207) PINOT, R. & TEISSIER, J. H. (1965). L'allaitement artificiel des agneaux. I. Comparaison entre differents laits de remplacement et le lait de brebis. *Annls Zootech.* 14 (3), 261–278.

(208) PINOT, R. & MAULEON, P. (1967). L'allaitement artificiel des agneaux. II. Comparaison entre trois modes d'elevage des agneaux. *Annls Zootech.* 16 (2), 151–164.

(209) POUT, D. D. (1965). Coccidiosis in lambs. *Vet. Rec.* 77, 887–888.

(210) POUT, D. D., OSTLER, D. C., JOYNER, L. P. & NORTON, C. C. (1966). The coccidial population in clinically normal sheep. *Vet. Rec.* 78, 455–460.

(211) ROSS, D. B. (1968). Success treatment of coccidiosis in lambs. *Vet. Rec.* 83, 189–190.

(212) SOUTHCOTT, W. H. & CORBETT, J. L. (1966). Age of weaning and parasitism in Merino lambs. *Proc. Aust. Soc. Anim. Prod.* 6, 194–197.

(213) SOUTHCOTT, W. H., HEATH, D. D. & LANGLANDS, J. P. (1967). Relationship of nematode infection to efficiency of wool production. *J. Brit. Grassld Soc.* 22 (2), 117–120.

(214) SPEDDING, C. R. W. (1968). The agricultural ecology of grassland. Sir Thomas Middleton Memorial Lecture, Dec. 1968. *Agric. Progress*. In press.

(215) SPEDDING, C. R. W., LARGE, R. V. & BROWN, T. H. (1961). The early weaning of lambs. *Vet. Rec.* 73 (51), 1428–1432.

(216) STAMPA, S. (1967). A contribution towards the influence of tapeworms on live-weights of lambs. *Vet. med. Rev.* 1, 81–85.

(217) Walker, D. M. (1950). Observations on behaviour in young calves. *Bull. Anim. Behav.* 8, 5–10.

(218) WALKER, D. M. (1959). The development of the digestive system of the young animal. *J. agric. Sci.* 53, 374.

(219) WALKER, D. M. (1964). Nutritional diarrhoea in the milk-fed lamb and its relation to the intake of sugar. *Brit. J. Nutr.* 18, 209.

(220) WELCH, J. C., VANDER NOOT, C. W. & GILBREATH, E. L. (1963). Effects of feeding milk replacers with varying amounts of fat for hothouse lamb production. *J. Anim. Sci.* 22, 155–158.

(221) WILLIAMS, H. L. (1967). *The Intensification of Sheep Production*. Ph.D. Thesis, University of London.

VI

THE PRODUCTION PROCESS

As mentioned earlier, the animal production process can be considered as independent of the particular nature of the food supply and the way in which this is produced; on the other hand, a production process is very incompletely described if it is not quantitatively related to the food input. It is important to appreciate that the quantity of food required for each unit of product is influenced by the rate of production. This in turn is influenced by individual food intake and therefore by both grazing conditions and the pattern of growth of the food supply.

The Main Products

The basic process of production may be conceived as the conversion of food, whether it is grazed herbage, hay or grain, into sheep products of agricultural value. There are, however, several different products with different values and costs, more than one of these often being produced simultaneously. The main products are meat, wool, store lambs for fattening, breeding stock of all ages and milk. Store lambs can be ignored as a separate item since change of ownership in the middle of a process does not, in principle, affect it. A similar view might be taken of breeding stock. It is as well, however, to bear in mind that, to the producer, the live animal is often a product, and that the condition appropriate to a product cannot always be described in terms of 'finish', or fatness, or even weight.

Meat, wool and milk are thus the chief products. In using terms like 'meat' and 'wool' it should not be forgotten that, whatever the main purpose of production, by-products may be of considerable value. Thus the wool of slaughtered sheep may constitute 30 per cent of the home-produced wool in Britain. The carcase may be only 45 to 50 per cent of the liveweight (this is known as the 'killing-out' or 'dressing-out' percentage) but the offal (the head, feet and some internal organs), may represent about 5 per cent of the value of the sheep and the skin anything from 10 to 20 per cent. The monetary value of each product is influenced in several ways by the quantity produced and is not necessarily related to the cost of production: within a product, however, there is some quantitative relationship between the food costs expressed in biological and in economic terms at any one time. If the kind of food used is changed, of course, then the same number of pounds of it may produce a different weight of product and one with a different unit cost.

Economic and Biological Relationships

The input–output relationship, expressed in biological terms, between one feed and one product, however, will remain unchanged by changes in the monetary cost of the one or in the value of the other. Thus the curve in Fig. 6.1, which illustrates one input–output relationship between grass eaten and liveweight produced, is independent of, and equally useful for, whatever costs and values we wish to allot.

This hypothetical curve is introduced for three reasons. First, the general shape is characteristic, with a steeply rising early part and a flattening off at the top. Secondly, although the curve starts at zero output this is associated with x units of input. This is because living animals always require a certain amount of food simply to remain alive and at the same weight: this is termed the maintenance requirement (see Chapter V) and no production occurs until more than this quantity is fed (2, 4). Thirdly, one object of obtaining biological relationships of this kind is to enable the economist to fit the appropriate monetary values. In this way, information of permanent biological value can be given its current economic interpretation at any future time. With this in mind, it is permissible to leave economic criteria out of the calculations to some extent and consider the production model as a biological construction. In doing so, it is necessary to recognize that a separate model is required for each process.

Consider, for example, meat and wool. If f lb of grass produce either m lb of meat or m lb of wool, the two processes cannot be said to be equally efficient from a biological point of view. In fact, a biological comparison can hardly be made, and any kind of comparison would require them to be expressed in comparable terms.

These might be the weight of the dry matter or the amount of energy contained in a given weight of dry matter (the usefulness of measuring the energy flow in various ecosystems has been discussed by Macfadyen (23)) but, in any event, the comparison would have no agricultural relevance since it is not entirely for their dry matter or their energy content that these products are required. Nor is there necessarily any relationship between the economic values of the two products and the quantities of food required to produce them. Not only is food

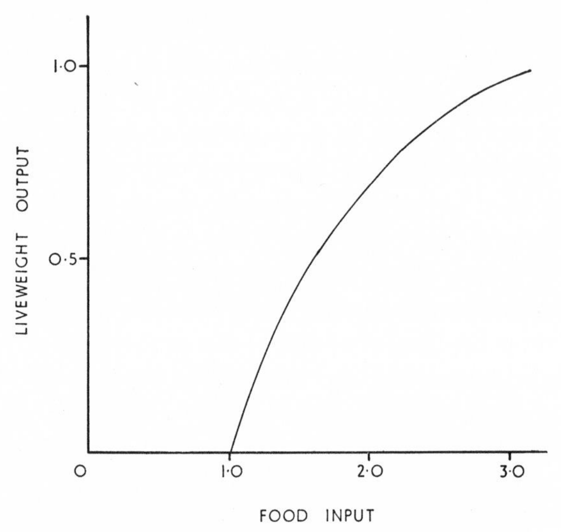

FIG. 6.1—*Hypothetical relationship between food input, (lb S.E./head/day) and output of liveweight (lb gain/head/day), for a sheep weighing approximately 80 lb*

only part of the total production cost but the value of the product is also affected by supply and demand relationships. Relevant comparisons between products can thus only be made in economic terms, and biological comparisons must be confined within product interactions. This is one of the limitations on the usefulness of the models considered here. Economic models would be extremely valuable, once it became possible to provide them: however, their provision rests to some extent on the construction of the intermediate, biological relationships with which we are concerned here. At present, the latter are essentially an aid to understanding but with more knowledge they could become something much more and have direct usefulness as a basis for practical action.

The study of input–output relationships

Although it is necessary to consider one product at a time, it is not always possible to study one without also varying another. Using mature sheep, the production of wool can be studied in relative isolation because, even if the animal's weight fluctuates, there is no other product involved. By employing full-grown wethers, i.e. castrated males, it is possible also to avoid all breeding complications. In milk production, however, the plane of nutrition may influence both lactation and wool growth. In general, ewes which lamb and lactate produce less wool than do those which are barren or lose their lambs at birth (44). Improved nutrition would normally be expected to increase the milk yield, but it is not entirely clear what the effect would be on wool growth in these circumstances. In meat production, many of the important factors will also affect the amount of wool produced. Larger litters would almost certainly impose a greater stress on the ewe and would therefore be expected to reduce the weight of fleece produced. At the same time, the greater number of lambs would increase the total wool yield. It may not be possible, therefore, to hold the level of output of one product, wool, for instance, constant whilst considering the relationship between food input and the output of another product, such as meat.

Just as, with different products, it may be necessary to consider production separately on different foods: in order to compare processes based on different foods, it is necessary to express them in the same terms. Terms like dry matter, energy or digestible organic matter have been used for various purposes and it may be that no one expression will prove appropriate for all productive processes. The level and efficiency of production of any animal depends upon the provision of a diet adequate in several respects (11). If a diet is relatively low in some essential component, this may prove limiting to production and the expression of dietary intake in terms of this component may be relevant. In practice a close relationship exists between the level of production attained by an animal and the intake of dietary energy (10).

The present purpose is the strictly limited one of clarifying the separate sheep production processes, to see which are the important factors affecting them and how they interact and, in consequence, to be better placed for the efficient application of such processes in any particular environment. The particular problems of applying the process within a grazing context are considered in subsequent chapters. Disease is left out of consideration at this point, being looked on as part of the problem of application to particular feeds and circumstances. For this reason, different diseases and disorders are dealt with in the context of application and at the appropriate points.

MEAT PRODUCTION

Production must ultimately be considered on a flock basis and over a substantial period of time. First, however, something must be said of the component processes.

The component processes

The output of the ewe (see Chapter IV), as far as meat production is concerned, is represented by new-born lambs and milk. The latter will be discussed in somewhat greater detail in a later section on milk production. The former is the first stage at which a sheep can possibly be regarded as a product, ignoring the rather remote possibility of transplanted eggs ever coming into this category on any large scale. Newly born lambs, by contrast, could do so (see Chapter XV) in the not-too-distant future.

Food Intake of the Pregnant Ewe

The total input of food during pregnancy is a substantial proportion of that required by the ewe for the year. In most cases, however, the requirement is not critical until the second half of gestation and is of major significance only during the last six weeks before lambing (see Fig. 6.2).

It is not certain that this will be true for large litters and one would expect the critical period to start much earlier in these cases.

As the number of foetuses increases, so the nutritional need might be expected to increase, but the situation is not quite like this, in fact.

The needs of the pregnant ewe are chiefly related to the total weight of the foetal burden. Russel *et al.* (39) found that the additional daily intake required for each 1 kg of foetus was 100 g D.O.M. Now the total foetal weight cannot be proportional to the foetal number because the total capacity of a ewe is limited. This can be illustrated by the birth weights of lambs born in different-sized litters (see Table XXIX).

The total birth weight levels out at the maximum weight the ewe can sustain: further increases in litter size would simply reduce the size of each lamb.

Of course, as mentioned in Chapter IV, there may be problems in achieving the necessary food intake by the ewe on certain feeds and more concentrated (or drier, ground) materials may be needed where the foetal burden is large.

The additional quantity required for larger litters is, however, a

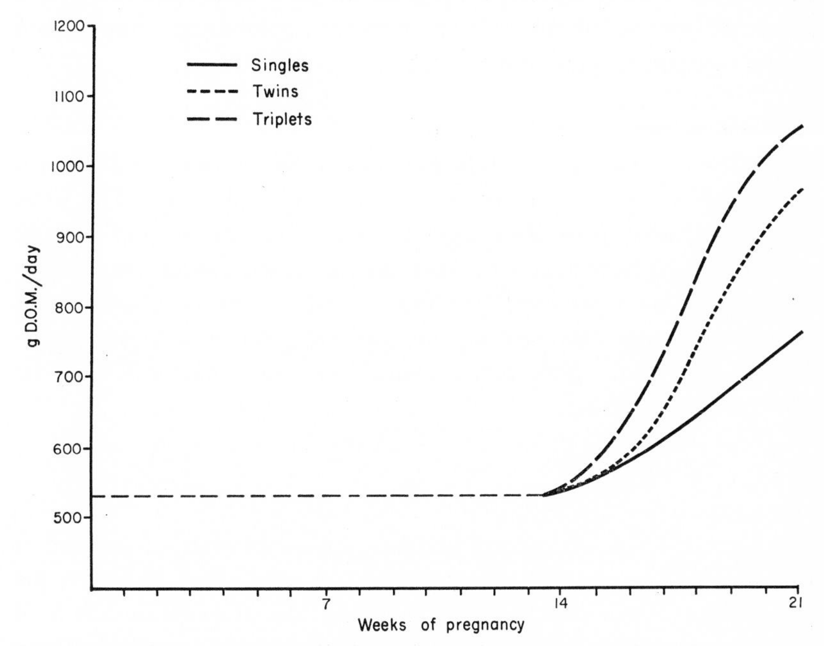

FIG. 6.2—*Intake of D.O.M. required during pregnancy for ewes weighing 55 kg at mating with singles, twins, or triplets*

(Source: data from experiments by R. V. Large)

relatively small proportion of that required in pregnancy and an almost negligible proportion of the annual total.

It may be objected that the requirement of a ewe cannot be stated as a single figure because reserve fat can be used and some imbalance between needs and current intake can be tolerated without detriment to subsequent production. If (or where) underfeeding is justified, it will of course reduce the quantity needed still further and overfeeding is only likely to be justified where the ewe was excessively thin.

TABLE XXIX

BIRTH WEIGHT OF LAMBS (KG) IN RELATION TO LITTER SIZE

	L/wt of ewe	Litter size	Birth wt	
			Mean	Total
Scottish Half-bred . .	78·0 (6)*	1	6·4	6·4
	75·2 (7)	2	5·1	10·2
	73·4 (10)	3	4·4	13·2
Welsh Mountain . .	34·7 (10)	1	3·8	3·8
	34·4 (6)	2	3·1	6·2
	35·3 (4)	3	2·5	7·5

* Number of ewes

TABLE XXX

FOOD REQUIRED IN PREGNANCY

	L/wt at mating (kg)	Litter size	Food (kg d.o.m.) (weeks)	
			0–15	16–21
Scottish Half-bred . .	(6)* 78·1	1	89·9	35·8
	(6) 74·5	2	85·8	47·4
	(3) 82·8	3	95·3	56·4
Kerry Hill . . .	(6) 53·1	1	53·4	27·4
	(6) 55·7	2	56·0	33·3
	(2) 67·5	3	67·4	47·1
Welsh Mountain . .	(6) 30·7	1	30·9	16·2
	(4) 34·4	2	34·6	22·1
	(2) 34·9	3	35·1	25·9

* Number of ewes

A reasonable estimate of needs during pregnancy can therefore be obtained by measuring the amount of food required to ensure that the weight gained above the initial (tupping) weight is equal to that lost at lambing. The food eaten has then supplied all that was needed for production of the new-born lambs and associated tissues. There are difficulties due to changes in body composition of the ewe but these should not greatly affect the estimates (Table XXX).

Lamb Growth and Food Intake Relationships

The growth of the lamb in relation to food intake must take into account the differential development of parts of the animal, the changes in voluntary intake with changes in size and the effect of rate of growth. This last was considered in Chapter V and the same point is implied in Fig. 6.1, that, in general, the faster the growth rate of any given animal, the higher is the proportion of food that is devoted to productive purposes. This remains broadly true for any one feed so that the production per unit of food tends to increase with food intake and rate of growth. If, however, we are concerned with different foods, then, just as with different products, the important differences may be economic. For example, when an artificially reared lamb is receiving mainly milk, its growth is directly related to its milk intake (Fig. 6.3) and the milk required per unit liveweight gain is greater at the lower levels of milk intake. After two to three weeks the lamb could live on a cheaper food, without prejudice to its final size and weight, but this would involve, initially, a considerable reduction in its rate of growth compared with that associated with continued milk feeding. So a lower growth rate might, in cases of this kind, be more profitable. It is worth noting that a reduction in early growth rate does not necessarily result in a smaller 'fat lamb': it may result in a different shape (conformation) but it has recently been suggested (32) that the shape of a sheep is of doubtful commercial importance. Permanent stunting, below the mature size, may occur if growth is retarded below what Dickinson (12) has termed the 'minimal growth path conforming to genetic growth competence'. Compared with that of the artificially reared lamb the situation with the ewe suckling lambs is more complicated and less easy to control, especially under grazing conditions. Before considering this total system of production by the ewe and lamb unit, however, it is necessary to consider the effect of growth on production efficiency.

The growth rate of the lamb rarely increases with time, but tends rather to decline with advancing age (Figs. 5.4 and 5.5, pp. 139, 140). Thus, apart from the growth of early weaned lambs and that obtained during 'compensatory' growth (34), absolute growth rates are, at best, maintained for a time before actually declining. Food intake, on the other hand, increases in some relation to the weight of the lamb (Fig. 5.15, p. 177) or to its size. The lamb must therefore become each day a less efficient converter of food to product (4). This applies to growth as measured by liveweight increments and ignores the

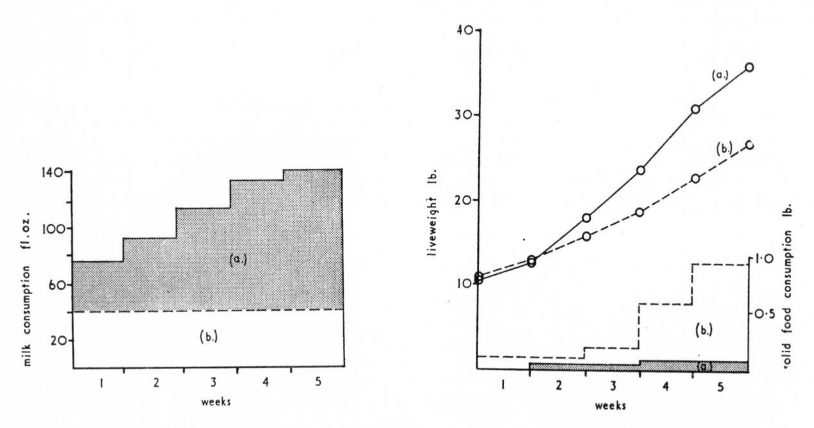

FIG. 6.3—*Liveweight gain and food consumption (daily means for each week) of Dorset Horn lambs fed concentrate pellets* ad lib. *and milk substitute (a)* ad lib. *four times a day, or (b) as a ration of ½ pint four times a day. (1 pint = 20 fluid ounces.)*
(Source : R. V. Large, M.Sc. Thesis, Reading)

calorific value of the weight gained. The weight increments of the *carcase* are likely to have a higher unit food requirement in the later (fattening) stages, simply because of the higher calorific value of fat (2); the efficiency with which food energy is converted is thus underestimated when the *weight* of the animal or its carcase is used to measure growth.

The important point is that, inevitably, as an animal grows bigger its maintenance requirement increases and, if it does not similarly increase its rate of growth, the proportion of food used for productive purposes must fall. So the efficiency of production tends to decline as the lamb grows. This is illustrated in Table XXXI by calculating,

TABLE XXXI

DATA FROM A TRIAL WITH LAMBS MILK FED FROM BIRTH TO SLAUGHTER

PERIOD	L-WT GAIN (kg/lamb)	MILK D.M. INTAKE (kg/lamb)	CONCENTRATE INTAKE (kg/lamb)	EFFICIENCY
1–7 days				
Group I* . .	1·80	2·084	0	0·86
Group II† . .	1·93	2·223	0·145	0·82
8–14 days				
Group I . .	2·40	3·469	0·116	0·67
Group II . .	2·81	3·080	0·081	0·89
15–21 days				
Group I . .	3·20	4·190	0·112	0·74
Group II . .	3·53	3·572	0·221	0·93
22–28 days				
Group I . .	3·57	4·739	0·284	0·84
Group II . .	3·37	4·602	0·581	0·64
29–35 days				
Group I . .	3·50	4·028	0·114	0·71
Group II . .	2·60	3·768	0·325	0·65
36–42 days				
Group I . .	3·90	6·082	0·535	0·59
Group II . .	3·70	4·976	0·645	0·66
43–49 days				
Group I . .	4·4	5·188	0·619	0·76
Group II . .	4·0	5·270	0·866	0·65
49–56 days				
Group I . .	3·0	5·585	1·124	0·45
Group II . .	2·4	5·054	1·023	0·39

* Group I 4 lambs Dorset Horn and Dorset × Scottish Half-bred
† Group II 5 lambs Dorset Horn and Dorset × Scottish Half-bred

Pritchitt's milk substitute fed cold from tanks
Commercial lamb creep pellet (18 per cent C.P.)
Hay offered
Efficiency calculated on total D.M. (milk + conc.) intake
Efficiency probably underestimated due to
1. Birds eating concentrate
2. Slight separation of milk

from an experiment relating milk intake and liveweight gain, the amount of food required to produce successive liveweight increments.

Efficiency of food conversion has been calculated as the weight of liveweight gained each week per unit of food (D.M.) eaten. The latter includes both the dry matter of the cold milk substitute and that of the supplementary concentrate (both fed *ad lib.*); a small amount of hay is also included. The maximum rate of liveweight gain, over a period of one week for a group of four lambs, was approx. 630 g/head/day. Maximum intakes of milk and concentrates were 870 g/day and 160 g/day, respectively.

Age of the lamb at slaughter

If the initial cost of a lamb be ignored, the sooner it is slaughtered the more efficient will the process have been, and it is worth noting that wherever the initial cost, i.e. the food requirement involved in producing the newly born animal, is small relative to the subsequent requirements for growth, slaughtering at an early age will be indicated. Obviously there is no point in this until the lamb has at least become a satisfactory product. After this time, the desirability of slaughtering will also depend on whether the extra food required to continue growth is being bought, e.g. grain, or provided as grazing. If it is bought, it might be more efficient to feed it to another small lamb and repeat the process, provided that another lamb is obtainable. If another lamb is not available immediately, it is necessary to take greater note of other resources invested, including time. Theoretically, there is no reason why grazing should be a special case, since grass also has a production cost. It is, however, a cheap food and, moreover, it may be more efficient to graze it with an inefficient animal than to conserve it, with current inefficiency. It goes without saying that it is more efficient to graze it in this way than not to do anything with it at all. The most efficient procedure for the grazing situation would also, theoretically, be to use another young lamb, but there are other factors to be considered in this (see subsequent chapters and Chapter V) in practice.

Production of the newly born lamb

The new-born lamb, or whatever stage is taken as the starting point in this discussion, however, always *has* a cost. Whether it be expressed in monetary terms or not, the cost of the lamb must reflect to a large extent the cost of keeping the ewe that produced it. Because the foods used by both ewe and lamb can be of the same kind, and can thus be expressed in the same terms, it is possible to assess the efficiency of the whole process.

There is one further point to consider, however, before this is done. Most of our information about lamb growth concerns liveweight and at times this may include atypical quantities of ingesta in the alimentary tract. It is necessary to use actual carcase weight for the final product, for although it is not entirely edible and, in any case, consists largely of water, it is the product that is bought and sold. The value of any intermediate body weight is primarily biological and depends on its capacity for progression to further stages. This further development is as much dependent on other parts of the body as it is on the carcase present at that time. Nevertheless, the production process must be concerned with the development of the product, however unlike the final version it may be at any particular stage. Since the carcase represents an almost constant proportion of the empty body weight as the lamb grows bigger (32), it is possible to calculate the intermediate points on the growth curve in terms of carcase weight, without incurring too large an error.

Importance of Ewe Maintenance

The largest single item in the food input to the ewe and lamb unit is that required for the maintenance of the ewe (5). This is so during the productive years, and it is relatively larger still if the whole life of the ewe is considered. The simpler situation will be discussed first, taking a year in the middle of a ewe's productive life.

Estimates of the maintenance requirements of ewes vary greatly. Kammlade & Esplin (18) summarized the position by giving a range of 2·6 to 3·8 lb food (of 90 per cent D.M.) required daily by dry ewes (gaining 0·07 lb/head/day) from 100 to 160 lb liveweight. Coop has emphasized the difference between the maintenance requirements of sheep under the conditions of pen feeding, 0·92 lb digestible organic matter (D.O.M.) for a sheep weighing 100 lb (6), and those associated with grazing conditions. The latter were found by Coop & Hill (9) to be much higher, three estimates being 1·48, 1·63 and 1·36 lb D.O.M. a day for a sheep of 100 lb liveweight. Other estimates are those of Langlands *et al.* (21) of 0·82 lb D.O.M. daily for a 100 lb adult sheep, and 10·25 lb starch equivalent (S.E.) for a 125 lb ewe (22) weekly.

The maintenance requirements of ewes, when neither lactating nor pregnant, have been summarized recently (37) as varying from 1·80 to 2·02 M cal metabolizable energy/day (for diets containing 3·0 to 1·8 M cal/kg dry matter). These values are for 70 kg ewes walking about 2 miles/day.

Measurements of the amount of food required by sheep of different sizes and breeds to remain at constant weight confirm that size is the major factor but suggest that other factors must have a considerable influence. Individual variation can be considerable even under very similar conditions.

Feeding the breeding ewe

The additional food required by the ewe during pregnancy is a relatively small proportion of the total annual intake: that required during lactation is substantially more. The magnitude of these additional needs varies with litter size and ewe size. Since ewes always have to be maintained at the same time as they are fed for any of these productive purposes, it is more useful to consider the proportions of the annual food requirement that are consumed in total in each of the main phases (Fig. 6.4).

Efficiency of Food Conversion for Meat Production

Consideration of the *proportions* of food devoted to different purposes should not obscure the fact that a different quality of food, and therefore a different cost, may be necessary at some times and for some purposes. It is true that a high proportion of the total food is consumed by the ewe but some of this can be of a lower quality and cost than any of the smaller total quantity required by the lambs.

The measurement of food requirement

The quantities of food referred to in this section are those measured in experiments at Hurley (38) in which ewes have been fed, in pens, to maintain constant weight, except during pregnancy, when they have been fed according to their needs. The latter have been judged (*a*) according to the number of foetuses, determined by X-ray examination, and the liveweight change necessary to ensure that the weight gained in pregnancy is approximately equal to that lost at parturition, or (*b*) by measuring the non-esterified fatty acid (N.E.F.A.) levels in the blood and feeding to prevent the level rising above that associated with adequate nutrition. The quantities of food required by lambs are

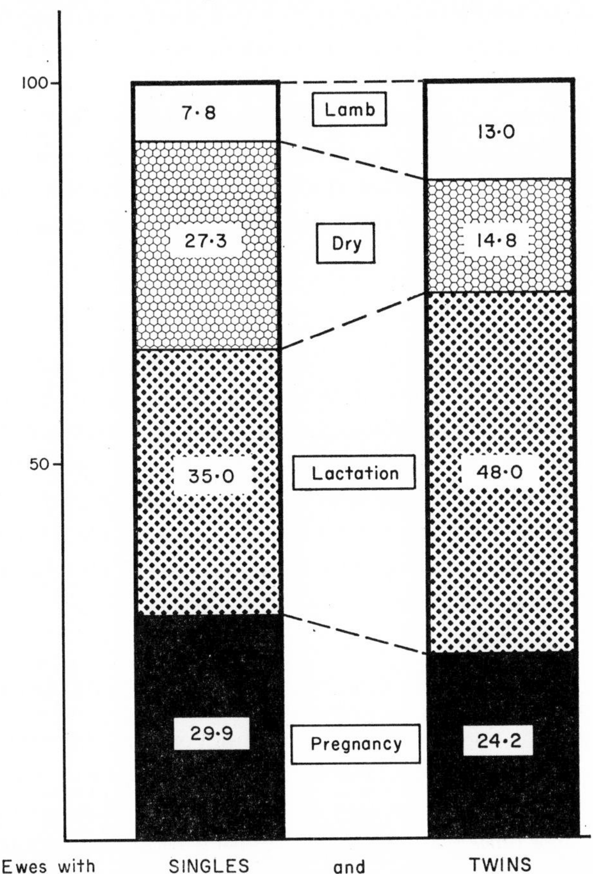

FIG. 6.4—*Percentage distribution of D.O.M. consumed by ewes with singles and twins* (*Suffolk* × *Half-bred*) *during a year*
(Source: Large, R. V. Expt. H. 650)

either the amounts of solid food consumed *ad lib.* by suckled lambs or the total amounts of milk substitute and other foods offered *ad lib.* to artificially reared lambs.

It may be that these quantities are not those necessarily associated with either the most efficient or the most economic food conversion. In the case of the lamb, *ad lib.* feeding is clearly not the most economic regime when expensive foods (such as milk substitute) are used: a very high plane of nutrition, based on a balanced diet, is likely to lead to the highest food conversion efficiency, however. For the ewe, it can be argued that, apart from the weight that must be gained during pregnancy, a constant weight will avoid the inefficiency of laying down and using up fat reserves. It is unlikely that the last word has been said on this topic, however, in terms of the efficiency of fat use in times of stress and when avoidance of fat loss (e.g. in lactation) may restrict production. It is obvious that variation in liveweight may be sensible in certain environments and it may often be cheaper to conserve herbage as animal fat than to do it by making hay or silage. Quite apart from all this, it is also arguable that it is inefficient to maintain a ewe for the whole year at a weight that may only be required at some critical time (e.g. mating).

It must be concluded that the optimum pattern of weight change cannot yet be specified, except perhaps for certain environments, and that used experimentally (i.e. constant liveweight) is the most satisfactory for the purpose at the present time. Unfortunately, a decision to maintain a constant weight does not automatically settle what weight should be maintained and the most satisfactory weight or size for a given breed cannot be specified either. The best that can currently be done is to aim at good condition without excessive fat. Most sheepkeepers know what is meant by this and probably aim to achieve it at mating (in fact, it is often referred to as the 'tupping weight'); nevertheless, it would be helpful to employ a simple scoring system to indicate degree of fatness, in order to reduce the subjectivity of the judgement.

The measurement of food conversion efficiency

Food conversion can be measured as the amount of product produced per unit of food consumed and can be most simply expressed by the following equation:

$$\text{Conversion efficiency } (E) = \frac{P}{F} \times 100$$

where P is the product (in kg) of carcase and F is the amount of food (also in kg) of D.O.M. required to produce it. Multiplying by 100 simply avoids very small fractions.

The main factors are thus determined, for a particular process, on logical grounds; they *must* be the main factors but many other things alter their values. The importance of a main factor must also be judged by the effect on E of variations in its value, within a practicable range.

The equation can be conveniently expanded as follows:

$$E = \frac{C \times N}{F_E + N(F_L)} \times 100$$

where C is the weight of one carcase, F_E is the food consumed by the ewe in one year, F_L is the total food eaten by one lamb and N is the number of lambs.

Looked at in this way, it is obvious that the major factors capable of influencing E are the size of the product (C), the size of the ewe (affecting F_E), and the number of lambs (N). The food eaten by the lamb (F_L), we already know to be a relatively small proportion of the total.

It is then easy to extend the list of factors to include all those that influence C, F_E, F_L and N. Some will be more important than others, and their importance can be tested by inserting changed values for them into the calculation and determining the effect on E. They must all operate, however, through the main factors. An economic calculation can be made on the same lines, for food conversion efficiency or, more usefully, for an expanded index that includes all costs (e.g. labour).

The general relationship, where lambs can be successfully reared, is likely to apply to all breeds (see next section), but the shape of the curve may be quite different.

The effect of the main factors

Any of the main factors can have a dominant effect: if $N=0$, there is no output and C must also be equal to 0. If $N=1$, the value of C becomes very important; the value of F_E must be important when $N \times F_L$ is small, but, if N is large, the importance of F_E diminishes. These main factors will be considered in turn.

The number of lambs (N). In an annual calculation, N can be affected by both litter size and lambing frequency. Their combined effect is largely a matter for the future and will be further considered in Chapter XV. At the present time, the number of lambs per ewe per year is determined by the lambing percentage. This is affected by barrenness and losses, but is chiefly influenced by the mean litter size.

The effect of litter size. This is illustrated in Fig. 6.5. Clearly, efficiency (E) increases with increasing litter size, simply because the effect on output is much greater than the effect on the extra food required to produce it. This will only be so whilst larger litters result

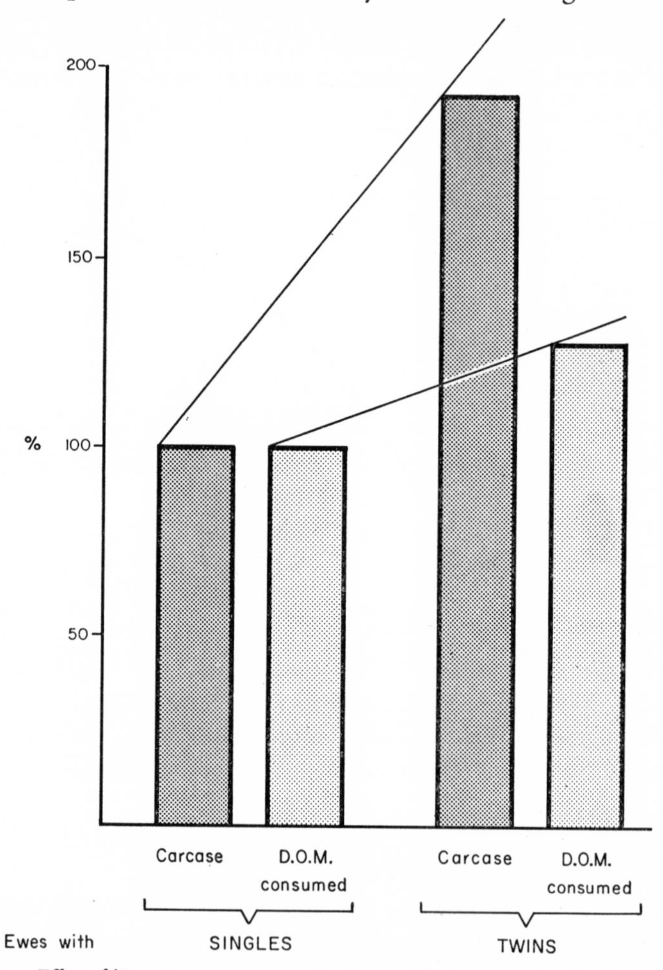

FIG. 6.5—*Effect of litter size on carcase production and food consumption* (both expressed for twins as percentages of the values for singles)

in higher output, of course; where they do not, because of increased mortality, for example, E may even be diminished.

The increase in E is not linear with litter size, however, since the relatively constant value of F_E represents an investment of food in the ewe, spread over an increasing number of lambs: the effect of 1 lamb instead of 0 is the greatest obtainable, and the effect of one more lamb gives a diminishing return.

This curve is illustrated in Fig. 6.6, in which it can be clearly seen that the biggest increases in E are to be obtained for increases in litter size within the normal range (0—3).

Such a ratio can be used, of course, for all products and for a variety of ways of expressing both product and the food used. For many

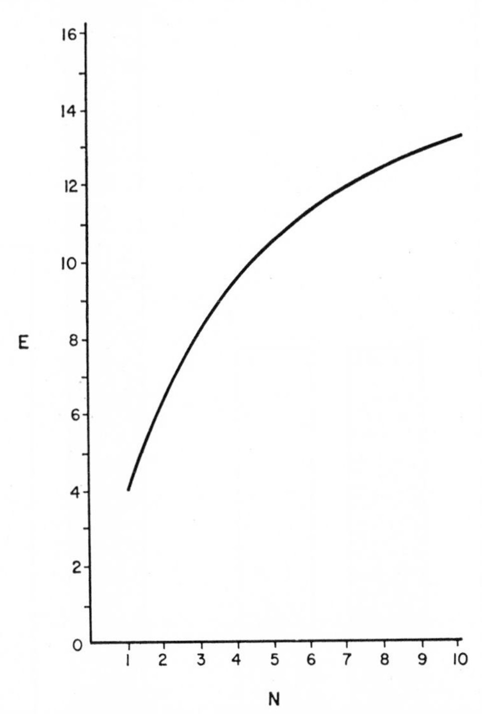

FIG. 6.6—*Effect of number of lambs per ewe per year* (*N*) *on E* (*for Suffolk × Scottish Half-bred*) (from a calculation by R. V. Large)

biological purposes, expression as energy or protein would be appropriate: for most agricultural purposes, however, the product is most usefully expressed as the saleable commodity (the carcase) and the food as digestible organic matter (D.O.M.).

The entire calculation can refer to different periods of time, according to the purpose for which it is made. The efficiency ratio (E) tends to be most useful, however, when the period of time is long enough to embrace a whole productive cycle. In frequently-breeding sheep, this could be from one conception to the next; for most sheep it is one year and, on a flock basis, this may even take account of age etc. A complete picture must include longevity and the food costs of rearing ewes to breeding age.

It is not yet possible to make such extended calculations with any confidence but data are increasingly available for calculations relating to a whole year.

The size of the ewe. This has a marked effect on F_E, the major food input. Large ewes eat more than small ones but the difference is not constant; it may vary with whether the ewe is lactating, pregnant or dry. Unfortunately, major differences in ewe size usually involve breed differences and it is difficult to dissociate the consequences of size alone (42).

In practice, the same applies and a large change in ewe size will probably require a change of breed or cross.

The relationship between total annual food intake (for a ewe and progeny) and ewe size is shown in Fig. 6.7, for ewes with singles and for ewes with twins.

An attempt to describe the effect of size on E, in these same situations, is illustrated in Fig. 6.8. The data refer to particular circumstances, however, and should not be regarded as a fair breed comparison. Quite apart from the omission of wool production, the method of assessment takes no account of the environment in which production can be sustained and the slaughter weights were selected somewhat arbitrarily. Clearly, if output were constant, the advantage would lie with the small ewe. A market demand for a low carcase weight tends to have this effect and reduces the chances of a large ewe operating efficiently. In general, however, the size of the ewe will have some effect on the size of the product, although the size of the ram is of about equal consequence in this.

In the same example given, it cannot be assumed that the maximum carcase weight was, in fact, obtained for any of the breeds used. The actual carcase weights will have influenced the calculation on E considerably.

It should be emphasized that the examples given are based on data derived from pen-fed ewes. Less information is available for the grazing situation, which could give different results: earlier calculations (30), based on data from the literature (19, 22, 25, 27, 31), gave results in the same direction, however.

The size of the carcase (C). If a lamb was slaughtered shortly after birth, the whole process would have been grossly inefficient—although the food conversion efficiency of the lamb itself would have been very high.

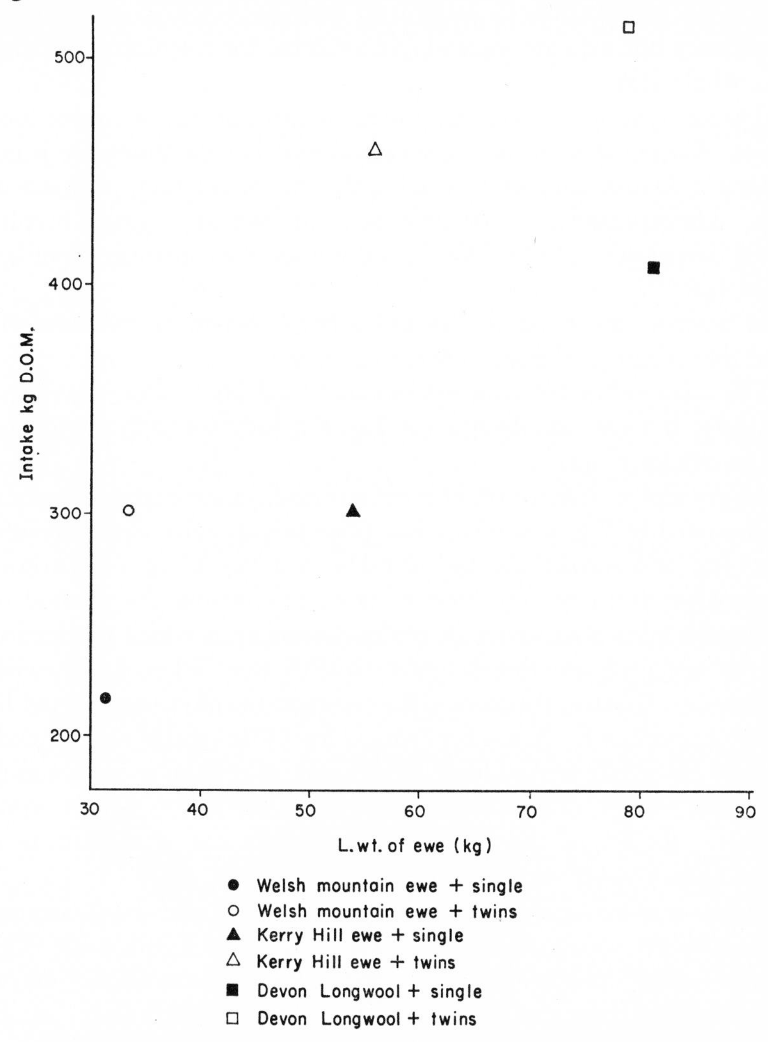

FIG. 6.7—*Effect of size of ewe on annual feed intake of the ewe/lamb unit*

Obviously, the total food input at birth has been high and, for a little more, a much greater output could have been obtained. It is not obvious, however, how far this argument can be pursued or at what case weight it ceases to be valid.

As the lamb grows larger, so its own efficiency declines but, whilst this remains higher than that of the whole family unit, the net effect is to increase *E*. It is therefore not so much a matter of size as of lamb

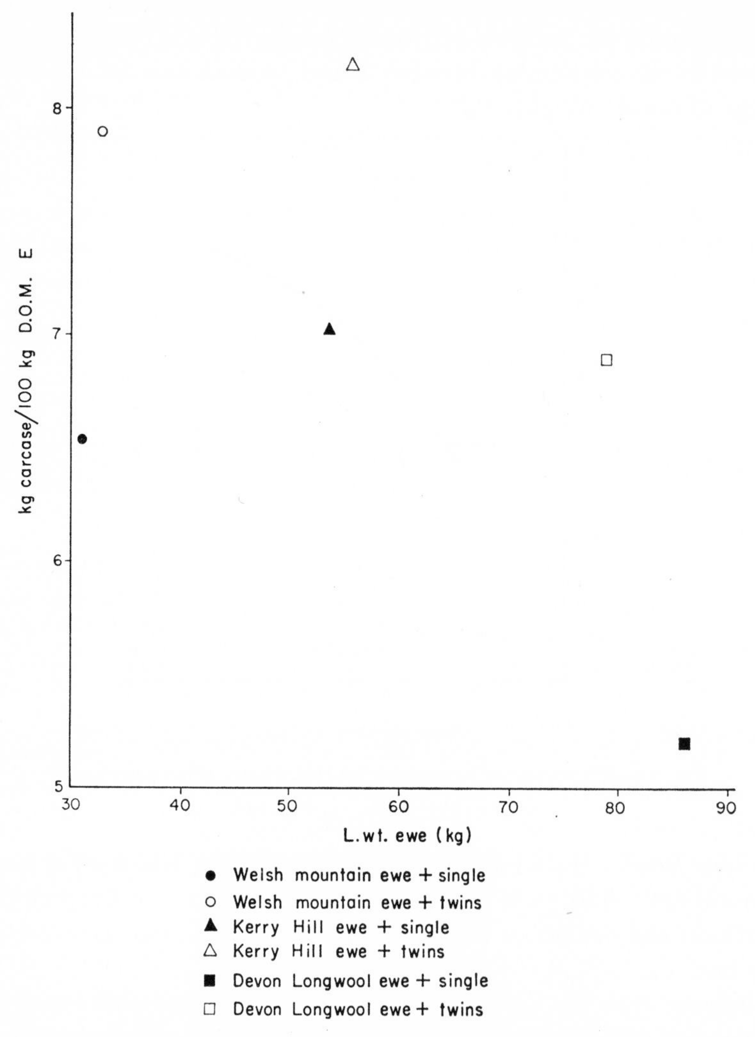

FIG. 6.8—*Effect of size of ewe on food conversion efficiency*

growth rate. The efficiency of food conversion by the lamb will be most influenced by its growth rate and, whilst this is high, it is likely to be higher than that of the family unit. If growth rate declines, for whatever reason (nutrition, disease or because the lamb is approaching its mature size), to the extent that food conversion efficiency of the lamb(s) falls below that of the unit (or, in practical terms, is not substantially above it), then the process should be terminated by slaughter.

Few figures are available yet for the whole relevant range: Fig. 6.9 illustrates the relationship between *E* and product size, for the lower range of carcase weights only.

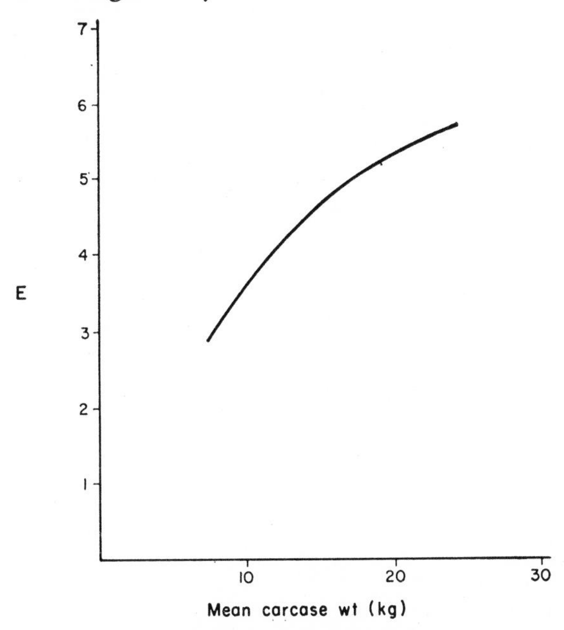

FIG. 6.9—*Effect of carcase weight on E (Suffolk × Kerry Hill = single lambs)* (data supplied by R. V. Large, 1969)

Other factors. Included here are all those factors which affect lamb growth rate: milk yield and prolificacy of the ewe (see Chapter IV), nutrition and diseases of the lamb (see Chapter V), and the effects of the ram.

Influence of the sire. The reason why the ram is largely left out of the calculation of food conversion efficiency is because it usually represents only one-thirtieth to one-fiftieth of the adult flock. Its contribution to food consumption is thus rather small. Perhaps it should not be dismissed quite so readily, even in these calculations; it should certainly figure in any economic calculations.

The importance of the ram, in relation to food conversion efficiency, is chiefly due to its influence on the growth potential of the lamb. In general, the potential growth rate of the progeny is intermediate between that of the two parents and the ram can therefore have a considerable effect in reducing any restriction on lamb growth rate and final size imposed by keeping a small ewe.

WOOL PRODUCTION

In many respects, the efficiency of wool production is by far the simplest relationship to determine, since wool can form the sole product. Similar arguments apply here, as they did to meat production, and the input of food and output of wool can be readily related (36), from a production point of view. Furthermore, the intermediate stages of growth of a fleece can be determined with relative ease and the wool grown in one month may be regarded as having a similar value, as it is the same kind of product, to that grown in any other month. There are considerations, such as of staple length, that make the whole fleece different from the simple sum of its monthly increments; but, whenever it is grown, and in whatever quantity, it is all usable.

As with meat, the more that can be produced from one animal in one year the more efficient the process is likely to be. There is one important difference, in that, as wool is grown, the resulting additional *weight* of the sheep is unlikely to affect significantly its maintenance requirement. Efficiency of conversion of food to wool, therefore, does not decline as the wool grows—unless, as in the lamb, the animal body is growing too.

Relationship with Body Size and Nutrition

This is a subject of some complexity, however, since wool production, unlike liveweight gain, is positively correlated with body size or, more accurately, surface area. As the animal grows larger, therefore, it eats more but it also grows more wool, and it has been suggested (8) that the efficiency of food conversion to wool may be independent of liveweight. In general, the bigger the sheep, the more wool it will produce, and the greater the ratio between surface area and body weight (a value deliberately sought for in the Merino, for example), the more efficient is production likely to be. The stimulation of wool growth by thyroxine implantation illustrates some of the important relationships which determine efficiency. Lambourne (20) found that thyroxine increased the length, but not the diameter, of wool fibres and resulted in up to 15 per cent more wool. At the same time, however, the maintenance requirement of the sheep was raised from 550 to 750 g D.O.M. per day. The increases of between 20 and 25 per cent in intake of D.O.M. per day had the effect of lowering the efficiency (E) of food conversion to wool. Lambourne expressed efficiency as follows:

$$\% E = \frac{\text{g wool}}{\text{g D.O.M. intake}} \times 100$$

and, over a six-month period, the untreated controls had values for percentage E of between 1·46 and 1·74. The thyroxine-treated sheep had values varying from 1·21 to 1·52.

Efficiency of food conversion in wool production

This subject has recently been discussed by Ryder & Stephenson (40), who have summarized the available experimental results, especially from Australia.

Within the Merino breed, fine-wools are least efficient and strong-wools most efficient.

In general, it seems that sheep producing more wool also eat more, but the increased wool production is due both to increased food intake and to greater efficiency of conversion of food to wool.

Sheep selected for high fleece weight tend to have high efficiencies and it appears that selection on one diet and in one environment (e.g. pen-feeding) results in the same order of ranking as selection on other diets and in other environments (e.g. under grazing conditions).

Efficiency of food conversion in an individual sheep is relatively unaffected by variations in energy intake but it may be decreased by pregnancy and lactation and by stress.

Under drought conditions, it has been found that intermittent feeding increases efficiency, compared with regular feeding.

Effect of poor nutrition

A further characteristic of wool growth, which affects the efficiency with which it may be produced, is its continued growth during periods of poor nutrition, even when this involves some loss of liveweight (17). This means that, except under conditions of great nutritional stress, wool once grown is not likely to be lost again. Thus the efficiency of wool production may be expected to be less affected than that of meat by an environment in which nutrition varies in an irregular manner. Indeed, since wool growth can continue whilst liveweight decreases, it is possible for the efficiency of wool production per acre to be increased, by restricting the food intake of the individual (24). Wool yield per sheep is decreased by rates of stocking which markedly affect liveweight (29) but yield per acre is higher. As Allden (1) has pointed out, however, it is necessary to take into account lifetime performance as well as current production. He found that only those factors which affected the growth of young sheep in the first few months of life had any long-term effects on their wool production. He also found, under Australian conditions, linear responses of wool growth to both energy and protein supplements when these were added to a diet obtained from summer pasture. Presumably the higher yields increased the over-all efficiency of food conversion, but supplementary feeding proved quite uneconomic.

Effects of specific components of the diet

Many studies have been conducted with the object of relating wool growth rate to various components of the diet. The literature has recently been reviewed by Schinckel (28), who considered it 'probable that wool growth in an animal in positive energy balance (or energy equilibrium) is primarily a function of the amount and composition of amino nitrogen at the absorptive level (that is, metabolizable nitrogen)'. Wool growth responds to increasing feed intake and the most marked variations in seasonal wool growth are considered to be found in those areas showing great variations in pasture growth (28). Schinckel concluded that during periods of weight gain a close association existed between rate of wool growth and rate of liveweight increase. The maximum rate of wool growth observed in Australia he considered to 'approach the limit of which sheep are capable (1·6 to 1·8 $mg/cm^2/day$)'.

It has already been pointed out that it is difficult to combine estimation of efficiency of both wool and meat production. In fat lamb production both occur together and, in some breeds, the fleece is the more important product. In comparing breeds, therefore, the most relevant terms for expressing both meat and wool will frequently be economic.

MILK PRODUCTION

In many parts of the world, for example in France and Israel, sheep are milked on a considerable scale and milk is a primary product. It is often made into cheese and, in this event, the first part of the lactation is often used to rear one or more lambs. In some parts of Eastern Europe sheep are triple purpose animals: lambs and humans share the milk, meat is produced and wool is grown. Here we shall consider the efficiency of the milk production process, irrespective of the use to which the milk is put. Much the same approach can be made as before. The major food input is again the requirement of the ewe for maintenance during one year. If additional feeding during pregnancy is debited to the production of a new-born lamb, we can then simply relate additional feeding during lactation to the milk produced.

Efficiency of Food Conversion to Milk

The efficiency with which the total food intake is used for milk production must vary with the level of yield, since the proportion of food converted into milk will be greatly influenced by this. Yield varies between breeds (see Chapters IV and XII) and often increases with the age of the ewe. Gruev (14), for example, found that milk production increased up to 6 or 7 years of age in the Précoce, Stara Zagora and Plovdiv breeds. It is interesting, when considering efficiency of production, that he found milk production to be lower in Précoce ewes, although their body weight was greater; the fleece weight was higher, however. The level of yield also varies with the stage of lactation and breed comparisons at one stage of lactation only may give misleading results. Hunter (16) has pointed out that, in his experiments, milk yield of the Border Leicester was higher than that of the Welsh Mountain ewes for most of the lactation but became very similar by 16 weeks. This was at a time when yields were not negligible and from ewes which at their peak of lactation were producing 121·6 lb a month (Border Leicester) and 92·8 lb a month (Welsh Mountain).

Clearly, where breeds differ markedly in milk output but not in weight, or in weight but not in milk output, the efficiency of the food to milk process must vary greatly. Brody (4) concluded that large and small animals produced the same amount of milk energy per unit of food energy and found that milk production varied with $W^{0 \cdot 7}$ (where W= liveweight of the animal), which he termed the 'lactationally effective body size'. These conclusions related, however, to the ratio of milk energy produced to food energy consumed by a wide range of species (cattle, goats, rats, etc.). Obviously one way of

TABLE XXXII

FOOD CONVERSION EFFICIENCY OF DORSET HORN EWES DURING LACTATION

MILK YIELD (kg)	FOOD CONSUMPTION (kg D.O.M.)	
	during 9 weeks' lactation	*annually*
80·7	53·7	252.7
Efficiency=	150·5	31·9

Efficiency has been calculated as $\frac{\text{Milk yield (kg)}}{\text{Food consumption (kg)}} \times 100$

Based on 46 ewes, covering 7 different nutritional treatments during lactation

(After T. T. Treacher, 1967)

describing the energy requirement for milk production is in terms of the starch equivalent (S.E.) (or total digestible nutrients (T.D.N.)) per lb of milk produced. Reid (26) has given values for the ewe of 0·42 lb S.E. or 0·48 lb T.D.N. and a gross energy value for ewe's milk of 536 kcal/lb. He points out, however, that 'even within species it appears probable that heredity determines the tendency to fatten and, therefore, influences the extent to which the lactational increments diminish as the level of dietary energy consumed increases'.

The efficiency of food conversion to milk can be calculated in the same way as for meat:

$$\frac{E}{F} = M \times 100$$

where *M* is the amount of milk produced (in kg) and *F* is the amount of food (kg D.O.M.) used (during the whole lactation, above maintenance or in total for the whole year).

An example of this calculation is given in Table XXXII for Dorset Horn ewes, machine-milked for a 9-week lactation (Treacher (43)).

Weight Changes During Lactation

It is well known that ewes may lose a considerable amount of weight during lactation, and, while this is often related to the number of lambs suckled, weight loss may vary markedly within groups with the same number of lambs and within a group on one plane of nutrition. The range of values for weight loss during the first four weeks of lactation for ewes with twins, all grazing together on the same occasion, is shown in Table XXXIII. Similar data are given in Table XXXIV for 6 ewes milked twice daily by machine, all offered dried grass nuts *ad lib.*

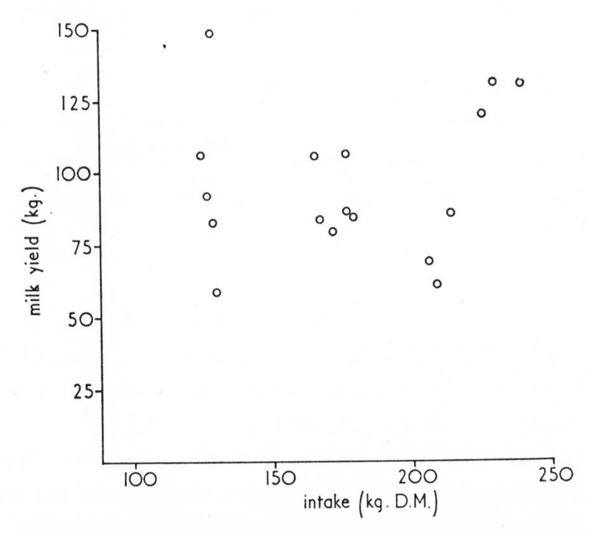

Fig. 6.10—*Total milk yield of machine milked, Dorset Horn ewes in relation to total D.M. intake, over a twelve-week period, of grass nuts (N.B. changes in digestibility with level of intake may have occurred). 1 kg of milk is approximately equivalent to 0·22 gal.*

(Source : T. T. Treacher, unpublished data from Expt. H. 367, the Grassland Research Institute, Hurley)

Quite apart from the variation between individuals, weight changes during lactation may obscure the relationship between food input and milk output. An example of the latter is shown in Fig. 6.10 for Dorset

TABLE XXXIII

LIVEWEIGHT LOSS DURING THE FIRST 4 WEEKS OF LACTATION IN HALF-BRED EWES AT PASTURE

Number of lambs born and suckled	Number of ewes	Liveweight change (lb)	
		Mean	*Range*
1	13	−10·0	+3·0 to −22·0
2	24	−15·0	−1·0 to −29·0
3	3	−20·0	−17·0 to −27·0

(From an experiment (H. 202) at the Grassland Research Institute, 1960)

TABLE XXXIV

CHANGES IN LIVEWEIGHT WITHIN THE FIRST 4 WEEKS OF LACTATION

(in 6 machine-milked Dorset Horns)

	Mean	Range
Maximum liveweight change (lb)	−4·2	−8·6 to +9·0
Liveweight change as a percentage of liveweight at parturition . . .	−4·6	−10·4 to +14·4
Number of weeks from parturition in which maximum weight change was reached	3	2 to 4

(Source : T. T. Treacher and C. R. W. Spedding, from an experiment (H. 367) at the Grassland Research Institute)

Horn ewes fed at 4 levels of food intake. In this case, the lack of any obvious relationship may have been due to coincident weight changes or to changes in the efficiency with which the food was digested.

Milk output per unit of food input could decline with increasing food intake if ewes lost a great deal of weight at the lower levels of intake. Unless an overweight ewe is being considered, however, this weight will have to be gained again and the use of fat reserves in this way may not normally be an efficient process. At the present time, it cannot be predicted with any confidence whether a particular ewe will respond to more food during lactation by producing more milk or by losing less weight. In some circumstances, ewes giving most milk are also found to be losing most weight. It is difficult to believe that an increase in their food intake would raise the milk yield but also increase the weight loss or that a decrease in their food intake would reduce milk output but also reduce the weight loss. It is probable that ewes which are producing less milk than their genetic potential would respond to an increase in food intake, provided that the extra milk produced was removed. For suckling conditions, this presupposes that the lambs would increase their intake of milk.

In other circumstances, more food would simply modify the pattern of weight change. There is a wide and profitable field for research on these problems.

Supposing that a ewe may produce at a level relatively independent of food intake, then the quantity of available reserves must be of critical importance. Milk cannot be produced from fat reserves if the animal has none; production at the expense of reserves may thus occur for a variable part of lactation, according to the condition of the animal immediately after lambing. Any measurement of productive efficiency must be greatly influenced by such considerations. As with meat production, the large investment of food simply to maintain a ewe for the year is almost bound to mean that higher milk yields will increase total efficiency. This is so whilst milk is regarded as the end-product, whatever the actual efficiency of the milk-producing process. A somewhat different situation may arise where the milk is converted into meat by the lamb; it is also worth noting that lactation may reduce wool growth (13).

Milk and Grass for Meat Production

Little is known of the feeding value of milk and herbage mixtures in relation to the development of the lamb. The double conversion, by the ewe and the lamb, of grass to milk and milk to meat, however, is often regarded as inefficient, once the lamb is capable of using grass directly, as a complete diet by itself. The ewe is considered, in this view, to be a transformer which is only necessary whilst the lamb is unable to live and grow on the raw materials directly. This is a central issue in questions of sheep management, particularly in relation to time of weaning. It was suggested by Brody (4) that the efficiency of early growth was similar to that of average milk production, but,

TABLE XXXV

INTAKE AND WEIGHT GAIN IN LAMBS

Total dry matter intake (of milk substitute and cold-stored ryegrass) and daily liveweight gain of lambs in successive weeks.

Age of lambs (weeks)	Number of lambs	Mean daily D.M. intake (g) (*I*)	Mean daily liveweight gain (g) (*G*)	D.M. required per unit of gain (*I*/*G*)
4–5	21	394	300	1·31
5–6	21	476	347	1·37
6–7	17	605	332	1·79

(Source : T. H. Brown and C. R. W. Spedding, from Expt. H. 286 at the Grassland Research Institute)

TABLE XXXVI

MEAN DAILY INTAKES OF MILK SUBSTITUTE AND GRASS BY LAMBS DURING THEIR SIXTH WEEK OF LIFE AND THEIR RELATED MEAN DAILY LIVEWEIGHT GAIN

The lambs were fed individually three levels of milk substitute (2½, 3½ and 4½ pints/head/day, respectively) and offered cold-stored ryegrass to appetite.

D.M. consumed (g.)			Liveweight increase (g) (*G*)	D.M. required per unit of gain (*I*/*G*)
Milk	*Grass*	*Total* (*I*)		
127	336	463	324	1·43
178	331	509	365	1·37
229	248	477	358	1·33

(Source : T. H. Brown and C. R. W. Spedding, from an experiment (H. 286) at the Grassland Research Institute)

of course, the former declines with increasing age and weight (Table XXXV). It is likely that if milk resulted in a higher growth rate than that obtained on herbage alone, milk feeding might well be the more efficient. Where milk and grass may be considered as alternatives in producing the same rate of growth in the lamb, this would not necessarily imply the same food intake. As shown in Table XXXVI, similar rates of liveweight increase can occur on widely differing ratios of milk substitute and cold-stored grass. In the experiment from which these figures are taken, the total dry matter intake required per unit of liveweight gain decreased with increasing proportions of 'milk' in the diet, however.

The production of milk can be viewed as superimposed on the ewe's maintenance. Whenever the lamb is weaned, the ewe still has to be maintained for the year. Some discussions of early weaning rather imply that the ewe disappears at weaning time. The information needed in order to compare the efficiency of the grass to milk to meat process with the direct grass to meat efficiency includes:

(*a*) the amount of grass required to produce one unit of milk (above maintenance),

(*b*) the amount of milk required to produce one unit of product (e.g. carcase),

(*c*) the amount of grass, directly consumed, required to produce one unit of product.

It must be emphasized that, during the time when there is any need to consider these relative efficiencies, the suckled lamb is also eating grass.

The suckled lamb is the more efficient, therefore, when $(a) \times (b) < (c)$. This could occur when either (*a*) or (*b*) was small. Such a situation might easily arise with a heavily milking ewe, losing weight. This is, of course, to over-simplify the problem greatly and a great deal more information is required before any general conclusions can result from this kind of calculation.

In most cases, the optimum time for weaning will be related to lamb growth rate. If this will be improved by weaning, food conversion efficiency will also be increased: conversely, a reduction in growth rate is certain to lead to reduced efficiency.

When there is no great change in performance, the question is more difficult to answer.

An attempt to calculate the consequences of weaning on annual food conversion efficiency (41), suggested that the effect might not be great, at least from birth to 4 months of age.

THE USE OF MODELS

There is little doubt, however, that much progress remains to be made in understanding the basic production processes and it appears more likely with the aid of the sort of model building described in this chapter. It is not without interest that such model-building is increasingly being used: recent examples include a general approach to the determination of alternatives in the conversion of food to animal products (3) and the clarification of problems of epidemiology (33). The really important animal factors in production efficiency are relatively few. It is true that they are not always simple factors, but are affected by many other things; growth rate, for example, has innumerable influences bearing upon it. Only this kind of reduction to essential components, however, offers a chance of assessing the importance of the many factors immediately involved in the translation of the model to the environment in which it has to work.

REFERENCES

(1) ALLDEN, W. G. (1961). Nutrition and the growth of young sheep. *Wool Tech. and Sheep Breeding 8* (2), 81–85.

(2) BLAXTER, K. L. (1962). *The Energy Metabolism of Ruminants.* London, Hutchinson.

(3) BLAXTER, K. L. & WILSON, R. S. (1963). The assessment of a crop husbandry technique in terms of animal production. *Anim. Prod. 5* (1), 27–42.

(4) BRODY, S. (1945). *Bioenergetics and Growth.* New York, Reinhold Pub. Co.

(5) COOP, I. E. (1961). The energy requirements of sheep. *Proc. N.Z. Soc. anim. Prod. 21*, 79–91.

(6) COOP, I. E. (1962 a). The energy requirements of sheep for maintenance and gain. I. Pen fed sheep. *J. agric. Sci. 58*, 179.

(7) COOP, I. E. (1962 b). Liveweight-productivity relationships in sheep. I. Liveweight and reproduction. *N.Z. J. agric. Res. 5* (3 & 4), 249–264.

(8) COOP, I. E. & HAYMAN, B. I. (1962). Liveweight-productivity relationships in sheep. II. Effect of liveweight on production and efficiency of production of lamb and wool. *N.Z. J. agric. Res. 5* (3 & 4), 265–277.

(9) COOP, I. E. & HILL, M. K. (1962). The energy requirements of sheep for maintenance and gain. II. Grazing sheep. *J. agric. Sci. 58*, 187.

(10) CORBETT, J. L. (1961). The feeding of livestock. *Jl R. agric. Soc. 122*, 175–186.

(11) DAVIDSON, J. (1961). Nutrition of livestock. I. Energy and protein in the productive efficiency of livestock. *J. Sci. Food Agric. 9*, 581–591.

(12) DICKINSON, A. G. (1960). Some genetic implications of maternal effects—an hypothesis of mammalian growth. *J. agric. Sci. 54* (3), 378–390.

(13) DONEY, J. M. & SMITH, W. F. (1961). The fleece of Blackface sheep. *Hill Farming Res., Org. 2nd Rep. 1958–61*, 34.
(14) GRUEV, V. (1959). Correlations between milk production, wool yield and body weight in sheep. *Mez. sel.-hoz. Z. 3* (2), 109–118.
(15) GUNN, R. G. & ROBINSON, J. F. (1963). Lamb mortality in Scottish Hill flocks. *Anim. Prod. 5* (1), 67–76.
(16) HUNTER, G. L. (1956). The maternal influence on size in sheep. *J. agric. Sci. 48*, 36–60.
(17) JORDAN, R. M. & WEDIN, W. F. (1961). Effect of grazing management on body weight and subsequent wool and lamb production of non-lactating ewes. *J. anim. Sci. 20* (4), 883–885.
(18) KAMMLADE, W. G. & ESPLIN, A. L. (1962). Sheep are efficient users of forage. Ch. 61 in *Forages*, St. Univ. Press, Iowa. 2nd Ed. Ed. Hughes, H. D., Heath, M. E. & Metcalfe, D. S.
(19) LAMBOURNE, L. J. (1955). Recent developments in the field study of sheep nutrition. *Proc. N.Z. Soc. anim. Prod. 15*, 36.
(20) LAMBOURNE, L. J. (1963). Stimulation of wool growth by thyroxine implantation. I and II. *Aust. J. agric. Res. 15*, 657, 676.
(21) LANGLANDS, J. P., CORBETT, J. L., McDONALD, I. & PULLAR, J. D. (1963). Estimates of the energy required for maintenance by adult sheep. *Anim. Prod. 5* (1), 1–9.
(22) LEITCH, I. & GODDEN, W. (1953). The efficiency of farm animals in the conversion of feeding stuffs to food for man. *Commonwealth Bur. Anim. Nutr., Tech. Commun.* No. 14.
(23) MACFADYEN, A. (1957). *Animal Ecology*. London, Pitman.
(24) MOLNAR, I. (1961). *A Manual of Australian Agriculture*. London, Heinemann.
(25) PHILLIPSON, A. T. (1959). Sheep. In *Scientific Principles of Feeding Farm Livestock*. London, Fmr. & Stockbreeder Pub. Ltd.
(26) REID, J. T. (1961). Nutrition of Lactating Farm Animals. Ch. 14 in *Milk, Vol. II*, 47. Ed. Kon, S. K. & Cowie, A. T. New York and London, Academic Press Inc.
(27) REID, R. L. & HINKS, N. T. (1962). Studies on the carbohydrate metabolism of sheep. XVII. *Aust. J. agric. Res. 13* (6), 1092–1111.
(28) SCHINCKEL, P. G. (1963). Nutrition and Sheep Production: A review. *I World Conf. Anim. Prod.*, Rome.
(29) SHARKEY, M. J., DAVIS, I. F. & KENNEY, P. A. (1962). The effect of previous and current nutrition on wool production in southern Victoria. *Aust. J. exp. agric. Anim. Husb. 2*, 160–169.
(30) SPEDDING, C. R. W. (1963). The efficiency of meat production in sheep. *I World Conf. Anim. Prod.*, Rome.
(31) THOMSON, W. & AITKEN, F. C. (1959). Diet in relation to production and the viability of the young. Part II. Sheep. *Commonwealth agric. Bur., Tech. Commun.* No. 20.
(32) TULLOH, N. M. (1963). Relation between carcase composition and liveweight in sheep. *Nature (Lond.) 197* (4869), 809–810.
(33) WHITLOCK, J. H. (1962). Bionics and experimental epidemiology. *Biological Prototypes and Synthetic Systems, Vol. 1*, 39–48. New York, Plenum Press Inc.
(34) WILSON, P. N. & OSBOURN, D. F. (1960). Compensatory growth after undernutrition in mammals and birds. *Biol. Rev. 35*, 324–363.
(35) WRIGHT, P. L. (1961). Some aspects of ewe and lamb nutrition. Ph.D. Thesis, University of Wisconsin.
(36) Rural Research in C.S.I.R.O. (1963). Merino strains compared in three environments. No. 44, 14–17.
(37) A.R.C. (1965). The nutrient requirements of farm livestock. No. 2, Ruminants. A.R.C., Lond.
(38) LARGE, R. V. (1970). Biological efficiency of meat production in sheep. *Anim. Prod.* In production.

(39) RUSSEL, A. J. F., DONEY, J. M. & REID, R. L. (1967). Energy requirements of the pregnant ewe. *J. agric. Sci., Camb. 68*, 359–363.

(40) RYDER, M. L. & STEPHENSON, S. K. (1968). Wool growth. London and New York, Academic Press.

(41) SPEDDING, C. R. W. (1968). Practical implications of genetic and environmental influences: Sheep. pp. 451–465 in Growth and development of Mammals, *Proc. 14th Easter Sch. Agric. 1967*. Ed. Lodge, G. A. and Lamming, G. E. London, Butterworth.

(42) SPEDDING, C. R. W., LARGE, R. V. & WALSINGHAM, J. M. (1969). The importance of the size of the female in sheep. *World Rev. Anim. Prod.* In Press.

(43) TREACHER, T. T. (1967). *Effects of Nutrition on Milk Production in the Ewe.* Ph.D. Thesis, Reading University.

(44) WILLIAMS, O. B. & SUIJDENDORP, H. (1968) Wool growth of wethers grazing *Acacia aneura-Triodia pungens* savanna and ewes grazing *Triodia pungens* hummock grass steppe in the Pilbara district, W. A. *Aust. J. exp. Agric. Anim. Husb. 8*, 653–660.

VII

UTILIZATION OF THE PASTURE

The previous chapter was primarily concerned with the principles underlying the production processes in sheep. These principles are unaffected by the kind of food used or the way in which it is produced; but the actual results may be greatly affected by food quality and quantity, and the successful application in practice of an efficient process based on sound principles may depend on many other factors as well. Where food is under close control, as in the use of a manufactured compound, the result might be expected to be predictable. This is in fact one argument in favour of close control; another is the possibility of improved efficiency.

The majority of sheep are pastured, however, a situation which allows only incomplete control of the food supply. In Chapter II the non-uniformity of pasture growth in Britain was stressed: it may be greater or less in other parts of the world, chiefly depending on the magnitude of the climatic variations. Some indication was given in Chapter III of the ways in which the demands of the sheep population also varied. The matching of these demands and supplies cannot be exact and a safety margin is always necessary. Efficient use of pasture thus requires, first of all, that this margin is available when needed, in suitable quantity and quality. Since the pasture does not remain static, a particular surplus of herbage has to be used within a given, comparatively short time, either because it is needed for grazing or because it must be conserved before it deteriorates. This problem of management, considering the whole year, is further dealt with in Chapter X.

Conservation is thus a concomitant of efficient pasture use, in the sense that without it much herbage may be grown, die and decay without having passed through a grazing animal at all. In other words, efficiency here refers to the proportion of the herbage grown that is consumed by ruminants directly or after processing; or, in the agronomic sense, to pasture utilization. There are, of course, other measures of efficiency and there are many circumstances where it would be uneconomic (involving, perhaps, inefficient use of labour or some other resource) to achieve a high degree of utilization. There is therefore nothing necessarily good or bad about any particular level and this chapter is simply concerned with the factors that influence the degree of utilization and the effect of this on animal production

It is always something of an over-simplification to say that herbage is grown for animals to eat, for some part of it is always grown in order that other parts may continue to grow and be eaten. Nor is this simply a reminder that important parts of the plant (such as roots) are not eaten at all; it is worth considering whether all the edible leaf that is grown can be consumed without reducing total herbage yield. Certainly the proportion of the sward removed at each harvest can influence the total amount grown. Although utilization frequently refers to the proportions removed and left behind during a short-term harvesting process, the results have ultimately to be assessed in terms of the annual yield. There is clearly no virtue in complete utilization if it results in a greatly diminished total production. For a given input of resources, it is often important to grow as much herbage as possible. Exceptions to this, such as giving priority to a particular pattern of growth, length of growing season or nutritive value, have been discussed in Chapter II. In general, however, a major attribute of any method of utilization is its effect on the total production of nutrients.

As has been pointed out in Chapter VI, the number of animals, between which the available nutrients are distributed, has a big influence on animal production. Under grazing conditions, the proportions of the pasture to be removed and left uneaten may affect both the stocking rate that can be sustained and the food intake of the individual animal. A further important attribute is therefore the effect of utilization on these two animal production factors. Before discussing these aspects separately, it is worth mentioning some of the factors which affect the process.

Factors Affecting Utilization

The proportion of herbage grown that is harvested over any prolonged period depends upon (*a*) the proportions removed and left at each harvest, (*b*) the frequency of defoliation, (*c*) the defoliating agent and any other consequences of its presence whilst defoliating, (*d*) the structure, physiological state and botanical composition of the pasture, (*e*) climatic conditions and (*f*) soil fertility. Although listed separately here, these factors all interact with each other and with others not mentioned.

Obviously, the proportion removed at one time represents the degree of utilization on that occasion and this must affect the over-all total harvested. The amount, or proportion, of herbage removed, however, may influence the structure of the sward as well as its productivity. Furthermore, changes in the structure of the pasture can affect the proportion that can be harvested subsequently. This is most easily illustrated with cattle. Very close grazing or cutting of an erect type of pasture can rapidly result in one so dense at the bottom that much of the herbage is no longer accessible to cattle, or at least not readily so. Frequency of defoliation matters greatly in the reaction of a sward to the removal of any particular proportion. In general, sward density is increased by frequent defoliation. The amount that can be removed is physically limited by the kind of grazing animal, since cattle and sheep differ considerably in closeness of normal grazing, and by the kind of machine employed for cutting. Machines have normally to work at an appreciable cutting height, unless on very even land. Other effects of the defoliating agent include the physical flattening of herbage, by feet or wheels, below the harvesting height and soiling of herbage with mud, faeces or urine. Mechanical damage can be of particular importance in wet weather or in long herbage. Soiling is most serious with cattle and considerable wastage can result later in the season when all harvesting is by cattle only. Naturally, all these reactions may be different on different pastures. Some plants react more swiftly in density and growth habit, some are more susceptible to mechanical damage. The weather has obvious effects in changing soil conditions and the external condition of the pasture plants. More importantly, climatic conditions and soil nutrient status govern the ability of the plant to grow, and only when it is growing can a plant change and react to any of the above factors, to any significant extent.

Effect of Method of Utilization on Pasture Production

Since herbage growth depends upon sunlight for photosynthesis, production from an area of soil depends upon the efficiency with which sunlight is trapped and used. This is influenced by sward structure, by the disposition of leaves on a plant, by light intensity and many other factors (11), but above all, perhaps, in grassland, by the quantity of herbage present. Because plants reflect light, i.e. they do not absorb all the light that falls on them, the optimum amount of herbage is far greater than that which would just cover the area of ground it occupies.

The optimal disposition of leaves varies with the growth form of the plant. For grasses, under British conditions, Wilson (19) has suggested that a vertical disposition would be most effective. It is essential that each leaf is exposed to a light intensity greater than that which its 'compensation point' requires. The latter is the intensity below which more energy is used than is assimilated and for grasses it is about 100 foot candles. There is no point, however, in exceeding an upper limit, called the 'saturation point', above which no more energy is gained.

The ratio of leaf area to ground area is termed the leaf area index (L.A.I.) and it has been suggested (see Chapter II) that, for grass-clover swards, it is at its optimum when the value is about 5. It has therefore been suggested that maximum plant growth rate would be sustained if pasture could be kept at this value. To do this, herbage would have to be removed daily at about the same rate as it grew. As shown in Chapter II, this rate varies markedly but, in any case, the daily increment is a rather small quantity to harvest, other than by grazing. It is sometimes claimed for set-stocking, where animals occupy the same area continuously for long periods, that only in this way could a sward be kept at its optimum L.A.I. value. Even then, of course, the stock numbers may have to be varied or the surplus grass cut frequently with, for example, a gangmower, in order to achieve the appropriate daily removal of herbage. In fact, this approach is only practicable where herbage is kept very short. Attempts to maintain swards at different L.A.I. values have demonstrated the difficulties (27). When a substantial quantity of herbage is maintained per unit area of land, the rate of senescence may so increase that any advantage of greater crop growth rate is lost. Furthermore, selective grazing by the animals presented with this amount of herbage may rapidly produce extreme non-uniformity of height, nutritive value and L.A.I.

Where a crop is grown for some time and then removed over a

short period, as in normal cutting or rotational grazing, the leaf area index must fluctuate between values much higher and much lower than any optimum value. The pattern of re-growth after cutting or grazing will then be affected by the amount left behind, because this influences how long it will be before the optimum L.A.I. is again reached. The more that is left behind the quicker this value will be attained. Cutting at a height well above the ground, however, rapidly influences the structure of the sward and may eventually result in little leaf being left below the cut level. Grazing to a height is only possible where animals graze uniformly. Obviously the optimum L.A.I. cannot be used both as a criterion of the amount to be left behind and also of the amount to be present at the beginning of defoliation; indeed, in a rotational system it is difficult to use it effectively for either purpose. In the first case, the herbage would always be growing at a higher than optimum L.A.I. value and never be utilized below it; in the second case, nearly all the growth would have to occur at less than the maximum rate. The conclusion must be that the maximum growth rate cannot be maintained, unless it be under a continuous grazing system, and that, for systems of alternate growth and harvest, only approximate upper and lower limits can be set for the heights or quantities to be taken or left.

The foregoing discussion has been based, however, on a rather narrow interpretation of the 'optimum' L.A.I. value and its relationship to sward growth rate. It may well be that, for practical purposes, it is sufficient to operate within quite a wide range of L.A.I. values, simply avoiding conditions where they fall outside certain upper and lower limits. It would be valuable to know what these are, for managements of a rotational kind could aim to defoliate in such a way that these limits were not exceeded. It must also be emphasized that any concept of optimum values in relation to light may only be fully applicable when no other factor is limiting. Clearly, when a pasture has been cut at a low level or grazed very closely, insufficient leaf may remain for any appreciable growth to take place as a direct consequence of photosynthesis. The fact that it does take place, even when all leaf has been removed, demonstrates that factors other than light have an important influence on re-growth immediately after defoliation. It is possible that the upper limit is the more important, because a lack of light will certainly limit plant growth. The effect of climatic conditions, particularly temperature and rainfall, on the ability of the pasture to react to different methods of utilization has been mentioned. Frequency of harvesting, by determining the interval for growth, can have a major effect on production, but if standards of the amount of herbage to be present at any one harvest are adopted the frequency of defoliation must vary accordingly. Advantages have been claimed for methods of utilization which allow the rest period to vary with the season (16) so that the amount on offer to the grazing animal is always approximately the same. It is always worth remembering, however, that, for the grass crop, time intervals vary in their significance according to the temperature, humidity, light intensity and so on. Different quantities of herbage will be produced in the same time-interval at different times of year but the optimum leaf area index may also vary with the season.

In general, infrequent harvesting results in greater total production of dry matter: it is often otherwise for the production of nutrients. Until the kind of plant material required for any particular purpose can be specified in greater detail, it is difficult to visualize specifications for methods of utilization in relation to herbage production.

The Intake of the Individual Animal

One of the most important effects of a system of utilization is that on the intake of the individual animal. The amount that a sheep needs to eat each day varies with its size and function; its need may also vary with temperature, wind velocity and other climatic conditions (3). At some stages of its life it is neither necessary nor desirable that the maximum voluntary intake should be achieved. Early pregnancy may, in many circumstances, be regarded as such a period. When production is required, however, an intake approaching the maximum is usually desirable (see Chapter V): it is therefore essential to try and understand the factors that affect the amount eaten.

Apart from the influence of the environment, intake may be visualized as being affected by the animal, by the food and by the way the food is presented.

The animal's capacity

Obviously, big animals eat more than small ones. This may be due to their greater need, or to their greater capacity, probably depending on which happens to be limiting. It is possible that, in animals whose genetic capacity for growth is not being exploited, an enlargement of the alimentary tract, relative to body size, could result in greater intake and, consequently, faster growth. This may well happen in an early weaned lamb whose rumen develops so that greater quantities of herbage can be dealt with; growth rate will then increase above that initially possible. This was illustrated in Chapter V. It does not mean, of course, that the growth rate of weaned lambs on poor pasture will exceed that of lambs still suckling (15), but that the growth rate of an early weaned lamb may increase with time provided that the quality of the diet does not decline.

Food attributes

The attributes of the food which influence intake may be classed as mechanical, i.e. structural, or chemical, according to the way in which the animal is supposed to appreciate them. Such categories probably over-simplify the problem. Certainly the structure of the plant may have an influence, but this may be due to the way in which it breaks down on digestion as well as to the possession of spines, hairs or a wax covering, etc. Foods may also be disliked because of their taste. There is no doubt that some grasses and clovers are eaten more readily than others (this is sometimes called 'preference' (6)) and that some are eaten in greater quantities. This is an important agronomic fact and requires elucidation. Blaxter, Wainman & Wilson (2) have shown that the intake of hay by sheep may be greatly influenced by the digestibility of the hay and there is general agreement that the rate at which food can be dealt with in the alimentary tract is one of the most important factors limiting intake in ruminants.

It is not the only one, however, and in some cases more will be eaten of a food of lower digestibility. Crampton, Donefer & Lloyd (7) endeavoured to incorporate daily forage intake, its digestibility and the size of the animal into one expression of food value, which they called the Nutritive Value Index (N.V.I.). This was calculated as follows:

$$\text{N.V.I.} = \frac{100 \times \text{daily forage intake (g)}}{80 \times \text{metabolic size (kg)}} \times \%D$$

Where D = digestibility, metabolic size = $W^{0.73}$, and W = liveweight. It is wise to bear in mind that a change in any part of a system as complex as that governing the intake of the grazing animal may result in a different factor becoming limiting. Indeed, a helpful way of looking at this whole question has been put forward by McClymont (8), based on the concept of a balance between facilitatory and inhibitory stimuli to the central nervous system resulting in the manifestation or inhibition of appetite. Thus the sheep may be thought of as having energy needs which it will satisfy unless any particular factor limits the amount it can eat.

The factors affecting food intake by ruminants have been discussed by Balch & Campling (21), Ulyatt (29) and Campling (23). With long roughages containing at least 8–10 per cent crude protein, intake appears to be limited by the capacity of the reticulo-rumen and the rate at which digesta leave this organ. The voluntary intake of low-protein diets (24) may be unaffected by the balance of concentrates to hay, unless the protein content is increased. Voluntary intakes of hays made from some herbage species appear not to be related primarily to digestibility (28) and even the time of year has been considered (26) to influence intake quite markedly.

Food presentation

One set of limiting factors is related to the way in which food is presented, offered or made available to the animal. Under housed or yarded conditions, with the supply of food under control, such things as frequency of feeding and size of pellets can influence intake. Other factors may be more important in particular situations. For example, Arnold (1), working with Merino and Merino × Border Leicester wethers fed on herbage cut daily from a mixed sward of *Phalaris tuberosa* and *Trifolium subterraneum* in Australia, found a relationship between the dry matter content of the pasture and the dry matter intake of the sheep. The voluntary dry matter intake was restricted when the dry matter content of the pasture was below about 25 per cent. Blaxter (2) has emphasized the need to present the animal with more than it requires, in such pen-feeding experiments, in order that maximal voluntary intake shall be expressed. This is probably of considerable importance under grazing conditions also. No doubt many other factors may be involved but, in the field, perhaps the most important are the quantity available to the animal and the extent of soiling. The latter may be due to faeces, urine, soil or a variety of other causes, any of which may result in lowered intakes.

Availability of Pasture

The quantity of herbage available to the grazing animal is an important but rather ill-defined concept. Much depends on what is meant by availability and clearly this must vary with the kind of animal being considered.

In general, the ruminant spends about nine out of every twenty-four hours grazing (see Chapter VIII). If it has difficulty in collecting its food it will graze for longer, but there is a limit to this. Similarly, animals can increase the number of bites they take per minute or the amount they take per bite in circumstances where either action will overcome difficulties in obtaining their requirements. At some point, however, a reduction in availability will result in a reduction in intake. What is needed is a measure of availability that will define this point.

Where herbage is plentiful per unit area of land only the total area allotted to the animal can be limiting, but where the amount of herbage per unit area is inadequate no amount of such area will greatly alter the situation. It is not always appreciated that when availability per unit area is limiting, increasing the available area, or reducing the stocking rate, may have no effect on food intake. The crux of this problem, then, is what constitutes an adequate quantity of herbage per unit area of land. Sheep can, of course, bite very close to the ground but, if they are forced to obtain a high proportion of their food in this way, their intake may decline. The amount that is adequate must therefore lie in a region that is readily accessible. Thus neither height nor density alone will suffice to describe the availability of pasture, but some combination of the two may do so.

A simple way of doing this in relevant terms could be very useful. Figs. 7.1 and 7.2 show how height and density determinations, made with a point-quadrat and recorded as a number of hits, each hit simply representing the presence of herbage at that height, can be used to give a picture of the way in which the herbage is distributed vertically (14). Fig. 7.1 shows the height-density relationships within a long and a short cocksfoot sward. The greater density close to the ground, found in the shorter sward, is typical. The increase in density in the bottom inch, in a ryegrass sward grazed continuously by ewes and lambs, is shown in Fig. 7.2. It is noticeable that swards of the same 'height' may differ markedly in the distribution of herbage within that 'height'. This is further illustrated in Fig. 7.3 by the height/density changes in an under-stocked ryegrass pasture. It can be seen at a glance that quite different proportions of such swards would be harvested by cutting at a standard height.

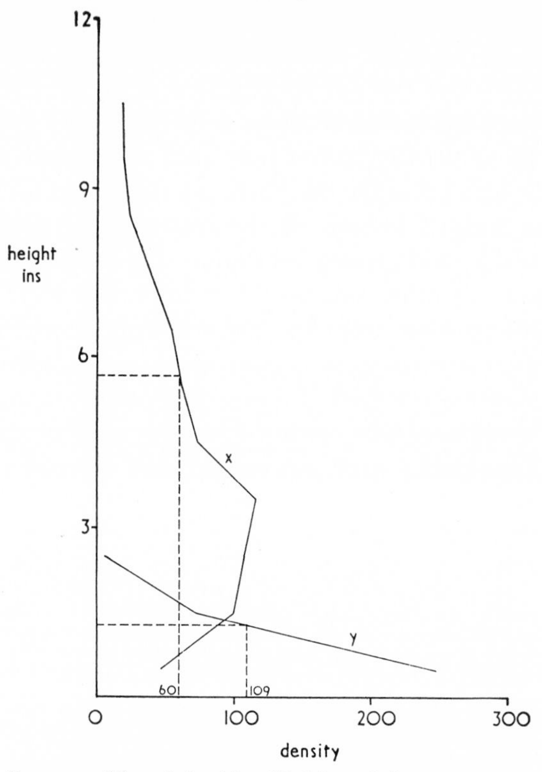

FIG. 7.1—*The relationship of height and density (expressed as the number of point quadrat hits per 100 points) in two cocksfoot swards*

The derivation of a 'height index' is shown on the figure and is based on the following calculation:

	Long sward X	Short sward Y
Total number of hits (T) . .	658	327
Range (in.) (R)	11	3
Mean number of hits per in. (T/R) .	60	109
Height Index (in.) . . . (defined as the height at which the mean density (T/R) occurs)	5·7	1·3

(Source : C. R. W. Spedding and R. V. Large (14))

In rotational grazing, the criterion required is that for the height or density of the sward to leave behind. The pasture condition when the animals arrive is unlikely to be limiting on intake: obviously it should not be so for productive stock. The question is: at what point during the grazing of a paddock or field does the amount per unit area begin to limit intake? Estimates have been made by Willoughby (18) that this can be expressed as about 1,400 lb of dry matter per acre (this would be about 0·03 lb or 14·5 g per ft²). As mentioned above, it may matter how such dry matter is distributed vertically. It is also possible that there is a maximum for some classes of stock but little is known of the interrelations between intake, quantity available, and grazing behaviour: the last is often influenced by the presence of other animals. It is important to recognize that such relationships must form the rational basis of grazing management. Rotational grazing, for example, may be a different management whenever it is practised, if these relationships are quite different. It is widely recognized that comparisons of grazing managements are often simply contrasting the relative skill of the operator at employing

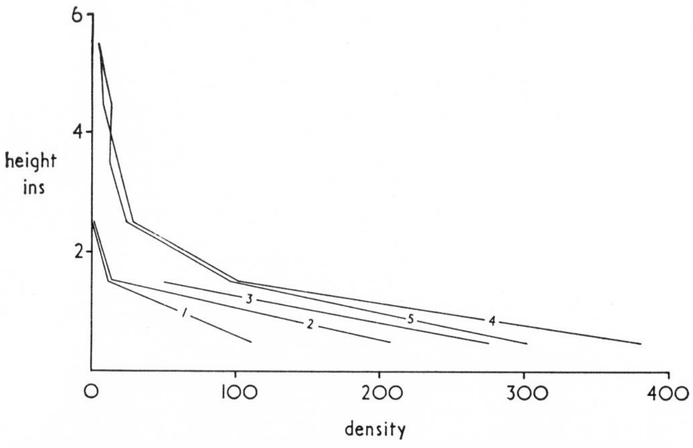

FIG. 7.2—*Development of a ryegrass sward, controlled by grazing, expressed as the change in the height/density relationship with time. Density is indicated by the number of hits per 100 points, using a point quadrat*

Dates of analysis: 1. 12 April
2. 15 May
3. 19 June
4. 18 July
5. 8 August

(Source: C. R. W. Spedding and R. V. Large (14))

each system. Where, over a season, a particular rotational system sometimes restricts intake and sometimes does not, it is difficult to

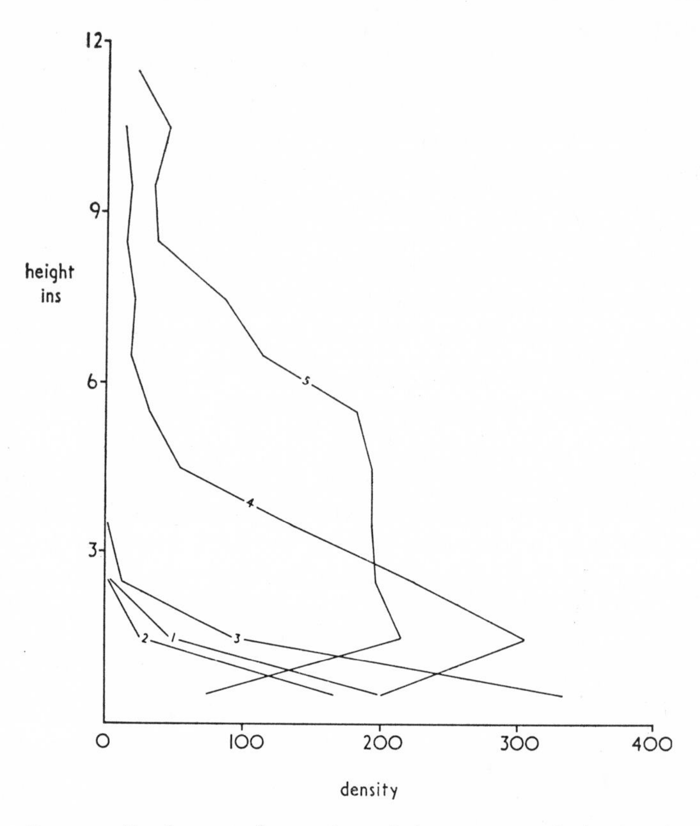

FIG. 7.3—*Development of an under-stocked ryegrass sward, showing the relatively small increase in the height at maximum density*

Dates of analysis: 1. 12 April
2. 15 May
3. 19 June
4. 18 July
5. 8 August

(Source : C. R. W. Spedding and R. V. Large (14))

interpret the meaning of the final result. It is a matter of urgency that biologically valid criteria be adopted in such studies. It is a matter of considerable practical difficulty, however, to obtain a rapid and accurate measure of any of the relevant quantities.

A new approach to the measurement of the quantity of herbage present was taken by Campbell *et al.* (22), who developed an electronic instrument, based on a suggestion by Fletcher & Robinson (25). The potential usefulness of such a capacitance meter is very great, since it does not involve destructive sampling. So far, however, the errors associated with its use are also considerable (20) and frequent calibration may be needed.

Output per Acre

To a very large extent production per individual reflects individual intake. It would also seem reasonable to suppose that the maximum output per acre would be obtained by stocking each acre with a number of sheep such that each received its maximum voluntary intake daily. In most experiments, however, this has not been the result.

In general, on a given pasture, the performance of each individual declines with increasing stocking rate. Obviously, the growth of individuals may be superior at high stocking rates on productive pasture to that of individuals at low stocking rates on poor pasture. In most comparisons of stocking rate on any one pasture, however, individual performance has been adversely affected at a lower stocking rate than that at which the maximum production per acre is obtained. Fig. 7.4 illustrates this for sheep: similar results have been obtained with cattle.

Importance of stocking rate

McMeekan (9, 10) has drawn attention to the importance of stocking rate in obtaining high production per acre and experience in Britain (4) has underlined the same point. Mott & Lucas (13) also noted the dominant effect of stocking rate and Mott (12) considered the relationship between production per individual and per acre. A curvilinear relationship between production per acre and stocking rate is almost bound to exist. Production per acre must increase with increasing stocking rate at least until each animal finds it difficult to obtain all the food it requires. Clearly, at some point, a stocking rate will be reached at which no animal obtains more than its maintenance requirement. Production per acre must be zero at this point. Grazing management can modify these relationships to some extent (17).

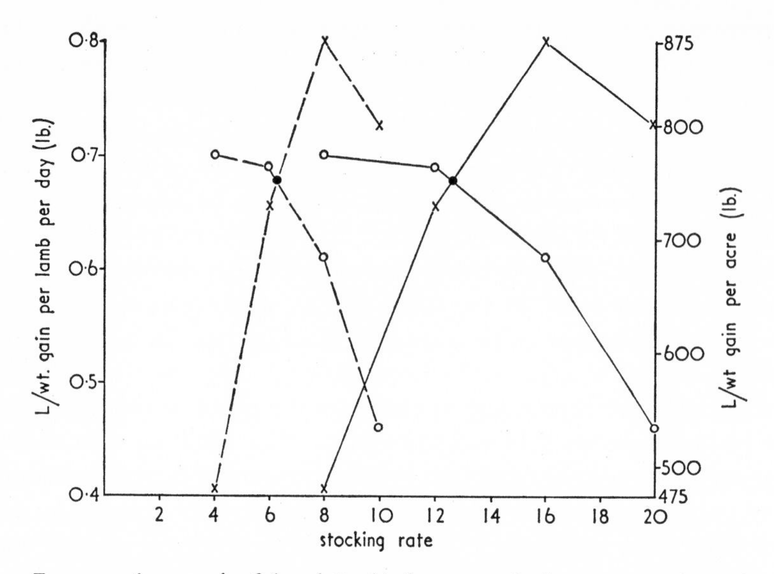

FIG. 7.4—*An example of the relationship between production per acre and growth rate of the individual animal. The data are derived from Half-bred ewes and their Suffolk cross twin lambs grazing a ryegrass pasture over a period of three months*

The figures have been plotted in four ways, as follows:

1. Mean L/wt. gain of lambs per acre against stocking rate of lambs . . . . ——×——
2. Mean L/wt. gain of lambs per acre against stocking rate of ewes . . . . – – × – –
3. Mean L/wt. gain per lamb per day against stocking rate of lambs . . . . ——○——
4. Mean L/wt. gain per lamb per day against stocking rate of ewes . . . . – – ○ – –

The vertical scales have been adjusted so that the maximum and minimum gains per head and per acre (for the whole experiment, which contained other groups of sheep also) coincide.

It has been suggested that the points of intersection (of 1 and 3 and 2 and 4) can be regarded as representing optima. It clearly does not matter whether the stocking rate of ewes or lambs is employed to determine them, but this is only true, of course, for a fixed ewe/lamb ratio.

It is helpful in considering grazing to compare the situation with one where grass is cut daily and fed to stock. A very low stocking rate, in either case, is clearly a situation at which individual performance should be the maximum possible on that particular feed. In other words, quantity is not a limiting factor. In the field, however, the quantity must not only be there, it must also be readily available to the animal. Furthermore, this situation will remain only as long as the consequences of undergrazing are prevented. Stocking rate experiments that do not take this into account may result in the association of a particular stocking rate with an over-all animal production which conceals a variety of rates of production within the period studied. Fixed stocking rates are apt to result in production figures which suffer in this way. Indeed, a series of stocking rates rigidly applied may not include one which can be regarded as anywhere near right for the whole period. In the lower range of a stocking rate series, therefore, individual performance may be near-constant, appearing as a plateau on the curve relating performance to stocking rate, or may decrease with increasing stocking rate in a near-linear fashion.

Attempts have been made to eliminate the effect of the animal on the pasture from such investigations (5). The resulting relationship suggested that even at quite low grazing pressures a further reduction gave greater growth per head. It is perhaps worth stressing at this point that some swards are extremely variable in species composition or stages of maturity or in other attributes affecting nutritive value. Where this is so, of course, the significant stocking rate or grazing pressure may be that at which not all the sheep can readily live on the highest quality fractions of the pasture. Selective grazing, on some swards, could mean that only at impracticably low stocking rates could high production per head be obtained. For most practical purposes, within a context of good grassland management, individual performance may be regarded as unaltered up to some stocking rate, beyond which it declines markedly. There may be a region within which the decline in individual performance is associated with increased efficiency. It has been suggested that this is so for dairy cows. Where milk production is the object and the main measure of production, more can be obtained per acre at a stocking rate at which no body weight is gained or at which body weight may be lost. This avoids what has been termed 'luxury consumption', a term that implies that, even over a long period, that part of the intake which occurs only at very low stocking rates can be safely or efficiently dispensed with.

Weight change in the ewe

Body changes in the ewe also complicate the relationship between fat lamb production and stocking rate. Higher meat production per acre is generally associated with some loss of weight in the ewes, at least during early lactation, and managements that succeed at high stocking rates frequently operate to prevent the ewe completely replacing this weight later in the season. This is done by restriction of ewe grazing so that the available grass can be reserved for fattening lambs. It is possible that loss of weight in the ewe during lactation is associated with some reduction in milk yield. Since the significance of milk yield varies with many other factors (see Chapters IV and V), it is not possible to generalize about the effect of weight changes in the ewe on lamb production. It is of interest, however, to examine the liveweight changes of ewes in relation to stocking rate and lambing percentage. The data given in Fig. 7.5 are for ewes grazing with the lambs referred to in Fig. 7.4. It is particularly important to note that the liveweight change over a long period, say 3 months, may obscure quite different changes that have occurred within that time.

Maximum output

The fact that the highest output per acre is usually associated with less than maximum performance per animal does not mean that this need necessarily be so. It is instructive to enquire why it should be so anyway. Sheep are rarely grown as fast as they are capable of growing and it is surprising that a reduction in growth rate should represent increased efficiency. If it does not, however, the benefit of a high stocking rate must be that more herbage is grown or that more is utilized. The most likely explanation, at present, is that the optimum degree of utilization cannot be achieved by the grazing animal without reducing the intake of the individual. Of course, according to economic circumstances, production per acre may be more or less important than production per individual. It should be remembered, however, that this is simply a matter of degree: production per acre is always dependent on performance of the individual. When the latter can be specified as an objective, the output per acre can only mean the required individual performance multiplied by the maximum number of animals that can achieve it.

In many circumstances the ewe with twins can be regarded as a more efficient production unit than the ewe with only one lamb (see Chapter VI). This is only true whilst the twin lambs grow at or above a certain rate. If the quality of the herbage is poor, the minimum milk production required from the ewe with twins may only be possible at a low stocking rate. This may be because only then can enough herbage be obtained by the ewe daily or because greater selection can take place. The relationship between the opportunity for selection and the grazing pressure depends enormously on the nature of the pasture. Selective grazing may be of great consequence on a pasture with a varied botanical composition: it may be negligible where a pasture consists, for example, of one species of grass most of which has been grown within a short period. Since the effects of

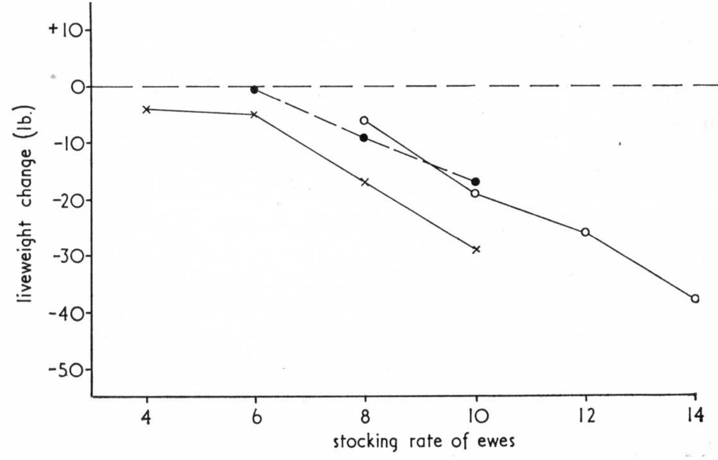

FIG. 7.5—*The liveweight change of ewes in relation to stocking rate and lambing percentage*

Lambing percentage = 100 (all singles) . . . . . ——O——
= 200 (all twins) . . . . . ——×——
= 150 (half singles and half twins) . – – ● – –

Data from Half-bred ewes with Suffolk cross lambs, set-stocked on S. 23 ryegrass receiving nitrogenous fertilizer in proportion to the stocking rate of sheep.

The weight change is that taking place between parturition (March 1962) and 17 June, a period of approximately 90 days.

stocking rate on plant growth may also vary with the species, it is obvious that the relationship between output per acre and per animal cannot be either simple or constant.

It is worth noting that, with increasing intensity, systems of utilization may have to be efficient and self-sufficient in the short term. An example may make this clear. At most stocking rates a ewe with one lamb, lambing only once a year, can safely lose a considerable amount of weight during lactation on pasture. It is likely that all this can be made up by grazing after the lamb has been sent for slaughter. If, however, more frequent lambing was required, or other crops of ewes and lambs needed the late summer grass, this could not be done so easily. It would then be necessary to decide whether the use of body reserves and their subsequent replenishment was an efficient process.

In summary of this chapter, utilization refers to the proportion of herbage grown that is eaten by ruminants. The methods employed to maximize this proportion depend chiefly on a high stocking rate and it is difficult to achieve both this and a high intake per individual at the same time. In the long term the methods adopted must increase, or at least not decrease, the total production of herbage in whatever form it is required. For simplicity, utilization has been discussed as if it was an almost mechanical process. In fact, of course, it combines living plants and animals whose individuality cannot be ignored. Some of the less tangible aspects of the combination are considered in the following chapter.

REFERENCES

(1) ARNOLD, G. W. (1962). Effects of pasture maturity on the diet of sheep. *Aust. J. agric. Res.* *13* (4), 701–706.

(2) BLAXTER, K. L., WAINMAN, F. W. & WILSON, R. S. (1961). The regulation of food intake by sheep. *Anim. Prod.* *3* (1), 51–61.

(3) BRODY, S. (1945). *Bioenergetics and Growth.* New York, Reinhold Pub. Co.

(4) COOPER, M. M. (1961). Grazing systems. *J. King's Coll. agric. Soc.* *15*, 5.

(5) COWLISHAW, S. J. (1962). The effect of stocking density on the productivity of yearling female sheep. *J. Brit. Grassld Soc.* *17* (1), 52–58.

(6) COWLISHAW, S. J. & ALDER, F. E. (1960). The grazing preferences of cattle and sheep. *J. agric. Sci.* *54* (2), 257–265.

(7) CRAMPTON, E. W., DONEFER, E. & LLOYD, L. E. (1960). A nutritive value index for forages. *Proc. VIII int. Grassld Congr.* (Reading), 462.

(8) MCCLYMONT, G. L. (1958) cited by MCDONALD, I. W. (1962). A review of nitrogen in the tropics, with particular reference to pastures. *Bull. Commonw. agric. Bur.* *46*.

(9) MCMEEKAN, C. P. (1956). Grazing management and animal production. *Proc. VII int. Grassld Congr.* (Palmerston North, N.Z.), 146.

(10) MCMEEKAN, C. P. (1960). Grazing management. *Proc. VIII int. Grassld Congr.* (Reading), 21.

(11) MITCHELL, K. J. (1956). The influence of light and temperature on the growth of pasture species. *Proc. VII int. Grassld Congr.* (Palmerston North, N.Z.), 1–12.

(12) MOTT, G. O. (1960). Grazing pressure and the measurement of pasture production. *Proc. VIII int. Grassld Congr.* (Reading), 606.

(13) MOTT, G. O. & LUCAS, H. L. (1952). The design, conduct and interpretation of grazing trials on cultivated and improved pastures. *Proc. VI int. Grassld Congr.* (Penn.), 1380–1385.

(14) SPEDDING, C. R. W. & LARGE, R. V. (1957). A point-quadrat method for the description of pasture in terms of height and density. *J. Brit. Grassld Soc.* *12* (4), 229–234.

(15) WATSON, R. H. & ELDER, E. M. (1960). Feed intake of suckling and weaning lambs on dry pasture. *Aust. vet. J. 36* (6), 266–270.

(16) VOISIN, A. (1959). *Grass Productivity*. pp. 353. London, Crosby Lockwood & Son Ltd.

(17) WHEELER, J. L. (1962). Experimentation in grazing management. *Herb. Abstr. 32*, 1.

(18) WILLOUGHBY, W. M. (1959). Limitations to animal production imposed by seasonal fluctuations in pasture and by management procedures. *Aust. J. agric. Res. 10*, 248–268.

(19) WILSON, J. W. (1960). Influence of spatial arrangement of foliage area on light interception and pasture growth. *Proc. VIII int. Grassld Congr.* (Reading), 275.

(20) BACK, H. L. (1968). An evaluation of an electronic instrument for pasture yield estimation. *J. Brit. Grassld Soc. 23* (3), 216–222.

(21) BALCH, C. C. & CAMPLING, R. C. (1962). Regulation of voluntary food intake in ruminants. *Nutr. Abstr. and Rev. 32*, 669–686.

(22) CAMPBELL, A. G., PHILLIPS, D. S. M. & O'REILLY, E. D. (1962). An electronic instrument for pasture yield estimation. *J. Brit. Grassld Soc. 17*, 89–100.

(23) CAMPLING, R. C. (1966). The control of voluntary intake of food in cattle. *Outlook on Agric. 5* (2), 74–79.

(24) ELLIOTT, R. C. (1967). Voluntary intake of low-protein diets by ruminants. *J. agric. Sci., Camb. 69*, 383–390.

(25) FLETCHER, J. E. & ROBINSON, M. E. (1956). A capacitance meter for estimating forage weight. *J. Range Mgmt. 9*, 96–97.

(26) GORDON, J. G. (1964). Effect of time of year on the roughage intake of housed sheep. *Nature (Lond.) 204* (4960), 798–799.

(27) MORRIS, R. M. (1967). *Pasture Growth in Relation to Pattern of Defoliation by Sheep.* Ph.D. Thesis, Reading University.

(28) OSBOURN, D. F., THOMSON, D. J. & TERRY, R. A. (1966). The relationship between voluntary intake and digestibility of forage crops, using sheep. *Proc. X int. Grassld Congr.* (Helsinki), 363–366.

(29) ULYATT, M. J. (1964). Studies on some factors influencing food intake in sheep. *Proc. N.Z. Soc. Anim. Prod. 24*, 43–56.

VIII

SHEEP BEHAVIOUR AND THE ART OF HUSBANDRY

It is necessary, in order to build a model of the production process, to reduce the input and output factors to almost mechanical terms. This simplifies the interactions somewhat and makes it easier to visualize the fundamental issues; but it tends to ignore the peculiar properties of living things. These properties may be very important, however, and, in the practical application of a production model, may set limits to what can and cannot be done. Knowing the kind of process (see Chapter VI) that one is trying to achieve in the field is a powerful first step in achieving it. The next step is to combine animals and plants in such a way as to maximize the efficiency of the process. This is the object of grazing management (Chapter IX) and it has to be based on an understanding of the effects that animals and plants have on each other (Chapter VII). It has also to take into account the more important of those attributes of living sheep, in particular those that make most difference to the grazing situation, that may be grouped under the heading of sheep behaviour.

A description of the way in which a sheep behaves is simply one way of referring to its reactions and responses to various stimuli. Behaviour could thus include efficiency of digestion or the relationship between growth and nutrient intake. To be useful, however, the term must be given a more restricted meaning. Sheep behaviour, as a subject, tends to conjure up visions of qualitative descriptions rather than quantitative expressions. This is partly due to the comparatively little attention it has received and partly to the tendency for anything that can be expressed in quantitative terms to be transplanted into some other category.

The main distinction that can be applied to behavioural reactions is that they tend to involve the whole animal and not just a part of it.

At the same time, they do not necessarily involve all parts of the animal. Perhaps this distinction cannot be strictly sustained: it is employed here simply for its usefulness.

FACTORS AFFECTING SHEEP BEHAVIOUR

There is some tendency for behavioural reactions, of the kind considered here, to be characterized by greater individual variation than occurs in other reactions. The behaviour of a sheep confronted by a bullock, another sheep or a man is much less predictable than the reactions of a sheep's digestive apparatus to a particular food. In addition, the reaction of one sheep to behavioural stimuli may well influence the reaction of others. Behaviour will therefore be discussed in relation to the main factors that influence it and which are important parts of the environment in which sheep have to live. Further details of all aspects of sheep behaviour can be found in an account by Hafez & Scott (15), which includes a full list of references.

THE INFLUENCE OF MAN

There is little point in generalizing about the influence of an average man on an average sheep. Obviously man can have a whole range of influences, whether he is accompanied by his dog or not. It will be noted again in Chapter IX, that grazing managements depend for their success not only on what is done but also on the way in which it is done. Any advantage of movement of sheep in, for example, a rotational grazing system, can be lost by careless movement from one paddock to another. It is hardly necessary to elaborate on this to those familiar with sheep and their management but the importance of the topic justifies a few examples.

It is well known that hungry sheep do not have to be driven into a fresh paddock: it is rapidly learned that the direction in which the gate is opened can assist or interfere with their smooth passage from one paddock to the other. In creep-grazing systems, movement of the ewes from one paddock to another should not begin until all lambs have returned from their creep areas, unless the ewes are moved to join the lambs. Otherwise lambs may damage themselves in their anxiety to get back through the creep in haste and all at the same time. Too frequent visiting of a flock expecting to be moved can have adverse effects on both the sheep and the areas of pasture near gateways that they crowd on to at such times. All this kind of thing is exaggerated by large numbers and has far more serious consequences when lambs are numerous and young. Mismothering is probably one of the more serious results of failure to understand sheep behaviour or to act upon the knowledge.

Little scientific study has been made of sheep behaviour but, of course, a wealth of information is stored in the minds of those who tend sheep. It can be learned but it is not easily taught and requires both patience and keen observation in application. Here it must suffice to recognize the existence and importance of these aspects of sheep husbandry. There are fascinating differences between individual sheep. Natural leaders often occur in a flock and such animals can be exploited to lead other sheep, for example, through muddy gateways. The same lambs frequently get left behind or get their heads stuck in wire-netting fences and some ewes are bad mothers, while some are excessively possessive. Sheep 'behave like sheep' in the face of a big enough stimulus—in this they probably differ less from other animals than is often thought to be the case. If the stimulus is alarming enough many animals will behave in a similar fashion. Nevertheless, this willingness to flock together is regarded as characteristic and a knowledge of the precise way in which it occurs can be tremendously useful in the management of sheep.

There are, no doubt, great differences between breeds (18) and some of these may make one breed more suitable for some purposes than for others. This subject will be discussed in Chapter XII.

Other Animals

The presence of other animals, in the same or adjacent paddocks, can have a marked influence on sheep behaviour. Within a field, the number of sheep present may have comparable effects to those which occur in poultry, for example. Little is known of this, but there may well be an optimum size for a group of sheep in a given situation. Apart from the numbers involved, sheep influence each other's grazing and other behaviour, a process known as social facilitation. Some aspects of this are very noticeable. Sheep may all get to their feet because one animal does so; although such observations have to be interpreted with caution and with, as it were, one eye on the observer.

It has been suggested that hungry sheep, such as heavily stocked ewes, can influence the grazing behaviour of less hungry ones, creep-grazed lambs, for example, and thus increase grazing time and intake. It is very difficult in most cases, however, to distinguish between temporary changes in rate of eating, changes in the number of hours spent grazing and changes in daily intake of nutrients. It seems more likely that behaviour could be influenced in such a way as to decrease intake but less likely that it could be influenced to result in greater intake than the normal voluntary level for the same animal.

In general, sheep tend to gravitate towards each other. This grouping is natural in cattle and sheep, but no doubt an upper limit exists to the group size. Thus, small numbers of animals in neighbouring paddocks, provided that they are aware of each other, tend to mass along the common fence (see Fig. 8.1). This may not occur in much

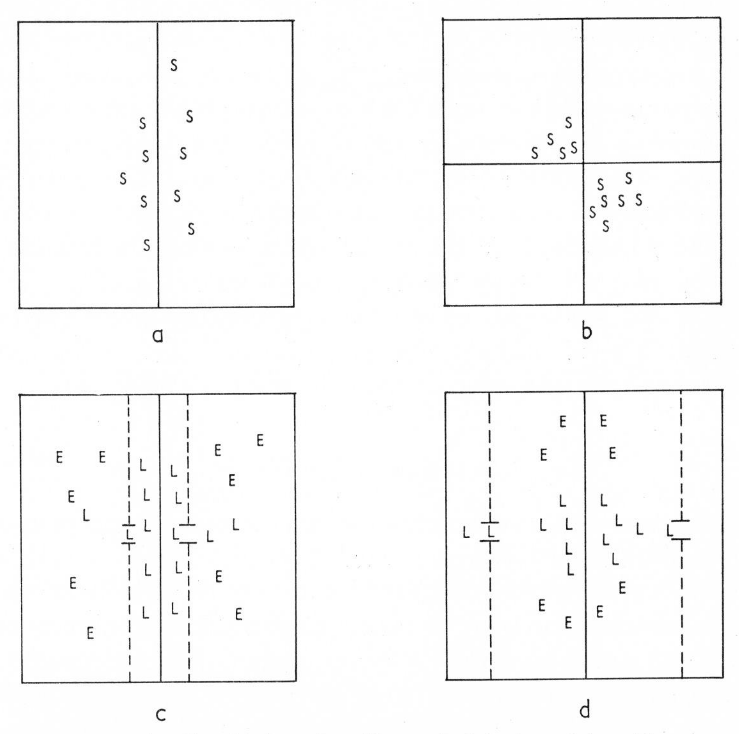

FIG. 8.1—*The effect of 'adjacent' stocking on the behaviour of sheep* (*S*)
(a) Congregation along a common fence.
(b) Concentration in corners.
(c) and (d) Possible effects of the proximity of other sheep on the behaviour of ewes (E) and lambs (L) in creep-grazing systems of management.

larger groups. Where it does happen, it can have adverse effects on the pasture. This is usually by 'poaching', a term used to describe the damage to herbage and soil caused by excessive treading in wet weather at high stocking rates, or, at the other extreme, neglect of the more distant parts of the pasture at low stocking rates.

Crofton (13), using aerial photography, was able to observe the distribution of sheep within a field. He found that the majority of grazing sheep were orientated so that two other sheep subtended an angle of approximately 110° to each one. This occurred only when the two sheep were in front of the individual, suggesting that a visual method of orientation might be involved. The angle of 110° was found to correspond to that between the optic axes of the sheep's eyes and Crofton suggested that the grazing sheep could most easily maintain its position in relation to others by using two other sheep as reference points, in such a way that they could be kept in view with the minimum of adaptation. The minimum number of sheep in a flock for all of them to do this is 5. Not all the sheep observed behaved in this way. Suckling lambs, weaned lambs and, to a lesser extent, ewes with lambs, departed somewhat from the pattern noted in mature sheep. In more than 90 per cent of the flocks considered, the mean distance between one sheep and the others in a flock was between 15 and 30 yards, even in fields varying in size from 12 to 55 acres. Although the mean distance between sheep varied between pastures, there was no relationship between this distance and the field size, the number of sheep in the flock or the stocking rate.

Size and Shape of Fields

The size of the field or paddock has an influence on the interaction between the stocking rate and the pasture. In general, small paddocks are more likely to be damaged by treading, for example, at any given stocking rate. The size of a field also affects the behaviour of the sheep and does so differently at different stocking rates. The information is not available to do more than state the existence of such effects. They can be illustrated by imagining extreme examples. If the area be very large indeed, an individual sheep may not be physically able to influence another particular animal simply because they will not meet: this is quite possible under range conditions and very sparse vegetation. If, on the other hand, ewes with young twins are crowded on to a small area, even with plenty to eat, they can distress each other by the constant need to sort out the lambs.

The effect of size of field, however, cannot be entirely divorced from that of shape. Long, corridor-like paddocks can produce temporary overcrowding effects. The nearer to a square or circle the shape is, the less boundary there is per acre. This may reduce poaching because there is less 'edge' to be damaged or increase it because all the animals still pound up and down a shorter route. As Thomas (26) has pointed out, however, sheep grazing over large areas will frequently make their own paths (this is most noticeable in snow: see Plates VII (a) and VII (b)). On the slopes of the Pewsey Vale in Wiltshire, he found them to be along the contours and to result in changes in the natural vegetation. The trooping and sitting of stock along fences is commonly due to the presence of stock on the other side or to some aspect of topography that may not be immediately obvious. Shelter to a sheep is not always a matter of hedges and walls: undulations of the ground may have the same effect in certain weather conditions. The dimensions of paddocks have direct economic importance if further subdivision is required. In general, shapes and sizes have most influence in controlled managements, like creep-grazing, where success depends, in fact, almost entirely on the behaviour of the sheep. It is essential, for example, that ewes get on with their own grazing and do not bother overmuch about their lambs. For this it may be desirable that they should be able to see them and certainly to hear them. Similarly, lambs should not be continually alarmed by finding that they have wandered too far away. Thus if areas are large, a long boundary fence between the ewes and lambs will help to maintain contact (Fig. 8.2). Where the lamb area is large because it also contains cattle, a central ewe paddock surrounded by lamb area would have the same effect (Fig. 8.3). Such managements may be quite unnecessary but serve to illustrate the point.

Considerations of animal behaviour find frequent expression in practice, however, and not least in the design of handling equipment. Intensive grazing managements have many of the same characteristics as such equipment, such as control of the animals, for example, and benefit from some thought given to their design. The design of creeps should not only encourage the lambs to go through and prevent the ewes doing so, it should, if possible, discourage a ewe from sitting in and blocking the entrance. This is really part of the problem of fencing materials. In relation to sheep behaviour, efficiency is the keynote for fencing. Once a sheep discovers it can do something that it is not intended to do, such as get under, over or through a fence, it becomes much more difficult to control; this applies especially

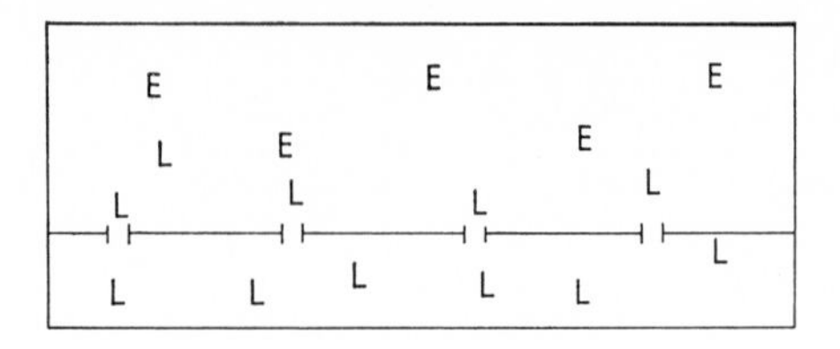

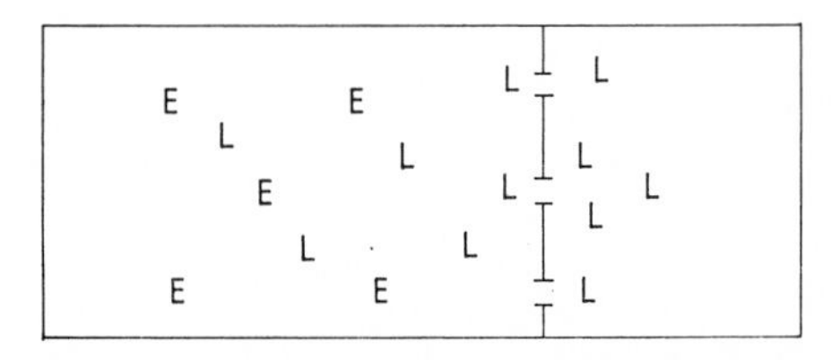

FIG. 8.2—*The effect of the shape of the lamb-creep area. In general, the longer the dividing fence the shorter the distance that lambs (L) need to travel from their ewes (E) in order to utilize the whole creep area*

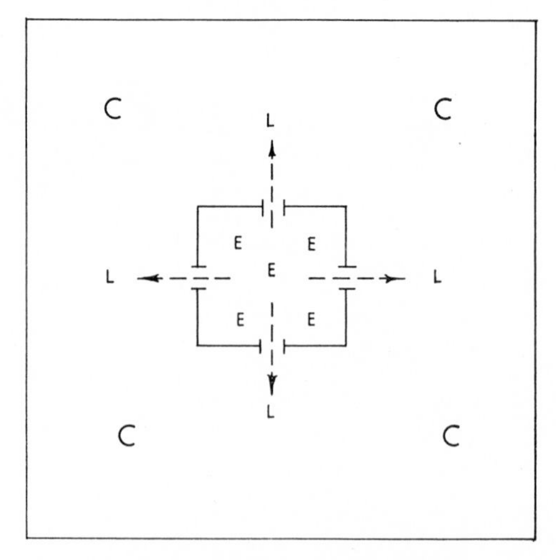

FIG. 8.3—*An arrangement in which lambs (L) would be able to utilize the maximum proportion of the area available to them. E=ewes; C=cattle*

(a)

Suffolk rams in snow demonstrating the way in which sheep keep to paths of their own making.

(b)

Hereford cross-calves will show the same path-following behaviour in the snow.

(a) Dorset Horn ewes seeking shade during hot weather.

(b) Lambing in covered yards: ewes and lambs in France.

(c) Half-bred ewes with their Suffolk cross lambs returning from a creep area.

to creep-grazing where ewes will try to squeeze through the creeps at first and, if this is not possible, leave them alone for the rest of the season. For this reason, it is a good idea to alter the dimensions of the creep as the lambs grow bigger. The proportion of the area devoted to lambs, in systems where separate ewe and lamb areas are allocated, must vary with the number of lambs relative to the number and size of ewes. The fence between them and the shapes and sizes of the resulting paddocks probably have a greater influence on sheep behaviour than in any other system.

Climate

Apart from the more obvious effects of climate on behaviour, there are others which, though less noticeable, may be more important. It is interesting that sheep may align themselves in a particular way to the direction of the wind but there is little fear that any management practice would prevent their doing so. Management could, however, interfere with the way a whole flock would normally react to adverse weather. Their behaviour may be a valuable guide to the need for or efficacy of shelter belts, but in many situations it represents a valuable reaction to unalterable circumstances. It is worth emphasizing that, although the need for shelter may be restricted to brief periods of the year, these may be of critical importance and influence, for example, the survival of new-born lambs. Hutchinson & Bennett (19) have pointed out that the kind of shelter belt required for sheep may be quite different from the kind used to protect a crop. Maximum protection is required for a small area and, because of the tendency for sheep to gravitate to the lee side of a paddock in windy weather, shelter belts may be best sited a short distance from the fence on the lee side. Other reactions to climatic conditions may have far-reaching consequences: only a few examples need be given here. Certain parts of the pasture may be frequented for shade (Plate VIII (a)) and shelter and thereby receive a disproportionate share of the animal's faeces and urine. It should be noted that shelter does not always improve performance, as where lambs may use it in such a way that their suckling time may be reduced (36).

In very hot climates, the time spent by sheep under scattered trees, or in woodlands associated with pastures, may be considerable. This can lead to a massive concentration of disease-producing organisms.

The significance of this depends on whether these organisms are likely to gain access to the sheep in these circumstances. It is possible that behaviour of this kind, well exemplified in 'night camping', could work in either direction. For example, internal parasites often depend upon faecal distribution for their dispersal. Where a high proportion of the faecal output is deposited in one place, under a dense shade tree, perhaps, this may effectively put it out of circulation if there is no herbage under the tree and therefore no grazing. If the result, however, is a concentration of faeces round a water hole, pond or stream, where herbage may be available when it is in short supply elsewhere, massive infestations may occur at certain times. This connection between sources of water and parasites has often had serious consequences in infections with, for example, liver fluke and lungworms.

Continuous heavy rain can reduce grazing time and, if prolonged, markedly reduce intake and liveweight gain. It is often suspected that water on the surface of the herbage adversely affects intake. Certainly young lambs often appear reluctant to graze very wet herbage, particularly when it is long. This can result in very patchy grazing caused by lambs standing on a short grass area and enlarging it at the edges. On the other hand, free water intake does not seem to affect the intake of herbage.

The need for water

This is, perhaps, a good point at which to emphasize the need for water, particularly in hot weather. Sheep need a considerable amount of water, but their actual requirement varies with their size, diet and physiological state. Weaned lambs naturally drink more water (see Fig. 8.4) than lambs still receiving milk. Milking ewes (see Table XXI) require the greatest quantities, but pregnant ewes on a dry diet may also drink a great deal. The water balance of a pregnant Dorset Horn ewe (given in Table XXXVII) illustrates the high proportion of the water requirement which may be derived from silage.

The temperature of the water may interact with that of the environment to affect the water consumption of sheep. Bailey, Hironaka & Slen (5) found that a reduction in environmental temperature from 15° C to 12° C reduced water intake from about 1,600 to about 800 ml a day in 2-year-old Cheviot cross wethers. At an environmental temperature of – 12° C, however, the temperature of the drinking water did not influence the amount of water consumed.

Grazing sheep normally receive a considerable amount of water in the herbage they eat, but a shortage can occur, and, when it does, it

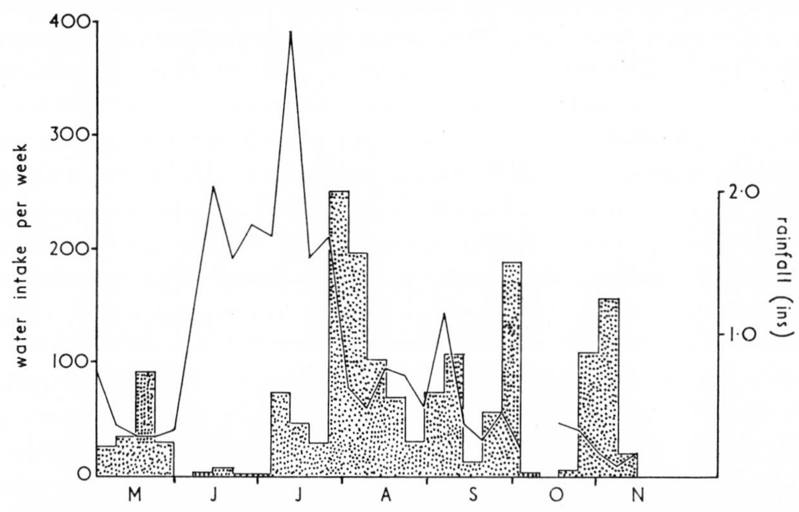

FIG. 8.4—*Mean water intake (ml of water drunk per kg of body weight) per week of weaned lambs from two months of age, grazing perennial ryegrass. Water intakes were calculated after allowing for rainfall (shown as a histogram) and evaporation; the latter was measured from similar containers under the same conditions as the water bowls available to the sheep*

(Source : data from Expt. H. 287, Grassland Research Institute, Hurley)

TABLE XXXVII

WATER BALANCE OF A DORSET HORN EWE

(150–160 lb liveweight), fed on silage in a digestibility crate, during mid-pregnancy. Lambing occurred on 5.4.61

Week		WATER (MEAN DAILY VALUE FOR EACH WEEK IN LB)			
		INTAKE		OUTPUT	
		In silage	*Drinking water*	*Urine*	*In faeces*
1	Beginning 3.1.61	10·89	0·66	7·03	0·74
2		9·27	0·66	4·56	0·78
3		10·54	0·11	5·60	1·23
4		10·31	—	5·31	0·97
5		10·04	0·02	4·56	0·95
6		9·45	0·66	4·76	0·96
7		10·72	0·02	5·02	1·21

(Source : J. Hodgson, unpublished data, Grassland Research Institute)

can have very serious consequences on production. Most serious is a shortage for the lactating ewe. Experimentally, the milk production of ewes can be terminated most readily by a cessation of withdrawal accompanied by restriction of water intake. In most cases, restriction for about a week after the removal of the lambs results in a dry ewe. Clearly, a few days' shortage in the field might have serious consequences for lamb growth that season. Supplying water afterwards may not restore the situation. It is important that the water provided should be in a suitable condition. Even at British summer temperatures, surface piping may deliver water that is far too hot for sheep to drink. If surface piping has to be used, therefore, it must be associated with a large trough rather than a small drinking bowl.

The mineral content of drinking water also influences whether sheep will take adequate amounts. Pierce (37) has summarized his findings for sheep in Australia as follows: pen-fed wethers tolerated drinking water with

1·3% sodium chloride
or 1·2% sodium chloride + 0·1% magnesium chloride
or 0·9% sodium chloride + 0·5% sodium sulphate
or 1·0% sodium chloride + 0·3% calcium chloride
or 0·9% sodium chloride + 0·4% of equal proportions of sodium carbonate and sodium bicarbonate

but ewes and lambs appeared somewhat less tolerant of saline waters.

Interactions of behaviour, weather and parasites

One of the concepts of behavioural interaction which has not survived detailed examination is that relating grazing time, dew formation and larval migration. It was commonly supposed that infective nematode larvae took advantage of the dew to wriggle up the grass so that they should be in position when the sheep began their early morning grazing. It is true that infective larvae require a water film in which to move (24) but it has been pointed out by Crofton (10) that they require a higher temperature than that frequently accompanying dew formation. In fact, under hill conditions in the North of England, Crofton (10) found that the majority of infective larvae were recovered from herbage samples in the middle of the day. It is probable that this would be different in other parts of the world. For example, in the southern part of Brazil, very heavy dew may remain on the pasture in the autumn until the middle of the day, for although the temperature may then be very high, so also is the relative humidity. A distinction has to be made, of course, between the conditions necessary for dew formation and those which can obtain whilst the dew is still present. Buckley (6) and Crofton (9) have shown that little or no migration of infective larvae occurs at temperatures below 10° C and, when it does occur, many factors, such as the density of the herbage, influence the numbers that reach the upper parts of the sward (24). At Hurley, the maximum numbers of larvae have

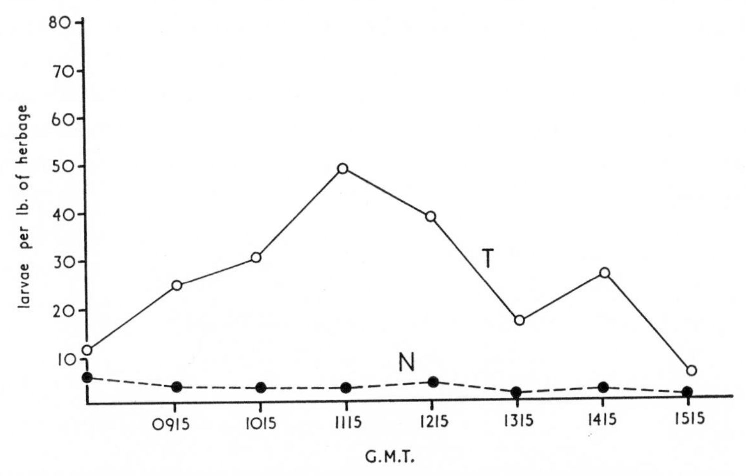

FIG. 8.5—*Variation in the number of infective larvae of* Nematodirus *spp. (N) and trichostrongyles other than* Nematodirus *(T), recovered from pasture sampled hourly during the day (25.9.63)*

(Source : data from the Grassland Research Institute, Hurley)

also frequently been recovered during the middle of the day (Fig. 8.5) but these reactions to temperature must be expected to vary with the species of parasite (28). If such patterns are of general occurrence, the relationship with grazing time might be of great importance. It is an interesting speculation that time of grazing might interact with other features of the herbage, such as the proportion of soluble carbohydrates, a quantity which may vary quite rapidly with the light intensity. There is a considerable difference, however, between grazing selectively in relation to parts of the plant or of the pasture and selection which results in fewer or more parasites. Preference for one species of grass compared with another (8, 23) cannot continue over a long period unless the sheep are moved frequently to a fresh pasture or are grossly under-stocked the whole time. It is possible to exploit selective grazing of those parts of the pasture that have the highest nutritive value (1, 4) by accepting, for one group of animals, a low level of efficiency of utilization (see Chapter VII). It is even possible to combine efficient utilization by ewes with lenient, and therefore selective, grazing by their lambs in systems involving a lamb 'creep' (see Chapter IX). Whatever the reasons for selective grazing, it may affect the number of infective stages of parasites that are ingested since these larvae are not randomly distributed on the pasture (11). Michel (21) observed that young cattle succeeded in avoiding infection with the free-living stages of the cattle lungworm (*Dictyocaulus viviparus*) by selective grazing. Cattle usually avoid herbage contaminated with faeces rather more than do sheep; nevertheless, it is quite possible that the grazing behaviour of sheep might have similar effects. Where this is based on time of grazing, whether time of day or in relation to when faeces were deposited, it is possible that the avoidance of infection may be consistent with eventual consumption of all the herbage and thus with a high degree of utilization. Even without selection between areas or species of herbage plants, the degree of utilization can affect the intake of parasite larvae. Crofton (12) and Silangwa and Todd (30) found that the majority of such larvae occur on the lower parts of the sward, so lenient grazing should result in a lower intake of infection.

Grazing Behaviour

The grazing behaviour of lowland sheep (Oxford Down crosses and Clun Forest) was studied in detail by Hughes & Reid (16). They found that sheep spent an average of 9 hours grazing (range: 7·80–10·47 hours) out of a 24-hour period, 95–100 per cent of this occurring during daylight in summer. In winter time, more grazing occurred at night and in December only 60 per cent during the day. Idling time, which included rumination, occupied 11·4 hours (range: 8·97–12·59 hours) and was mostly spent lying down. Average standing time was 3·4 hours (range: 0·98–5·57 hours) and the average distance walked was estimated to be 0·92–1·70 miles in winter and 0·44–1·00 mile in summer.

Marked periodicity was exhibited: an early morning grazing period appeared to be related to sunrise and a late afternoon grazing was related to sunset (see Figs. 8.6–8.9). It has been suggested (14) that the time required to consume an optimum amount of food may affect the rate of liveweight gain, and that the time required is influenced by sward conditions, such as height, density, fibre and dry-matter content. Certainly an animal cannot spend all its time grazing and the range of grazing periods recorded in the literature suggest that, whilst the period can be extended, it could also be a limiting factor on food intake under conditions of restricted pasture availability (2, 3).

Rutter (38) has demonstrated the usefulness of time-lapse photography, using a cine camera and colour reversal film in daylight, for recording group behaviour, for example in relation to weather conditions. He recorded the effects of wind and rain and noted marked behavioural responses, such as huddling near a hedge, that affect the pattern of grazing.

Sharafeldin & Shafie (39) found differences in grazing activity between four breeds (Ossimi, Texel, Caucasian Merino and Fleisch Merino), including the ability to cover distances to reach pasture, in the subtropics. Clearly, in difficult environments behavioural adaptation may be of the utmost importance.

Behaviour and Disease

The importance of early diagnosis of disease is probably greatest in metabolic disorders, such as hypomagnesaemia and hypocalcaemia. Scientific assessment of the mineral status of sheep requires measurements which cannot always be taken on a routine scale. Diagnosis of many diseases, therefore, must depend upon observation and knowledge of sheep behaviour. This is an art in that it relies, and probably will continue to rely, on visual judgement. Its worth can be very high.

It is well known that sick sheep tend to be sluggish and hide themselves away. Lagging behind during movement and other evidence of reluctance to be active can all be used to judge the fitness of sheep. Even terms such as 'carriage', 'brightness of eye' and 'erectness of ears', often associated with agriculturally irrelevant show standards, may represent useful indices of health. A compromise between the art of 'looking over' sheep and assessments such as blood analysis, can be obtained by physical examination of every individual. This is occasionally essential, as for early detection of footrot, often desirable,

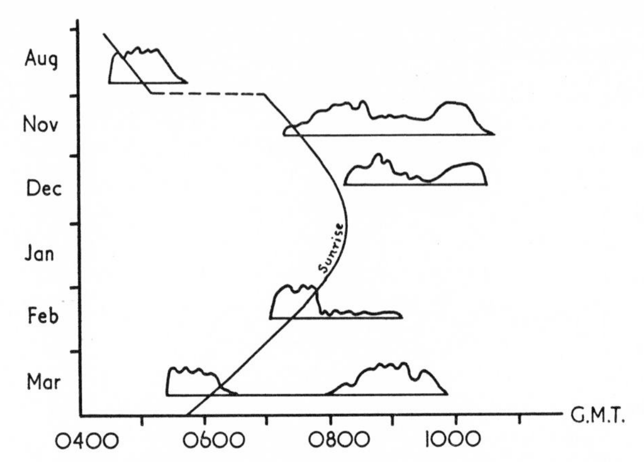

FIG. 8.6—*Relation of first daylight grazing period to time of sunrise (G.M.T.)*

10 sheep. August 1948 to March 1949

(Source : G. P. Hughes and D. Reid (16))

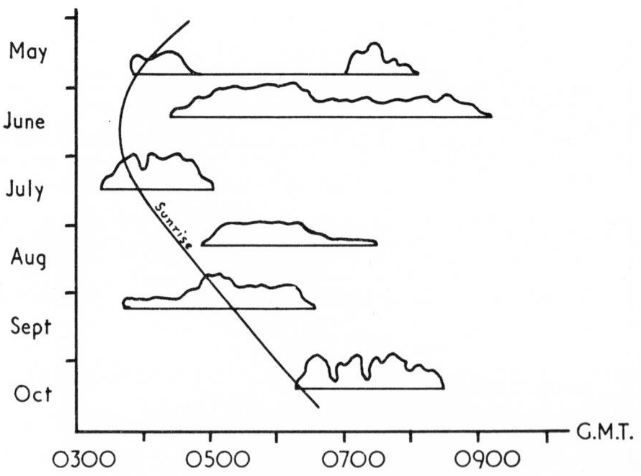

FIG. 8.7—*Relation of first daylight grazing period to time of sunrise*

10 sheep. May to October 1949

(Source : G. P. Hughes and D. Reid (16))

when checking for fly strike, frequently impracticable and sometimes dangerous because clinical symptoms of some metabolic disorders occur in response to the excitement involved in such examinations. Quiet observation is ideal in these latter circumstances, provided that the indicative muscular tremblings or stiff-legged gait are being looked for. Bloat is another disorder which can be costly if not treated

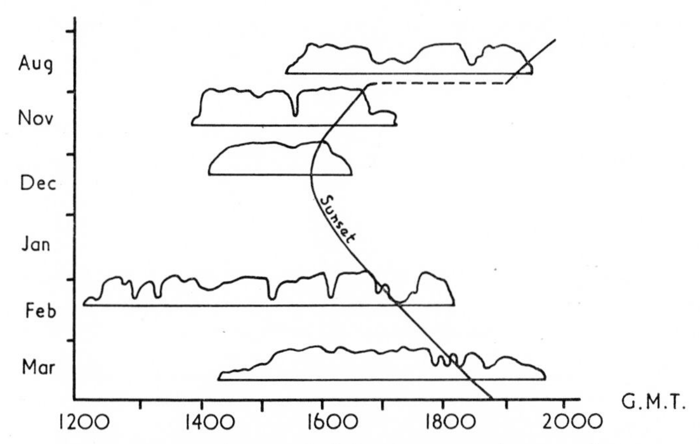

FIG. 8.8—*Relation of last daylight grazing period to time of sunset*
10 sheep. August 1948 to March 1949
(Source : G. P. Hughes and D. Reid (16))

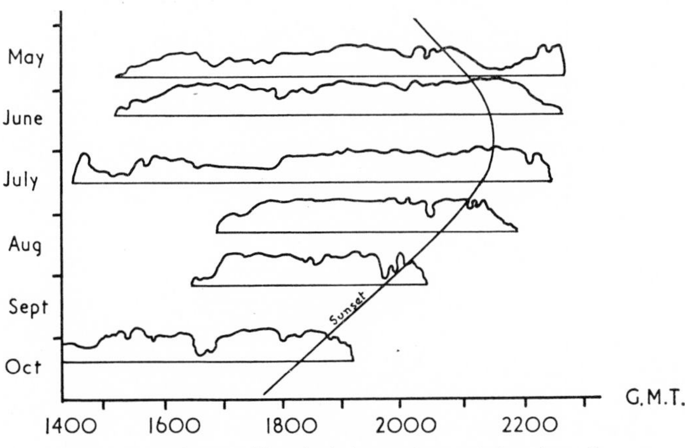

FIG. 8.9—*Relation of last daylight grazing period to time of sunset*
10 sheep. May to October 1949
(Source : G. P. Hughes and D. Reid (16))

quickly. Judgement is required to know when a much inflated rumen indicates bloat and when it does not. In sheep under British conditions it most frequently does not.

Observation of scouring and the resultant dirty hind-quarters of lambs is often part of the necessary information on which to base a decision about anthelmintic dosing. Observations of tail-wagging and a straining of the head sideways and back may indicate maggots (fly strike) of the greenbottle or European green blowfly (*Lucilia sericata*) or the secondary bluebottles (*Calliphora* spp.*). These flies lay their eggs on the fleece, particularly when it is wet or soiled. The eggs are commonly called 'blow'. When the eggs hatch, the maggots burrow under the skin and this is described as 'fly strike'. If these eggs and maggots are not treated promptly great loss and appalling suffering may result. It is not unknown for animals to be eaten alive, particularly where it is possible for them to hide, amongst bracken for example, and avoid observation.

Behaviour studies of sheep have tended to concentrate on the more easily measurable aspects. The knowledge of behaviour which will form part of the art of husbandry for a long time, however, is concerned precisely with those things which cannot be easily measured. Some of them do not call for immediate action but add up as a kind of background information. The characteristic head-jerking that indicates the presence of sheep nostril flies (*Oestrus ovis*) is an example of this. Sheep will also push their noses into loose earth to prevent the fly from laying its eggs on their nostrils. In some parts of the world the quite large maggots of this fly may be numerous and cause considerable trouble. Sheep do not 'gad', rush wildly about, like cattle but they are not unaffected by flies. The importance of such irritants and other influences on behaviour may be more or less serious according to the time of year or stage in the sheep's life.

Lancashire & Keogh (35) have emphasized the need to include studies of grazing behaviour in investigations of certain diseases and have noted different effects due to facial eczema and ill-thrift on the grazing behaviour of lambs.

BEHAVIOUR AT CRITICAL TIMES

Some of these times are of great importance and the impact of animal behaviour during them is worth separate consideration.

* Different species of both these blowflies may be chiefly involved in fly strikes in other parts of the world. Descriptions and distributions of these and other parasites may be found in parasitology textbooks (Cameron (7), Lapage (20)).

Mating

This is one time, more than any other, when behavioural patterns matter. The conception rate is a major determinant of productive potential and mismanagement during the mating period can adversely affect it to a marked extent. At this time, the ram is of obvious importance, but to ensure that he is effective it is necessary to give some thought to management throughout the year.

Anything which limits the ability of the ram to find and serve each ewe at the optimum time is worth eliminating. Impediments such as footrot are so obvious that they would not be worth mentioning if they were not quite common faults in rams. The ways in which rams distinguished between oestrous and non-oestrous ewes have been investigated (34) and the results suggest that sight, smell and hearing are important, in that order.

The crucial factor in many cases is the number of ewes per ram (25). This normally varies from 25 to 50, 40 being a common figure. The correct number depends on age, breed, dispersal over the area and so on. The average number of ewes that one ram can serve is small in relation to the ram's reproductive potential. Artificial insemination is relatively little used in sheep breeding but, where it is, for example, in Brazil, Spain and Russia, the semen from one ram may be used to fertilize 400–500 ewes.

In normal circumstances, however, the lambing period may be considerably extended unless the ram can cover all the ewes that may come on heat at the beginning of the mating season. Fig. 8.10 illustrates the pattern for the number of ewes mated each day in a small Dorset Horn flock. The average pattern over 4 years in a flock of Half-breds is shown in Fig. 8.11. The considerable daily variation in the number of ewes marked would be quite possible as a chance result. If, however, in a flock of 100 ewes, the ewes' heat periods were distributed in a random manner, the mean number of ewes on heat, i.e. in the middle of their periods of heat, each day would only be 100/17 or about 6. After seventeen days the number would be proportionately reduced by the number of ewes which had already conceived. Hulet *et al.* (17) have noted that ewes may initiate contact with the rams and, indeed, may compete for their attention. It is often sensible to start with a small number of ewes per ram and increase the number, i.e. take rams out, after about half of the ewes are marked.

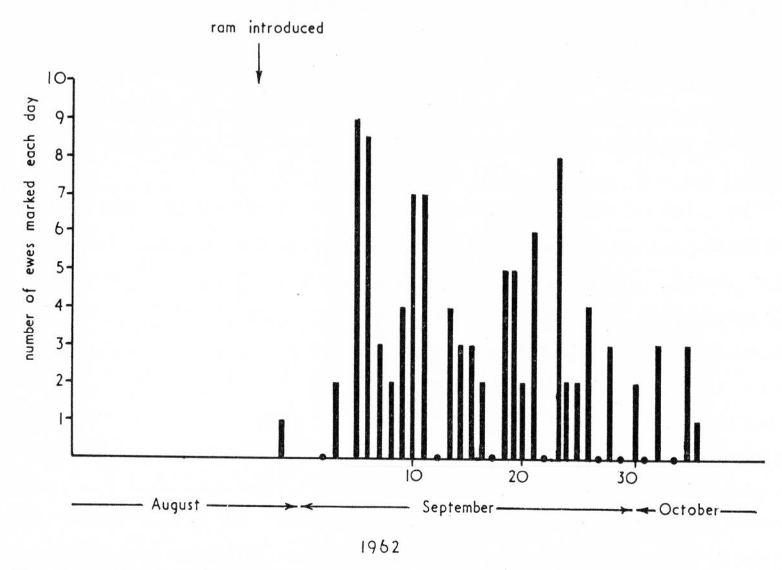

FIG. 8.10—*Pattern of mating. The number of Dorset Horn ewes (in a flock of 96) marked each day has been plotted from the day the ram was introduced*

(Source : data from a flock at the Grassland Research Institute, Hurley)

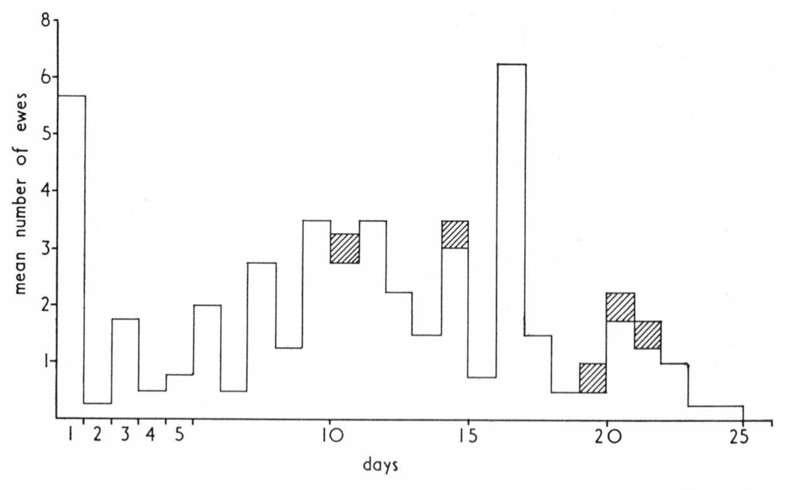

FIG. 8.11—*Mean number of ewes marked daily, from the time the rams were introduced (mean of 48 ewes over 4 years: 1959–1962). Day 1 varied from 10 Oct. to 16 Oct. The number of ewes returning to first service is indicated by shading*

(Source : Observations on a flock of Half-bred ewes at the Grassland Research Institute, Hurley)

It is interesting that traditional recognition of the importance of behaviour at this time has resulted in rams being 'keeled', i.e. plastered with coloured paste on their chests, or, more recently, fitted with harnesses carrying coloured crayons. Periodic changes in colour make it possible to distinguish which ram served which ewe and to record dates of mating and returns to first service.

The onset of the breeding season in sheep is sometimes characterized by ineffective matings which unnecessarily work the rams and prolong the period over which lambs are born. Mention was made in Chapter IV of the possibilities of using hormone injections to facilitate more 'compact' lambing. There is, however, some evidence of the same kind of effect being achieved by the use of 'teasers'. These are vasectomized rams, which behave in the same way as a normal ram but which cannot cause conception. Their presence can be used to detect which ewes are on heat but it is also claimed that they have a stimulatory effect on the ewes. It has been suggested that putting rams in an adjacent paddock, so that rams and ewes are only separated by a fence, may have similar effects. Watson & Radford (27) found that the presence of rams had a stimulating influence on the onset of oestrus in Merino ewes but that the smell or sound of them was a sufficient stimulus without any physical contact.

Behaviour may become spectacular if rams are put together during the mating season. Fighting is not only an unnecessary dissipation of energy but may result in serious damage to the contestants. It can usually be avoided by making sure that rams are accustomed to one another before use. This does not always solve the problem and the mating of ewes can be a very inefficient affair where two or more rams follow each other about instead of attending to their essential functions.

The mating behaviour of both the ewe and the ram is quite characteristic and a relationship may be formed between the ram and a ewe about to be mated, some time before the event. Unnecessary disturbance can upset such behaviour. It is often thought advisable to put young rams with older ewes and older, experienced rams with young ewes. Very young rams are usually allotted fewer ewes. Probably the most important behavioural feature to observe in the rams is failure to feed. This varies greatly between individuals, and perhaps between breeds, but can be very serious. Some rams are so active and single-minded that they simply do not eat enough to maintain their condition. This is exacerbated by bad feet and by fighting. Supplementary feeding is often necessary but even this is sometimes ignored unless the ram is penned daily for it.

Needless to say, many of these problems are more acute or more difficult to solve under range or mountain conditions.

Lambing

The subject of maternal care was discussed in Chapter IV as one of the essential contributions of the ewe to the production process. Mention was made of the effect of nutrition during late pregnancy on the onset of lactation. The behaviour patterns of the ewe and lamb shortly after parturition may be upset by mismanagement, by prolonged or difficult lambing, or interference by other sheep. Aberrant behaviour was also mentioned in Chapter IV.

Commonly ewes select suitable places, away from other sheep, in which to lamb. Some ewes, on the other hand, will lamb in such unsuitable sites as muddy ruts, often with adverse consequences to both ewe and progeny. If ewes are housed or yarded (see Plate VI (b)), it is important to organize the ewes and new-born lambs so that confusion is minimized. The time taken over lambing is very variable and it is one of the more difficult practical problems to decide when assistance, or interference, is justified. One thing is clear, that skill and care at this time can significantly affect productivity and that neither can be fully applied without knowledge of and some sympathy towards the instinctive behaviour of the lambing ewe.

Suckling

Strictly speaking, the ewe suckles the lamb—the lamb sucks the ewe. As with the cow, a 'milk-ejection' mechanism operates. It involves the release of a hormone called oxytocin, which can be injected for experimental purposes, and this occurs in response to the stimulus of sucking. The lamb not only sucks but also 'bunts' or buffets the ewe's udder quite vigorously and this may play its stimulatory part. In practice, unless udders or teats are physically damaged, little difficulty is encountered in lambs sucking, as long as they are the ewe's own lambs. Many methods have been evolved to persuade a ewe to adopt orphan lambs. These methods are based on trying to disguise the smell of the new lamb or to make it smell like the ewe's own, by rubbing it in the fluids ejected at lambing or coating it in a dead lamb's skin, for example, or simply to confuse the ewe. Ewes are sometimes tied up, confronted with dogs to arouse maternal instinct and so on. Many of these methods work, especially within 4 hours of lambing (29), and a satisfactory balance of live ewes and lambs created. They can be time consuming, however, and if orphans are numerous artificial rearing may be justified (see Chapter V).

There is much to be learned about the behaviour of the sucking lamb. Single lambs change sides and suck both quarters on any one suckling occasion. This may not be observed because, although some singles suck one side and then rush round to the other, other singles simply reach over whilst remaining on the same side of the ewe. Twins may soon settle down so that each knows, uses and confines itself to its own side. Indeed after a few weeks, about 4 in our own observations, so firmly is this habit fixed that when one lamb is weaned the other may continue to suck one side only. Before about 4 weeks of age, when one lamb of a twin pair is removed, the other will usually start to suck both quarters.

With triplets and quadruplets this kind of arrangement is not possible, but little is known about the normal behaviour of such lambs. Probably, it is extremely variable since the possibility of variation in lamb size increases with the size of the litter. In early weaning procedures (see Chapter XV) observation of suckling behaviour might be necessary to determine which lamb(s) to remove and which to leave on the ewe. It is most common for twins to be suckled together, the arrival of one lamb being a signal for the other to join it. Ewbank (31) found that the sucking behaviour of lambs was very variable, however, and later studies (32, 33) suggested that after 5 weeks about half of the twin pairs seemed to have a fixed preference as to which side each lamb should suck. About half the singles also showed a preference by this time. It also appeared that lambs which had already shown a preference for one side of the udder, modified their behaviour and used both sides when the other lamb was removed. Ewes similarly modified any behaviour they had established and allowed the remaining lamb to suck.

The frequency of suckling varies chiefly with the age of the lambs (see Table XXXVIII), tending to decrease with increasing age. It is much lower in ewes with a relatively poor milk yield. The majority

TABLE XXXVIII

NUMBER OF TIMES THAT LAMBS WERE OBSERVED TO SUCK, BETWEEN DAWN AND DUSK (4.4.58–5.7.58)

Age of lambs (days)	Suckling frequency of lambs on ewes with good or poor milk supply	
	Good	*Poor*
19	26	12
26	22	4
40	10	8
54	15	6
103	10	4

(Source : T. H. Brown, unpublished data from the Grassland Research Institute)

TABLE XXXIX

PATTERN OF SUCKLING BY TWIN LAMBS

(3–4 weeks old from dawn to dusk (19.4.58))

Time of suckling (G.M.T.)	Number of seconds spent sucking
0440	20
0516	14
0536	8
0620	19
0636	5
0736	20
0756	17
0820	14
0952	8
1040	7
1212	18
1324	17
1404	16
1504	22
1548	18
1648	19
1720	18
1744	15
1800	12
1824	11
1852	10
1900	15

(Source : observations by T. H. Brown at the Grassland Research Institute)

of suckling appears to be done during the daylight hours and an example of the distribution of suckling by Half-bred ewes and their Suffolk cross lambs is shown in Table XXXIX. Whether the time (number of seconds) spent suckling was indicative of the amount of milk received by the lambs is not known. Moule & Young (22) recorded quite large variations in milk intake by single Merino lambs, both daily and at different times of the day. Ewbank (32) found that up to the 4th week of life, twins sucked more frequently than singles but thereafter the rate fell at the same speed for both. The duration of suckling was about the same for single and twin lambs of the same age and dropped as they grew older.

Finally, it cannot be emphasized too strongly that, although it is essential to have a clear model of the productive process in mind, success in practical application depends upon recognition of the importance and relevance of the behaviour patterns of the living animals involved.

REFERENCES

(1) Arnold, G. W. (1960). Selective grazing by sheep of two forage species at different stages of growth. *Aust. J. agric. Res. 11* (6), 1026–1033.

(2) Arnold, G. W. (1960). The effect of the quantity and quality of pasture available to sheep on their grazing behaviour. *Aust. J. agric. Res. 11* (6), 1034–1043.

(3) Arnold, G. W. (1962). The influence of several factors in determining the grazing behaviour of Border Leicester × Merino Sheep. *J. Brit. Grassld Soc. 17* (1), 41–51.

(4) Arnold, G. W. (1962). Effects of pasture maturity on the diet of sheep. *Aust. J. agric. Res. 13* (4), 701–706.

(5) Bailey, C. B., Hironaka, R. & Slen, S. B. (1962). Effects of the temperature of the environment and the drinking water on the body temperature and water consumption of sheep. *Can. J. anim. Sci. 42*, 1–8.

(6) Buckley, J. J. C. (1940). Observations on the vertical migration of infective larvae of certain bursate nematodes. *J. Helminth. 18* (4), 173.

(7) Cameron, T. W. M. (1951). *The Parasites of Domestic Animals.* 2nd ed. London, Adam & Charles Black.

(8) Cowlishaw, S. J. & Alder, F. E. (1960). The grazing preferences of cattle and sheep. *J. agric. Sci. 54* (2), 257–265.

(9) Crofton, H. D. (1948). The ecology of immature phases of trichostrongyle nematodes. II. *Parasitology 39* (1 & 2), 26–37.

(10) Crofton, H. D. (1949). The ecology of immature phases of trichostrongyle nematodes. III. *Parasitology 39* (3 & 4), 274–280.

(11) Crofton, H. D. (1952). The ecology of the immature phases of trichostrongyle nematodes. IV. *Parasitology 42* (1 & 2), 77–84.

(12) Crofton, H. D. (1954). The ecology of the immature phases of trichostrongyle nematodes. V. *Parasitology 44* (3 & 4), 313–324.

(13) Crofton, H. D. (1958). Nematode parasite populations in sheep on lowland farms. VI. *Parasitology 48* (3 & 4), 251–260.

(14) DUCKWORTH, J. E. & SHIRLAW, D. W. (1958). The value of animal behaviour records in pasture evaluation studies. *Anim. Behav.* 6 (3 & 4), 139–146.

(15) HAFEZ, E. S. E. & SCOTT, J. P. (1962). The behaviour of sheep and goats. In *The Behaviour of Domestic Animals.* Ed. Hafez, E. S. E. London, Baillière, Tindall & Cox.

(16) HUGHES, G. P. & REID, D. (1952). Studies on the behaviour of cattle and sheep in relation to the utilization of grass. *J. agric. Sci.* 41 (4), 350–366.

(17) HULET, C. V., BLACKWELL, R. L., ERCANBRACK, S. K., PRICE, D. A. & WILSON, L. O. (1962). Mating behaviour of the ewe. *J. anim. Sci.* 21 (4), 870–874.

(18) HUNTER, R. F. (1960). Aims and methods in grazing-behaviour studies on hill pastures. *Proc. VIII int. Grassld Congr.* (Reading), 454.

(19) HUTCHINSON, J. C. D. & BENNET, J. W. (1962). The effect of cold on sheep. *Wool Tech. and Sheep Breeding* 9 (1), 11–16.

(20) LAPAGE, G. (1956). *Veterinary Parasitology.* London, Oliver & Boyd.

(21) MICHEL, J. F. (1955). Parasitological significance of bovine grazing behaviour. *Nature (Lond.)* 175, 1088.

(22) MOULE, G. R. & YOUNG, R. B. (1961). Field observations on the daily milk intake of Merino lambs in semi-arid tropical Queensland. *Qd. J. agric. Sci.* 18 (2), 221–229.

(23) REID, D. (1951). A quantitative method for determining palatability of pasture plants. *J. Brit. Grassld Soc.* 6 (4), 187–195.

(24) ROGERS, W. P. (1940). The effects of environmental conditions on the accessibility of 3rd stage trichostrongyle larvae to grazing animals. *Parasitology* 32 (2), 208.

(25) TERRILL, C. E. (1962). The reproduction of sheep. In *Reproduction in Farm Animals.* Ed. Hafez, E. S. E. London, Baillière, Tindall & Cox.

(26) THOMAS, A. S. (1959). Sheep paths. Observations on the variability of chalk pastures. *J. Brit. Grassld Soc.* 14 (3), 157–164.

(27) WATSON, R. H. & RADFORD, H. M. (1960). The influence of rams on onset of oestrus in Merino ewes in the spring. *Aust. J. agric. Res.* 2, 65.

(28) ROSE, J. H. (1964). Relationship between environment and the development and migration of the free-living stages of *Haemonchus contortus.* *J. Comp. Path.* 74 (2), 163–172.

(29) SCOTT, J. (1964). Critical periods for social behaviour. *Discovery* 25 (6), 20–24.

(30) SILANGWA, S. M. & TODD, A. C. (1964). Vertical migration of trichostrongylid larvae on grasses. *J. Parasit.* 50 (2), 278–285.

(31) EWBANK, R. (1964). Observations on the suckling habits of twin lambs. *Anim. Behav.* 12, 34–37.

(32) EWBANK, R. (1967). Nursing and suckling behaviour amongst Clun Forest ewes and lambs. *Anim. Behav.* 15 (2–3), 251–258.

(33) EWBANK, R. & MASON, AGNES C. (1967). A note on the sucking behaviour of twin lambs reared as singles. *Anim. Prod.* 9 (3), 417–420.

(34) FLETCHER, I. C. & LINDSAY, D. R. (1968). Sensory involvement in the mating behaviour of domestic sheep. *Anim. Behav.* 16, 410–414.

(35) LANCASHIRE, J. A. & KEOGH, R. G. (1966). Some aspects of the behaviour of grazing sheep. *Proc. N.Z. Soc. Anim. Prod.* 26, 22–35.

(36) MILLER, G. (1968). Some responses of hill ewes and lambs to artificial shelter. *Anim. Prod.* 10 (1), 59–66.

(37) PEIRCE, A. W. (1968). Studies on salt tolerance of sheep. VII and VIII. *Aust. J. agric. Res.* 19, 577–587 and 589–595.

(38) RUTTER, N. (1968). Time lapse photographic studies of livestock behaviour outdoors on the College Farm, Aberystwyth. *J. agric. Sci., Camb.* 71, 257–265.

(39) SHARAFELDIN, M. A. & SHAFIE, M. M. (1965). Animal behaviour in the subtropics. II. Grazing behaviour of sheep. *Neth. J. agric. Sci.* 13 (3), 239–247.

IX

GRAZING MANAGEMENT

Grazing management concerns the short-term disposition of animals on pasture. It deals with questions such as how long they should stay on one area and how long they should stay off it, which animals should graze what pasture and with what other stock; how many should graze together and what other activities should be integrated with their grazing. Other topics, already discussed, have posed similar questions: the difference is that grazing management must take all relevant aspects into account at the same time, with a few limited objects in view. It is a problem, or, more accurately, a series of problems, in applied ecology.

Factors Affecting Grazing Management

Clearly, many factors must be considered and some have to be given greater weight than others. The kind of sward and the kind of animals that have to be combined must both influence the choice of management, for any particular purpose. In addition, the weather can have a very marked effect.

Climate and weather

In many parts of the world climatic conditions are sufficiently predictable for managements to be based on assumptions as to future temperature, rainfall and so on. In other parts this is not the case, and the British Isles probably represents the extreme of irregularity though not of range. It should not be supposed, however, that this unpredictability is confined to any one part of the world. It applies, for example, to Uruguay in respect of severe drought, which may last for several weeks and can occur quite unexpectedly. In Chapters II and III it was pointed out that, where conservation is required, it adds tremendous flexibility because it can be used both to add food to the system and also to subtract it. Indeed, it is a common feature of grazing management in relation to different kinds of grazing animal, that conservation of surplus herbage growth should be an integral part of it (21). Where the purpose of grazing is simply to exploit an environment with a minimum input of resources, however, conservation may be omitted, in which case large safety margins must be allowed between pasture supply and animal demand, or occasional crises tolerated. Where a high degree of intensification of production per acre is practised, margins may be so small that even conservation provides an insufficient buffer against exceptional climatic conditions. In this event, supplementary feeding must supply the missing food—but only when it is required.

In most circumstances, therefore, it is not unrealistic to consider grazing management as if it can be applied with sufficient flexibility to cope with unexpected variations in the weather.

Productive purpose

The productive purpose must dominate the issue and different grazing systems are appropriate for different purposes. The following discussion is based on the needs of a lamb-producing ewe flock from a spring lambing to the end of the grazing season. It will be evident from Chapters IV, V and VI that the major components of the production process under grazing conditions are the stocking rate, the lambing percentage, the milk supply, the incidence of disease and the supply of nutrients from the pasture. All these interact with each other. Stocking rate in relation to pasture supply represents grazing pressure, but stocking rate and the incidence of disease may have a quite different relationship, unrelated to the pasture supply. The milk yield of the ewe is influenced by grazing pressure and may affect worm-infestation in the lamb. The lambing percentage affects the significance of any stocking rate of ewes and also affects the milk supply to each lamb. Many of these interactions have been considered in earlier chapters; this chapter is concerned with the influence of grazing management upon them. The influence of management is most conveniently discussed by considering the important categories of management in turn, but this does not mean that the distinctions are necessarily so sharp or that several managements cannot or should not be integrated within one grazing season. Indeed, as with stocking rates, the rigid application of one management over a whole season may simply ensure that there are times when it is inappropriate.

Unfortunately, much of the available evidence is derived from comparisons of two or more rigidly applied systems. It is necessary, of course, that experimental evidence should come from the consistent application of managements. This is relevant to the testing, in an agricultural environment, of managements considered to be useful over the period studied. Such hypotheses, however, should be based on information as to the period for which a management is suitable and this cannot necessarily be obtained in an experiment where it is applied over a longer period. For example, if the result of management A is worse than that of management B when compared over a period of 4 months, it could be due to the fact that A was totally unsuitable during the first month only. The result does not say anything about A versus B in the fourth month, except for the case where both A and B have been preceded by 3 months of the treatments A and B, respectively.

This is an important argument with a relevance outside the realm of grassland experimentation. It must be borne in mind in the interpretation of past practical experience and in the planning of future application. There is nothing new in this; in fact, it is simply a restatement of the difficulty of attempting to test an hypothesis in the same experiment in which one is obtaining the facts on which the hypothesis is based. What is most required is elucidation of the principles on which grazing management should be based. Towards this end, a description is needed of the attributes or characteristics of managements, and an understanding obtained of what they do in relation to the important parts of the sheep/grass situation. It is from this point of view that the kinds of grazing management will be considered here.

SET-STOCKING

It is often thought that set-stocking is the management that 'the farmer' traditionally employed: in this sense it is often thought of as opposed to whatever recent managements are being advocated. This attitude obscures several important issues. Set-stocking simply means that the stock are not moved about but remain in one field or paddock for a prolonged period. It does not mean for their lives, of course, but it is only a useful term if it implies a substantial period. Otherwise all sheep can be regarded as set-stocked for varying periods whatever the management. It is commonly used as synonymous with continuous grazing.

Planning of management

There is no reason why set-stocking should not be as carefully planned as any of the more complicated systems. It is not helpful to put all unplanned grazing managements into a category of this kind. There is, of course, no virtue in planning without adequate information and the dangers of a lack of flexibility have already been mentioned. What is necessary is that the terms in which plans are expressed should be both relevant and recognizable in practice. Unplanned movement of sheep among several fields, in such a way that they always have enough to eat, is obviously superior to any plan under which the animals go short of food, whilst proceeding from field to field at specified intervals, because grass has not grown as expected. Planned managements, including set-stocking, offer greater control: this is useful in proportion to the knowledge available on how to exercise it. Set-stocking offers less control than other forms of management, however; it also calls for less interference. It is not necessarily inflexible and there is no reason why stock numbers should not be adjusted (7) or, alternatively, the area adjusted to suit a fixed head of stock (see Fig. 9.1). These topics have recently been reviewed and discussed by Wheeler (50, 51).

Control of sheep

Two facets of set-stocking should be recognized at the outset. If, for any reason, it is desired that the sheep should graze one part of their area and not another, this cannot be done. At least, it cannot be done absolutely or cannot be guaranteed. There have been attempts to control the movement of sheep without fences, but where these are highly successful they may be regarded as doing the same job as fences and the management is no longer set-stocking. Strict shepherding with the constant attendance of dogs, as practised in France, cannot reasonably be viewed as other than the same kind of movement organized by fences. The occasional movement of sheep up and down the hill, as practised in parts of Scotland, however, is rather different. Gangmowing different parts of an area in rotation has also been used to influence the grazing behaviour of sheep; in general they prefer to graze on the recently cut areas (8).

This raises the second problem. Although movement cannot be controlled it cannot be prevented either. It is doubtless an exaggeration to suggest that any system of movement for which fences might be used could, theoretically, be practised by unrestrained sheep. Sheep, it would be felt, are incapable of such organization even if they could perceive reasons for it. Before dismissing the whole idea, however, it should be remembered that sheep do move about within a set-stocked area, that they cannot physically occupy it all at once, that only the amount of herbage required for one day will be removed

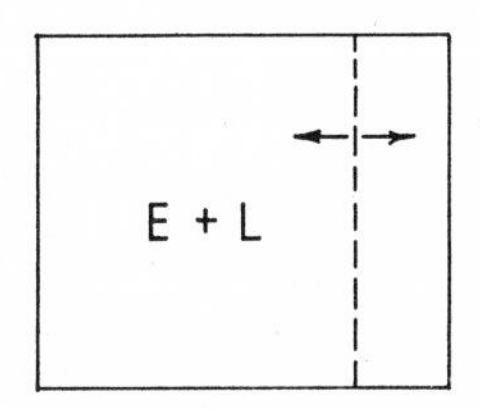

FIG. 9.1 — *Set-stocking with a movable fence to adjust the stocking rate of ewes (E) and lambs (L)*

each day—whether it is all from a concentrated area or whether it is from a large number of scattered small sites.

In recent years, more detailed studies have been carried out on the way in which sheep graze within continuously stocked lowland pasture.

Morris (66) studied lambs grazing S.37 cocksfoot, maintained at 3 levels of L.A.I. by adjusting the rate of stocking. In all cases, the lambs removed about 25 per cent of the leaf present but the grazing interval (i.e. the mean number of days between two grazings of the same plant or small area of sward) was related to leaf area, as follows:

L.A.I.	*Interval (days)*
3·0	19
4·5	24
6·0	36

These intervals are comparable to those which might well be employed in rotational grazing systems.

The proportion removed on any one day has been found to vary from 20 per cent to 1 per cent (62), under Australian conditions of set-stocking, with sheep showing a preference for the younger grass leaves. The same preference finding by Hodgson (63), was associated with removal of 27 per cent and 40 per cent at moderate and high levels of stocking, respectively. The intervals between defoliation of the same tillers were 11–14 days at the low and 7–8 days at the high stocking rate.

There are interesting implications in these results for nematode parasites. Clearly, in a continuously stocked situation (and in contrast to a rotationally grazed one), there is no reason to suppose that the patterns of defoliation and of dung deposition will be similar. Thus the relationship between the intervals between one grazing and the next and between dung deposition and the development of infective larvae may be much more complex.

Furthermore, breeds of sheep differ markedly in the way they move about and graze within an area. Some breeds, such as the Dorset Horn, will frequently graze in line abreast. Others, such as the mountain breeds, tend to spread out and 'occupy' the whole area, sometimes by the establishment of 'territories'. Still others prefer to graze in groups but without further organization. Within a breed there may, in addition, be strains which dominate or otherwise influence the rest. Thus, within a flock on the open hill, Hunter (22) has shown that some sheep always tend to occupy the best grazing while others are virtually confined to the worst. Clearly, these aspects of animal behaviour (considered in Chapter VIII) can influence the effect and significance of set-stocking.

Disturbance of sheep

The advantages or, more accurately, the positive attributes of set-stocking are lack of interference with the sheep and the possibility of the achievement of a natural balance between all the important populations involved. If sheep are continually disturbed (44), grazing time may be reduced and performance may suffer. They do, on the other hand, respond favourably to habitual activity and regular movement from plot to plot has an adverse effect only if too frequently done or if done irregularly. There is, however, one time when lack of interference may be of enormous value and that is in the first week or two after lambing. Disturbance at this time can result in a separation of ewes and their lambs which may prove permanent. Movement of a flock at critical times, of which lambing is by far the most important (though mating can also be affected), can result in unprofitable confusion. This is perhaps a good point at which to qualify all discussions on management: it is not only what is done that is important but also how and on what scale. Movement of a pregnant flock can be good or bad according to the care with which it is carried out. Movements of ewes and very young lambs may have no ill-effects on small numbers but could have quite serious consequences for a large flock. These are the more obvious effects of the scale of a sheep enterprise. There may well be others less noticeable but no less important and they may not be the same for different managements.

Balance between sheep and pasture

The achievement of a natural balance between sheep and pasture is unlikely without adjustment of the numbers or size of the one or the area or quantity grown of the other. As mentioned above, set-stocking does not preclude this. It does, however, impose some limitations to the extent that adjustments can be made and these may result in a reduced amount of herbage growth and poorer results at high stocking rates. In comparisons of one management with another, set-stocking has infrequently given better, and has frequently given worse results, than other managements at the same stocking rates. This is particularly true of high stocking rates.

Resistance to parasites

A natural balance between the sheep and its parasites is sometimes thought to be more likely under a set-stocking management. The basis for this is the idea that resistance to parasite infestation is dependent to some degree on continual ingestion of infective larvae. There is no question of the value to the lamb, from an age when it is immunologically competent, of acquired resistance to nematode infestations (16) and of the potential value of a knowledge of the immune mechanisms involved (29, 43), to the research worker and, ultimately, the farmer. The possibility of stimulating immunity by vaccination, using X-irradiated larvae, has been investigated by Jarrett *et al.* (23). If successful, such a procedure might entirely change some aspects of the interaction of the development of resistance and management, by putting the process under some control. At present, if continuous contact with infection is required, it cannot be guaranteed by any management, including set-stocking. Climatic conditions, notably drought, can break this continuity quite as effectively as rotational forms of management may do. Thus hot dry conditions followed by rain can give rise to a situation in which sheep are suddenly exposed to large numbers of infective larvae of roundworms on every part of the sward. High stocking rates may not only make this more likely but also result in the pasture carrying a relatively small quantity of herbage. If this occurs when lambs are dependent to a large extent on grass, competition for the available herbage cannot be avoided between ewes and lambs. Indeed, no differential control of ewes and lambs is possible in a simple set-stocking management, for any purpose whatever. It is perhaps not surprising that set-stocking has appeared least satisfactory in circumstances where worm-infestation has constituted a serious risk. These may be summarized, from Chapters IV and V, as (*a*) where the general level of infestation is high, in the ewes or on the pasture, at the beginning of the season, (*b*) where a high stocking rate of lambs coincides with a less than adequate milk supply to each and (*c*) where lambs are dependent on herbage for a high proportion of their diet (37, 39).

The attributes of set-stocking therefore suggest that it would be inappropriate in these circumstances unless supplementary feeding increased the milk output of the ewe or decreased the dependence of the lamb on grass.

Control of parasites

Control of parasites by other means, such as anthelmintic dosing (see Chapter X) could greatly influence the relevance of set-stocking. It should be noted that alternation of drought and rainfall may result in a breakdown of resistance to parasites, even in mature sheep, and is often associated with species such as *Haemonchus contortus*, which are capable of a high rate of reproduction. Crofton (13) has suggested that species with a short generation time and a high rate of egg production are likely to be of economic importance because they are likely to be most affected by climatic changes and to give rise to the extreme variations in parasite populations characteristic of epidemics.

Where climatic conditions can be predicted with reasonable accuracy and regular seasonal patterns of change in worm-infestation can be associated with them, it is possible to plan an anthelmintic dosing programme in advance. Gordon (20) has proposed a programme of this kind, for the conditions of Western Australia, on the basis that 'while the complexity of the problem must be recognized, it is essential to arrive at some simplification if the results of research are to be applied effectively in preventive veterinary medicine'. This statement, of course, has a relevance outside the confines of applied parasitology. The application of veterinary science within animal husbandry (15) is increasingly directed to prevention of disease and it is essential in this to recognize such biological patterns as can be discerned. Walker (70) suggested that, in New Zealand farm practice, 'strategic' dosing of ewes and suckling lambs on irrigated pasture may be wasteful and should be replaced by treatment based on the level of infestation of parasites. Naturally, the need for a dosing programme has to be related to an assessment of the damage, done by parasites, that can be prevented by dosing (2) and thus depends upon both the pathogenicity of the parasites and their numbers and on the efficacy of the available drugs. The need may be expected to change, therefore, with changes in systems of husbandry and with the development of new anthelmintics.

FOLDING

Strictly speaking, folding means the enclosure of sheep in pens or 'folds'. In the past, these were often hurdles made of wood or wattle and one of the reasons for the decline in the practice of arable folding was the labour involved in moving folds. The object of such close confinement of sheep was and is to achieve a high degree of managemental control. This may be used for various purposes, such as uniform manuring and treading of land, rationing of the food supply or a high degree of utilization. The permanent subdivision of an area by conventional fences cannot conceivably be used to achieve the fineness of control typical of folding. Temporary fencing of any kind may do exactly the same job as the hurdles did. The main point is that quite a small amount of fencing can be used to subdivide a large area into a multiplicity of small plots, simply because the same fencing is used over and over again. Strip-grazing with electric fencing is based on exactly the same approach, but it is as well to recognize that when a fence becomes simple and cheap enough it can be used so liberally that this may be more economic than the frequent moving of small quantities of it from one place to another.

Scale of operation

An interesting aspect of the influence of 'scale of operation' emerges at this point. It is true that folding usually involves small areas occupied by animals for only a few days or less, but the size of the area is not characteristic in itself. The area of the fold unit is simply that which is required to feed the stock for a desired time: it thus depends on the amount of feed and the number of stock to be fed. In short, a very large flock, particularly on a sparse crop, may require a large area each day and, in such circumstances, moving from field to field could be regarded as folding. Clearly, each individual sheep, however, is not in the same situation in these different cases and this influence of 'scale' must be borne in mind in any discussion of grazing management.

Rapid movement

The main characteristic of folding then is rapid movement of animals and close control within temporary fences. This implies movement over fresh ground, because the use of temporary, movable fencing hardly makes sense if the same fences are to be re-erected in the same place within a short time. Historically, folding has been most closely associated with arable crops, such as roots. Specially sown crops, like rape* or rye, are utilized by folding either to ration them or to prevent a large proportion being wasted by trampling, soiling and so on. There are obviously good reasons for folding over a large, unfenced crop from which only one harvest is to be taken. Grass is characterized, on the other hand, by the large number of harvests which can be taken and the relatively short intervals in between that it can tolerate. There are often, however, very good reasons why sheep should only take one harvest from any crop, grass included, within a given period. These reasons are mainly concerned with the prevention of disease.

* Enlarged thyroids have been reported (68) in lambs grazing rape, suggesting the presence of a goitrogenic factor of the thiocyanate type known to occur in other *Brassicas*.

Control of the Herbage

The chief importance of folding, therefore, lies in its ability to present to the sheep at frequent intervals a 'clean' crop, free from certain infestations. This presents no difficulty with arable crops. Nor does it with the grass crop in certain circumstances. If, as illustrated in Fig. 9.2, folding is visualized as a sheep unit passing over any point once only, then grass has to be harvested, in some other way, both in front of and behind the stock. The manner in which the grass is controlled in front of the sheep is crucial to their performance: the control of it after they have gone influences the efficiency of the whole procedure but matters not at all to the sheep.

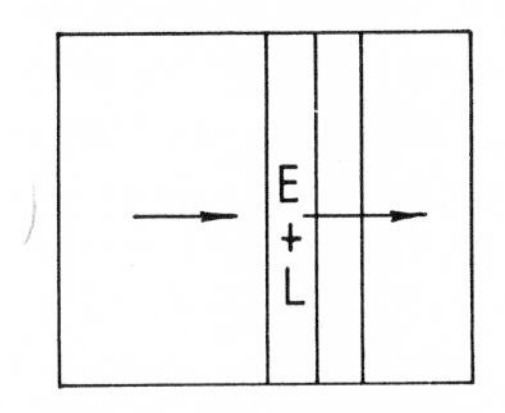

Fig. 9.2 — *Continuous folding of ewes (E) and lambs (L)*

Folding presents a considerable challenge precisely because it makes it possible to offer to the sheep almost any kind of grass, in almost any kind of quantity and of whatever quality is required. All this can only be done within certain limits, nevertheless it raises questions which can only be answered in very general terms, at present.

The kind of grass required and the stage of growth at which it is wanted for folding can differ in one respect only from that needed in all other grazing situations. It does not have to support the same sheep population again, or at the same stocking rate. Grass that might not survive if the same treatment were to be frequently repeated, can be satisfactorily folded once. It may suffer greatly in patches, nevertheless, from a combination of wet weather and dense stocking. Theoretically, the amount of herbage allowed each day can be adjusted, by adjusting the area, so that it is exactly what is required. In practice, this cannot be done without risk of reducing intake or, if this is to be avoided, without wastage, for what is not consumed that day cannot be eaten later on, at any rate by the same sheep. In fact, folding has normally to be undertaken deliberately either to ration food or in order to exploit its ability to feed each individual to capacity. For the latter purpose, a surplus must be offered every day to avoid any risk of quantitative shortage and to allow selective grazing so that only the higher quality part of the herbage is taken.

If the pasture growth in front of the sheep is controlled by cutting, no disease problems may arise. Similarly, the greater the difference between the kind of stock that graze it and the sheep that are going to do so, the less the disease risk. The only case for following sheep with sheep in these circumstances is where productive, young or susceptible ones precede those that are unproductive, old or resistant.

Clearly the greatest advantage is to be derived from folding where sheep are not preceded by anything that can leave behind disease agents or that can adversely affect grass intake, as, for example, by fouling. In these circumstances, the only possible source of trouble is from the sheep themselves. Anything that can be passed directly from one sheep to another, such as orf, may well be spread more rapidly by the crowding of sheep in a small space. With appropriate utilization of the herbage left behind, this aspect can be covered by avoiding any overcrowding. This obviously becomes more difficult to achieve when herbage production is high per acre, yet it is in these conditions that folding appears physically more attractive, since less ground has to be covered to obtain any given quantity of feed.

Control of Disease

Any disease which requires a time interval between one sheep and another can virtually be eliminated by adjusting the frequency with which sheep are moved, assuming, of course, that one can start with a pasture which is itself free from the disease-producing organism.

The most important in this category are the stomach, intestinal and lung parasites. The lungworm (*Dictyocaulus filaria*) tends to be short-lived on pasture in summer, though small numbers may survive from one year to another. Tapeworms and some roundworms, e.g. *Nematodirus* spp., are normally troublesome only where sheep graze year after year on the same ground. This longer-term aspect will be further considered in Chapter XI. It is worth noting here, however, that the numbers of parasites such as *Nematodirus* spp., which develop over a longer period (18), may still be influenced by short-term grazing management on pastures that have been grazed in previous years. Parasites with intermediate hosts may similarly be affected by management, although normally entirely different considerations seem to be more important. For example, Jensen, Mapes & Whitlock (24) reported that manipulation of the microclimate, by destruction of plant cover, failed to reduce the population of snails (*Cionella lubrica*) which are one of the intermediate hosts of the lancet fluke (*Dicrocoelium dendriticum*). Ploughing, however, had a very marked effect, because this snail is apparently incapable of returning to the surface through even an inch of superimposed soil.

Many species of nematode parasites can survive ploughing, but the important question would be whether they do so in significant numbers. The definition of what is a significant number, however, must vary with the reproductive potential of the species. Some roundworms, such as *Haemonchus contortus*, are notable for the speed with which one sheep can infect another with them—theoretically, in less than one week; it is in the control of this kind of parasite that folding should be outstanding. Nematode eggs reaching the pasture in the faeces of the sheep must have, even under optimum conditions of temperature, a day or so before they are likely to hatch. After this the developing larvae must reach the third or infective stage before they can infect another sheep. This, again under optimum conditions, must take several days. It follows that by moving sheep at shorter intervals, e.g. daily, all such infective larvae would develop after the animals had moved on to a fresh area. In this way, reinfection can be prevented in sheep that are already parasitized; it might take a very long period of such folding, however, before a worm-free sheep could be obtained in this way. By folding alone it might take more than two years.

Although little work has been done on the effect of different species of plants on the worm-burden of sheep, there is some evidence that grazing on certain crops results in the expulsion of certain species of worm. This has been demonstrated on green oats in relation to the intestinal roundworm, *Oesophagostomum columbianum*, for example (30). Equally little work has been devoted to a combination of anthelmintic dosing and folding to eliminate worms. The reasons for this are chiefly concerned with whether it is desirable, or not, to have sheep completely free from worms anyway, a topic further discussed in Chapter XIV.

The most important aspect of folding and the control of parasites relates to ewes and lambs grazing together. Rapid folding can prevent the infection of lambs whilst suckling and grazing with normally infected ewes (31, 32, 33). This it does simply because the lamb is normally born free from worms and remains so if the infective larval stages are never available to it. Complete freedom from parasites must depend upon the pasture being absolutely free. This might require several years and only be justified in experimental work (39) or where a completely worm-free sheep enterprise was contemplated (see Chapter XIV).

In practice, at present, one complete year without ruminant grazing would probably be as much as was warranted. For ewe and lamb folding, the pasture would be further infected by the ewes only and the number of larvae on the pasture offered to the lamb, after one year's rest, would probably be negligible for a folding system where reinfection could not occur.

Even assuming that there were no parasites on the pasture, it is likely that small numbers would reach the lambs from their dams. In practical terms, this is unimportant, again where reinfection cannot occur. Occasional worms of two relatively unimportant species frequently defeat attempts to rear lambs with no worms at all. These species are *Trichuris ovis*, which is most unlikely to cause trouble, and *Strongyloides papillosus*, only troublesome in very large numbers. The second species is interesting because it can gain access to the host by penetrating the skin and because it can actually reproduce and multiply in the free-living state on the pasture at least for one generation. It has been shown experimentally that invasion by *S. papillosus* larvae through the feet of sheep can enable the organism (*Fusiformis nodosus*) that causes footrot to gain entrance.

Experimentally, rapid folding of ewes and lambs has made it possible to eliminate worms as a problem (40) and has resulted in very high performance of lambs. It can only be practised where the number of sheep is small relative to the total number of acres; this implies other forms of grass usage, however, rather than extensive farming.

As has been said, there is no need to visualize any management having to be applied at all times. In relation to the control of parasites, the marked seasonality of incidence of many species should be noted. These rhythms in nematode parasitism have been studied under New Zealand conditions by Tetley (47) and, in Australia, Gordon (19) has constructed bioclimatographs to 'delineate the seasonal occurrence and regional distribution of some of the trichostrongyloidoses of sheep'. In Britain, the general increase in the numbers of infective larvae on the pasture towards the late summer and the autumn (17) is probably associated with the pattern of lamb rearing. The need to control parasites may thus be related to both season and husbandry contexts.

In many circumstances in Britain, the level of infestation on the pasture suddenly increases in July or early August and Gibson & Everett (61) have suggested that moving ewes and lambs to clean pasture just before this time might form the basis of a grazing system that would greatly reduce the risk of severe parasitism.

Folding has advantages in certain circumstances, and these advantages are probably greatest with ewes fattening lambs. The management, and the clean ground to go with it, then only has to supply 3–4 months of grazing. This period may be further shortened if, as was implied in the preceding section, folding is not used immediately after lambing. Even where folding is intended primarily to control worm-infestation, it can successfully do so by starting when lambs are up to one month old. Lambs do not generally appear to acquire a significant number of worms in the first 3–4 weeks, at least with a normal milk supply and spring lambing.

ROTATIONAL GRAZING

Rotational grazing is characterized by permanent* subdivision of the total available area into a number of fields or paddocks (Fig. 9.3). In systems of rotational grazing the number varies from 2 up to about 20. A small number allows a change of area for stock but makes little contribution to the control of grazing. Change alone, as is possible with 2 paddocks, simply means that stock can move to an area which is to some extent fresh, fertilizer can be applied on one while the other is grazed and one paddock can be rested prior to cutting for conservation. It is rarely that exactly 50 per cent of the area needs to be cut at any one time, however. With a small number of fields, rest periods can only be long if grazing periods are also long. It is somewhat inconsistent to allow both since, within a long grazing period, plants can be cropped at successive short intervals, and one object of a long rest period is to prevent this.

* This term is, of course, used to indicate division by so-called 'permanent' fences, as distinct from 'temporary' fencing.

Interaction with stocking rate

Rotational grazing has sometimes enabled a higher stocking rate to be carried (25) but in these cases the number of paddocks has generally been 3 or more. The number required depends chiefly on how long each one is to be occupied for any one grazing, since grass has to be harvested at intervals related to its growth rate and such intervals are often approximately 3 weeks. If sheep stay 3 days in each paddock, about 8 are needed. If they stay for only one day then about 20 are required; where they stay for a week, 4 will suffice. It is nearly always advantageous to have more rather than less paddocks, unless the area of each thereby becomes too small for farm operations. Morley (65) has suggested that the optimum number of subdivisions in a rotational grazing system is probably less than 10, unless considerations other than pasture growth are important.

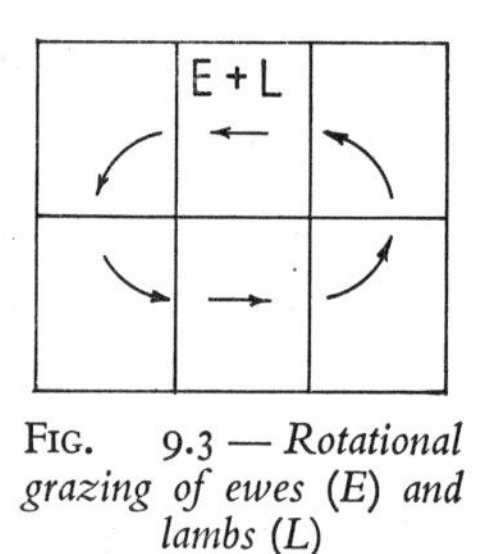

FIG. 9.3 — *Rotational grazing of ewes (E) and lambs (L)*

The control of grazing, inherent in rotational management, is characteristic of all systems that have a rotational element. This includes practically everything except set-stocking: continuous folding is not, in this sense, a system, since it does not occupy a given area for the period during which it is operated. There is relatively little value in rotational grazing as such, but a particular form of rotation may be of the utmost value for a particular purpose. The inclusion in a rotational system of the whole area available for a year, gives the maximum chance of conserving just the herbage that is surplus to the requirements of the flock. It also means, since any paddock can be used for either grazing or cutting, maximum choice, within the system, as to where sheep graze.

The interaction between method of grazing management and stocking rate has been stressed by McMeekan (26): he has also pointed out that damage to the sward may occur under any system and, if it does so at particular, critical times, this may seriously reduce pasture yield. McMeekan suggested that destruction of the apical meristem during the reproductive phase of grass growth could easily have this effect and considered it 'probable that the importance of the reproductive phase of herbage grasses has not been sufficiently recognized'. Wheeler (51) has pointed out that there has been 'some tendency for the experiments in which intermittent grazing was shown to be advantageous to have been conducted on pastures of lucerne/cocksfoot or other lucerne mixture rather than on grass/clover pasture'. Clearly, the appropriateness of a management to the particular species present in the pasture, over a prolonged period, must be added to the considerations of intake, grass growth and efficient utilization discussed in Chapter VII.

Disease Control

In addition, as indicated in the previous sections, disease control must be taken into account. The frequency with which sheep are moved carries the same implications in rotational grazing as it does in folding. The critical factor in the rotational system, however, is the rest period between grazings. From the point of view of grass production and utilization, the optimum rest period is likely to be related to the rate of grass growth and thus to temperature, water supply, fertilizer application and so on.

For some diseases a critical rest period can be stated. The footrot organism, for example, appears to be unable to survive on pasture for more than a fortnight, whatever the weather. The survival of parasitic worms, on the other hand, depends greatly on climatic conditions. Hot, dry conditions are unfavourable to larval survival; low temperatures delay their development and, if low enough, may also destroy them; warm, humid weather assists development and, although it may not prolong individual survival, usually results in the maintenance of a high level of infection on the pasture.

Silverman & Campbell (28) found that, under laboratory conditions, the eggs of *Haemonchus contortus* took at least 5 days, at a constant temperature of 21·7° C, to reach the infective larval stage. At 14·4° C they took 9 days and at 11° C at least 15 days were required before third-stage larvae first appeared. They also found that embryonated eggs survived at 7·2° C for up to 4 months and that rapid desiccation of faecal pellets resulted in the destruction of the contained nematode eggs, unless these were in an advanced stage of embryonation.

Todd, Kelley & Hansen (48), studying the survival of the infective larvae of *H. contortus*, *Ostertagia circumcincta*, *O. trifurcata*, *Chabertia ovina* and *Nematodirus spathiger* on pasture in central Kentucky, found that all species were represented at the end of a winter with temperatures down to 23·4° F (mean for January, 1948; mean minimum temperature= 14·7° F). It is worth noting that these authors also recorded the survival of the infective larvae of the tapeworm *Moniezia expansa*, within its intermediate host, under these conditions. Parasite-free lambs were used as a grazing test in this work. The development by Taylor (45) of a method for the estimation of pasture infestation enabled this kind of study to be greatly extended. Crofton (12) further studied the pasture sampling technique and the distribution of infective larvae on the herbage. He found, for example, that the greater numbers of larvae were within the bottom inch of the pasture. Crofton (11) also pointed out the need to take into account the length of the grass in these studies. Donald (58) has developed a technique for the recovery of infective larvae from small units of pasture (4 in./sq. quadrats yielding about 25 g of fresh herbage): this should make it possible to study the bionomics of these larvae in greater detail in the field.

The effect of temperature on larval development (57) and survival (52) makes it unlikely that the same management will have the same effect in different parts of the world.

The carry-over of larvae from one season to another varies with species (53, 67) and their activation in the rainy season of Rhodesia appears to depend on atmospheric humidity also (54).

Interactions involving the pasture

It is necessary to take into account the whole dynamic pasture complex when attempting to relate the mortality, hatching and developmental rates of parasites to the avoidance of disease in the grazing sheep.

It is a gross over-simplification of the problem of parasite control by management to suppose that the survival or mortality rate at any particular time is necessarily the major factor. When sheep leave a paddock the infective larvae present at that time die at a rate greatly affected by the microclimate in which they live. The latter changes with the weather and because the herbage continues to grow. At the same time, parasite eggs deposited during any or all previous grazings may be hatching and the emergent larvae developing also at rates affected by the microclimate. The proportions dying and surviving are less important than the absolute numbers present and the latter only matter to the sheep during the next period of grazing: it is then the number present on each unit weight of the herbage eaten that is of major consequence.

The severity of grazing thus affects the numbers ingested, the numbers left behind and the microclimate in which they are, initially, left. The infective larvae are thus a changing population, dying and being replenished in a changing environment. The influence of a particular length of rest period is understandably difficult to predict, but it may have little to do with the rate of development of the eggs deposited at the end of the last grazing period.

Characteristics of Rotational Grazing

Several features are worth noting, however. First, in a closed system without conservation, rotational grazing tends to involve more rapid movement when grass growth is slow, because each paddock supports the flock for a shorter time. When grass growth is rapid, the rate of rotation tends to slow down. It will be clear that one cannot generalize about the precise effects of this. It does, however, represent a situation where drought followed by rain and warmth can result in a combination of short grass and weather suitable for a rapid increase in the numbers of parasites. This can also occur in set-stocking but is less likely in a rotational system which incorporates an area for conservation. Where this is done, the amount of grass in each paddock when it is grazed can be kept more constant by using conservation paddocks for grazing: thus the longer the grass takes to grow the longer the rest period that is allowed.

Secondly, in a sheep-only system the rest period between grazings is unlikely to exceed 4 weeks, except for the paddocks cut for conservation: the rest period is likely to be shorter than this under intensive stocking. Under British conditions such a period cannot be relied upon, by itself, to ensure a significant reduction in the infective population. Better control under these circumstances appears to result, however, when each paddock is severely defoliated, leaving a very short pasture behind each time. In other parts of the world, particularly where very high temperatures and intense insolation occur, a rest period satisfactory for grass growth might also be satisfactory for the control of parasites. Little benefit from simple rotational grazing, in terms of parasite control, appears to have been found in Australia (59), however, and worm-burdens have recently been reported to be similar for lambs set-stocked and for those rotationally grazed under British conditions (60).

Thirdly, ewes and lambs, when kept under these conditions, compete with one another for herbage and food intake may be reduced for both. This is most serious in the case of the lamb and particularly, of course, in relation to lambs that are receiving relatively little milk.

Mixed grazing and rotational management

Intensive rotational grazing is probably not suitable for mixed stocking, whether of ewes and lambs or sheep and cattle, for it is difficult to satisfy two different needs at once in a system of this kind. Where one group of animals can be regarded as of less importance than another, it is more sensible to rotate in such a way that these follow behind the others. It may be noted at this point that there is a time at which a ewe and its lambs may be regarded as a single unit: this is where their interests coincide because the lambs are virtually milk fed.

Restricted grazing

A further possibility, little explored as yet (64) but relevant to intensive grazing, is that of restricting the grazing time to that necessary for feeding activities, with confinement to special areas at other times.

CREEP-GRAZING

Creep-grazing (Plate VI (c)) involves a fence through which only lambs can pass. This is arranged by the insertion of a 'creep' or hole in the fence big enough for lambs but too small for ewes (34).

Design of creeps

Such creeps vary in design (see Fig. 9.4) from temporary, home-made, modifications to an existing fence, to specially designed structures with rollers to prevent damage to the sheep. The need for a carefully thought-out design, though not necessarily for elaborate or complicated models, is greater when the lambs and ewes to be separated are similar in size. This occurs when creep-grazing systems take unweaned lambs on to heavy weights, but it can occur at an early age where light ewes are crossed with heavy rams. The progeny may then weigh as much as their dams long before weaning. Since the width of a sheep is largely wool and abdomen, both of which are readily compressible, creeps which attempt to distinguish between ewes and lambs, using such methods as vertical bars, tend to be less efficient when ewes and lambs are not so very different in size (9). The greater depth of chest of the ewe and its unwillingness to lower its rear end, compared with the lamb and its capacity for 'crawling' under things, make it possible to distinguish between them effectively on a height basis. Horizontal bars or rollers then appear very effective but the height of the top roller from the ground may be important and not merely the distance between top and bottom bars.

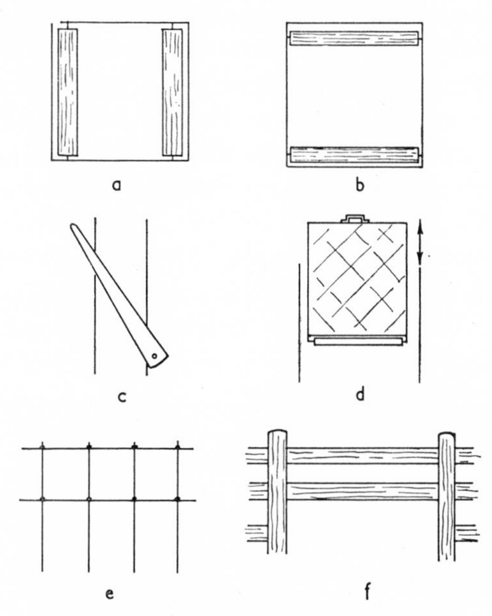

FIG. 9.4—*Types of 'creep' for use with lambs*

(a) With vertical rollers
(b) With horizontal rollers
(c) With adjustable diagonal lever
(d) Adjustable gate with horizontal roller
(e) Spaced vertical palings
(f) Simple gap beneath horizontal fence rail

The object of such arrangements is simply to allow the lambs to go where the ewes cannot: it is not a complete separation because the lambs can return to the ewes. Indeed, they are not obliged to leave them at all and it is a feature of all creep-grazing systems that lambs can only be offered the facility—it is a different and sometimes difficult matter to persuade them to use it.

Creeps have been used for a very long time for creep-feeding, where lambs have had access, sometimes on an *ad lib.* basis, to supplementary food in a trough surrounded by a fence with a creep in it. Creep-grazing does not imply creep-feeding, although obviously it can be combined with it (5). It was often associated with arable folding but it is only in recent years, from 1954 onwards, that it has been applied to pastured sheep. For this purpose it has been embodied in several different management systems. These will be considered separately but, first, the many features that they have in common will be mentioned.

Features common to creep-grazing systems

They all aim to improve the growth of the lamb, generally within systems which have other main advantages. Wherever heavy stocking rates result in undesirable competition between ewes and lambs, creep-grazing can be used to relieve this by giving the lambs a grazing area unavailable to the ewe. In the same way, creep-grazing protects the lamb from most of the ill consequences, such as a reduced intake, of a closeness of defoliation which may be desired in order to achieve a high degree of utilization. Wherever parasites become too numerous within a ewe-grazing system, creep-grazing can offer the lamb an outlet to a less infected environment.

Clearly, in almost any system of grazing management, creeps can be inserted in one or more of the fences and there are thus as many creep-grazing systems as there are non-creep-grazing ones. Only the more important will be mentioned here. These will be discussed in two main categories: (*a*) those in which ewes and lambs successively occupy the same land and (*b*) those in which lambs creep on to an area never occupied by the ewes.

FORWARDS CREEP-GRAZING

Although, when left together, ewes and lambs may compete for grass later in the season, when the lambs' grass intake has increased, their needs are not irreconcilable. As the lamb grows, it gets less milk and eats more grass, so the ewe's need for nutrients generally decreases as that of the lamb increases. Forwards creep-grazing has as its major aim the partition of the grass available between ewes and lambs in the most efficient manner. It is essentially a rotational grazing system with a creep between every pair of paddocks (9, 10, 14). About 6 paddocks are generally used but these must be arranged in such a way that when the ewes occupy one, lambs can creep ahead into the one the ewes will occupy next (see Fig. 9.5). Since the lambs can graze very selectively from a surplus of herbage, their intake should be maximal for quality and quantity. The ewes, on the other hand, can be restricted to whatever extent may be required, without restricting the lambs. By adjusting the rate of movement the needs of both can be satisfied.

There is little doubt that this system of management enables good use to be made of the herbage grown and removes some of the worst features of rotational grazing as applied to ewes and lambs at high stocking rates. Nevertheless, both ewes and lambs in fact graze the same pasture and which ones may be regarded as being in front of the others depends on the grazing interval considered.

Parasite control

Parasite control thus depends on two things. First, selective grazing by the lamb means that it need never ingest the more infected parts of the sward. These are usually nearer to the ground and this herbage is avoided if lambs graze at a higher level. Unfortunately they do not always do so. Lambs on occasion graze right down in patches, leaving other patches untouched, and little is yet known about the reasons for these differences in grazing behaviour. Secondly, the hard grazing that can be safely obtained with the ewe later in the season means the maximum ingestion of larvae by older animals and maximum exposure of those uneaten. This should result in a generally low level of infection on the pasture. The weather may greatly affect all this, of course.

Practical application

The disadvantages of forwards creep-grazing in practice are chiefly those of inefficient application. Ewes may be restricted too much or too soon: this can result in a reduction in the milk yield. Unless integrated with conservation, the system may be applied too long after grass growth has failed to keep pace with the needs of the stocking rate employed.

In the absence of parasites, forwards creep-grazing is an almost ideal way of utilizing one sward by ewes suckling lambs. Cooper (9) has emphasized that forwards creep-grazing was devised to minimize ewe/lamb competition for food and to avoid the fluctuations in the plane of nutrition which occur with rotational grazing. He has pointed out that many failures of the system are attributable to a delayed start resulting in lambs first meeting 'creeps' at the age of 4 to 6 weeks. Cooper (10) regards the early introduction of lambs to creep-grazing as an essential element in its successful use. At this

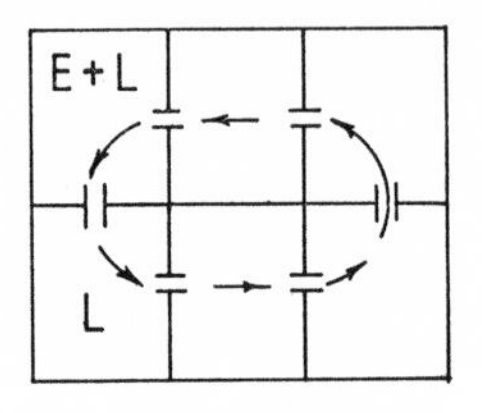

⊣⊢ fence with a creep in it

FIG. 9.5 — *Forwards creep-grazing of lambs (L) ahead of their ewes (E)*

stage, i.e. when lambs are under 3 weeks of age, it is chiefly curiosity that causes the lamb to investigate the creep area. It is essential that the pasture on offer should be such as to encourage grazing. During this time, when lactation is of great importance and the lamb is not consuming a great deal of grass, there is no point at all in risking a shortage of food for the ewe. It is later on, when lambs are over 6 weeks at least, that a high grazing pressure in the ewe area can be used to encourage lambs to graze ahead. Dickson (14) reported not only high output per acre using this system, but also the successful application of it on a large scale, using 150 Half-bred ewes and 220 Down cross lambs on 24 acres.

Creep-grazing of lambs on to an area not grazed by the ewes at all has been incorporated into a variety of systems: two different groups may be distinguished.

Set-stocking with a Creep

Sometimes called 'lateral' creep-grazing, this is simply a set-stocked system with an area fenced off for the lambs (3). Ewes cannot be restricted in a very positive sense, unless stock are added, but they may become so as the season progresses, either because less grass is grown or because the lambs take more.

In common with all methods in this group, entirely different swards can be grown for ewes and lambs, if desired. One difficulty about a separate lamb grazing area is that, in the absence of the ewe, it may prove difficult to control the pasture. It has often appeared so in practice. In principle, it is no different from the general difficulty which is met by conservation and there is no reason why conservation should not take place in the lamb area. Naturally it must be taken

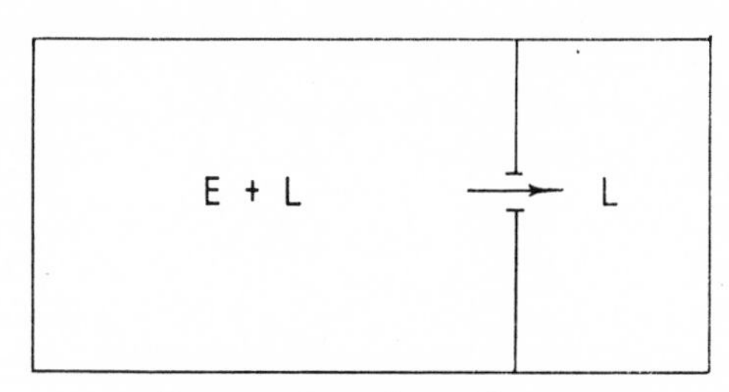

Fig. 9.6—*Set-stocking with a creep*

E = Ewes; L = Lambs; ——| |—— with a creep in it = a fence

at an early stage or the aftermath may be unsuitable for lambs. A different pasture, or another crop, offers a different solution, however, and speculation about this (little more has yet been done) has centred on clovers. These have the merit that their growth starts later in the season and their pattern of production more nearly fits the changing requirement of spring-born lambs. In addition, if growth does exceed the animals' needs, the quality of the herbage may not decline as rapidly as does that of grass. None of this would necessarily represent sufficient reason for providing it, unless it was also of high, or higher, nutritive value or was particularly attractive to the lamb. The latter point is made because, more than in other creep-grazing systems, it is necessary to persuade the lambs to use the area set aside (see Fig. 9.6). This is more necessary because, with no movement of ewes, there can be no shortage of herbage where the ewes are until a general shortage exists over the whole ewe area. If this occurs early enough to influence the behaviour of the young lamb, milk production is also likely to be affected. Whether the lamb creeps early or not, it is essential to control the lamb pasture. In addition to conservation, other stock can also be used.

Parasite control here depends on the proportion of grazing done by the lamb in its own area. It is also affected by how soon this occurs. If lambs are infected early by grazing with the ewes, they may then infect their own area to a greater extent. The lack of flexibility in the management suggests that it would be most useful when integrated into a larger system, where cattle and conservation, at least, could provide greater flexibility. It has, however, the great merit of simplicity and can be combined simply with supplementary feeding.

Sideways Creep-grazing

Like set-stocking with a creep area, sideways creep-grazing includes lamb areas which are not grazed by the ewes. The same possibilities exist, therefore, for using entirely different kinds of pasture for ewes and lambs, respectively. As a matter of fact, even when an initially uniform pasture is used, once it is divided for sideways creep-grazing differences in botanical composition rapidly develop. These are partly due to the different grazing pressures applied to the two areas and they are greatly exaggerated by any differential use of nitrogenous fertilizer. At high stocking rates, nitrogen is normally required in the early spring to produce herbage for the ewe during the peak of lactation but there is likely to be more growth than is needed on the lamb area. As with the management previously described, this problem can be overcome by the use of cattle or conservation.

Sideways creep-grazing (35, 41) is a rotational system in which each paddock has a creep area (Fig. 9.7). Paddocks need not be adjacent; theoretically, separate fields can be used. Whether it is practicable to use separate fields depends on the distance between them and also on how long sheep occupy each paddock. In intensive systems, sheep usually stay between 2 and 4 days in each paddock or field but, as mentioned earlier, these intervals can be varied considerably. Probably all systems are made more flexible by integrating conservation and grazing areas. Where creep-grazing involves separate lamb areas, however, it is usually advantageous to conserve in the lamb

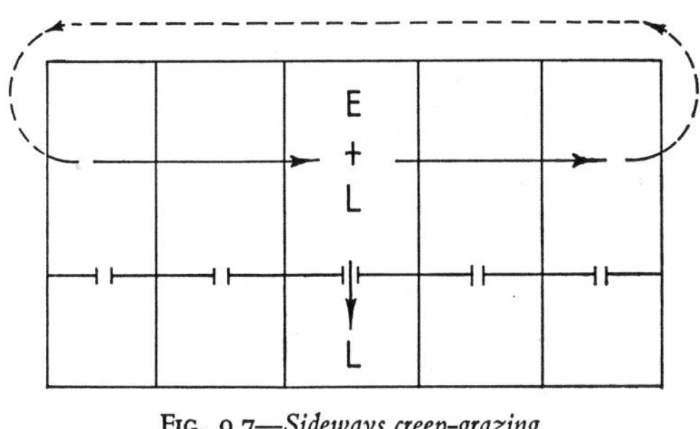

FIG. 9.7—*Sideways creep-grazing*

E = Ewes; L = Lambs; ——| |—— = a fence with a creep in it

pastures. There is little point in having to control the latter with no return for the surplus grass and there is often difficulty in cutting and using the small surpluses that occur in an all-grazing system.

Practical adaptation

Sideways creep-grazing has potential advantages in that lambs can graze a relatively clean area; or, at least, one in which the ewes do not graze in the same season. This advantage can be exploited by allowing lambs access to a large area. This means that they are always at a low stocking rate and always have plenty of herbage, not only to eat but also to select from. Such a large area, however, requires an additional form of utilization and this is more easily arranged if the area has few internal fences. A variant on a fully rotational sideways creep-grazing system, therefore, dispenses with fences within the lamb area (Fig. 9.8). The lower stocking rate is of greater value than the fences for purposes of control of worm-infestation.

Theoretically, of course, the ewe area could be placed almost anywhere in relation to a large unfenced lamb area. It could, for example, be in the middle of it, or occupy a short or long side of the field. Similar considerations apply to the set-stocked area with a creep. The amount and cost of fencing is a major factor in this but there are also important considerations of animal behaviour. These have been discussed in Chapter VIII and will not be further elaborated here, except to stress that the greater the time (or more accurately, perhaps, the greater the proportion of its grazing time) that the lamb spends in a creep area, the more influence the management may be expected to have. In relation to nutrition this will matter most where a more nutritive feed is available in the lamb area. In most cases, lambs creep when they need to from a feed intake point of view. The exceptions occur when the lamb pasture has become unattractive, usually because it is too mature. Over-mature herbage can have marked effects on the diet of sheep (1) and this is most marked with young lambs. In relation to parasites, use of the creep area has to be fostered. The whole pattern of infestation in the lamb and on its pasture can be altered by a different use of the lamb creep area.

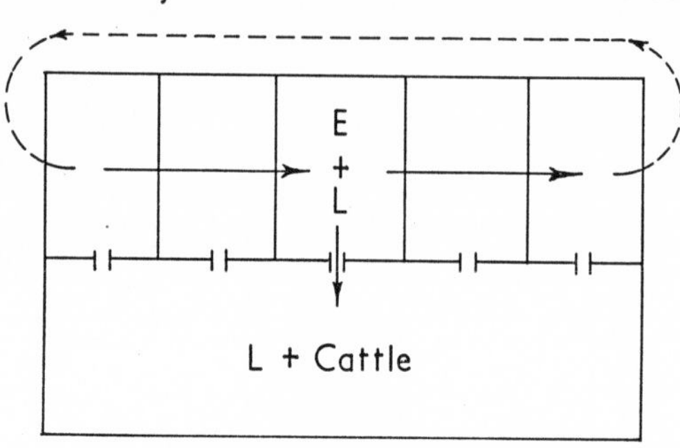

Fig. 9.8—*Sideways creep-grazing with cattle in addition*

E = Ewes; L = Lambs; ——| | —— = a fence with a creep in it

Consideration of the long-term consequences of management are deferred until Chapter XI but one point is worth making here. If lambs graze sufficiently, in a ewe-free area which starts the season worm-free, to result in extremely low worm-infestation in one year, there is no reason why the procedure should not be repeated on the same land in succeeding years. This, of course, can only apply to creep-grazing systems where the ewe does not graze the lamb areas at all.

MIXED GRAZING

The grazing together, at the same place and time, of different kinds of stock has advantages and disadvantages. The latter, related chiefly to the difficulty of satisfying two or more different needs at once, outweigh the former at high stocking rates and high grazing pressures. Intensity per acre, in short, tends to lead towards specialization of enterprise: this is exemplified by some parts of New Zealand (6, 49). Intensity of production per individual, on the other hand, is often associated with a lower intensity per acre. In these circumstances, the advantages of mixed grazing may outweigh the disadvantages.

For sheep, mixed grazing with cattle must mean a lower stocking rate of sheep per acre and most of the effects of mixed stocking on, for example, parasite control, can be attributed to this. Adverse consequences of a high stocking rate of one kind of animal can therefore be avoided, to a greater or lesser extent, by mixed grazing. Cattle, particularly, can be mixed with sheep in most of the managements described (see Fig. 9.9). Much remains to be learned about the interactions involved in mixed grazing but there is little evidence as yet to

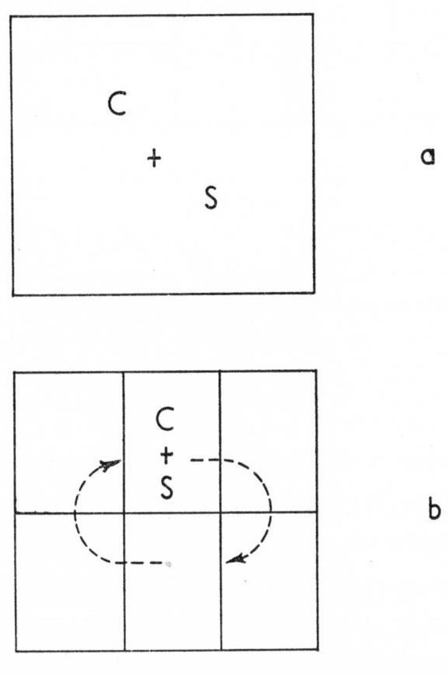

FIG. 9.9—*Mixed grazing with cattle (C) and sheep (S):*
(a) Set-stocked
(b) Rotationally grazed

support some of the often-repeated claims that the presence of cattle has a beneficial effect on the sheep. There is no doubt that rapid growth of lambs is often associated with mixed stocking at a low stocking rate, but it is probable that the same number of lambs per acre would give the same result with different forms of additional defoliation.

This does not rule out the possibility of a higher production per acre resulting from mixed grazing at a very high total stocking rate. There are limiting factors to animal production per acre with any one class of stock, e.g. worms in the case of triplet lambs, factors which increase in effect with increasing stocking rate and which are not necessarily related to the amount of herbage grown per acre. There may be disease risks in a high concentration per acre of young susceptible lambs (46), particularly early-weaned animals which are totally dependent on pasture for their nutrient intake (4, 38, 42). Mixed stocking in circumstances of this kind could lead to higher total production per acre by using two kinds of animal, each at less than its critical stocking rate. The need for this kind of situation might arise if highly productive individuals were combined with very highly productive acres.

Mixed grazing may be employed at a high grazing pressure where one of the kinds of animal is not in a productive state. Otherwise it is most likely to be efficient at a low grazing pressure, independent of stocking rate.

Alternate Grazing

In rotational grazing systems different classes of stock can be alternated (Fig. 9.10). If one class contains animals with a low order of requirement, that is, for food of low nutritive value or where requirements are not critical, such alternation can be used to benefit the other. In general, alternate grazing is a means of lengthening the period between two successive defoliations by any one kind of animal. This may have beneficial effects in reducing the influence of soiling, for example, but it is chiefly of significance in the avoidance of disease. The significance is not, however, very well expressed as a time interval. As mentioned earlier, periods cannot simply be doubled in these biological interactions. Grass continues to grow, it is harvested and, as a consequence, its structure and microclimate are suddenly altered, and it grows again, changing all the time. The pasture may be 'rested' for twice as long from sheep, but few of the original leaves may remain at the end (36). The effect of alternate grazing on disease control is most marked where organisms are involved that are highly host-specific. For example, it has been assumed for a long time that the lungworms of sheep would not infect cattle and *vice versa*. Thus each class of stock could be safely grazed after the other and to some extent this could be exploited to reduce the risk to each. Parfitt (27) has recently pointed out, however, that the sheep lungworm (*Dictyocaulus filaria*) *can* infect calves, with serious consequences, and has suggested that this may be a relatively common occurrence. The principle is still a sound one, but this particular example may have to be reconsidered, should this kind of cross-infection prove common. Cross-transmission of *Cooperia oncophora*, *Trichuris ovis* and *Moniezia benedeni* from calves to lambs has also been recorded (69) but it was thought that the role of cattle in perpetuating parasitism in sheep was insignificant in the Maritime provinces of Canada.

From the point of view of grassland utilization, in both mixed and alternate stocking much depends on the proportions of one class of stock to the other. If their total demands are unequal, they can only rotate at the same speed if one group grazes more intensively than the other. This suggests that the group which requires the lower grazing pressure should also have the smaller total demand. If, however, the two groups differ in demand and a lower demand group (e.g. cattle) is

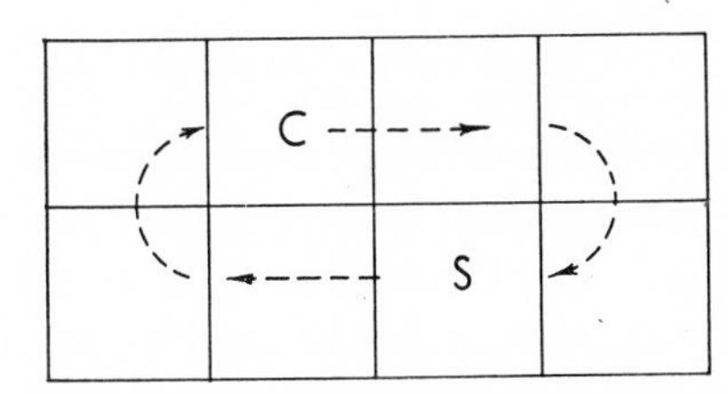

FIG. 9.10—*Alternate grazing of cattle (C) and sheep (S)*

grazed for the same time as a higher demand group (e.g. sheep), they should each rotate at the same speed but leave different quantities of herbage behind. It seems then that the intervals should also be adjusted in such a way that the smaller demand group is further behind the higher demand group. In the example used, sheep would be closer behind the cattle than the cattle would be behind the sheep.

Alternate stocking with beef cattle and breeding ewes has been investigated by Conway (55) in Eire. The results generally support the view that, in some circumstances, greater production can be obtained by combining the two classes of stock than can be obtained with either alone.

One of the most useful applications of the idea of alternate stocking, however, is the alternation of sheep and cattle in years. This has been used as a major method of preventing disease due to *Nematodirus* spp. infestation in lambs (56).

REFERENCES

(1) ARNOLD, G. W. (1962). Effects of pasture maturity on the diet of sheep. *Aust. J. agric. Res. 13* (4), 701–706.

(2) BANKS, A. W. (1958). Epidemiology of helminth infestation in sheep. South Australian aspects. *Aust. vet. J. 34* (1), 20–26.

(3) BOAZ, T. G. (1959). Sheep set-stocking experiment III. *Pub. AH/ 5/1/1959*, Dept. agric. Univ. Leeds.

(4) BROWN, T. H. & SPEDDING, C. R. W. (1962). *Exp. Grassld Res. Inst.* (Hurley), 14.

(5) CALDER, F. W., NICHOLSON, J. W. G. & CUNNINGHAM, H. M. (1962). Grazing systems for ewes and lambs. *Can. J. anim. Sci. 42*, 139–144.

(6) CALDER, J. W. (1961). Fat lamb farming in New Zealand. *N.S.B.A. Yearbk.*, 11.

(7) CAVE, W. E. (1961). N.Z. methods in mid-Wales. *N.S.B.A. Yearbk.*, 13–15.

(8) CHIPPENDALE, H. G. & MERRICKS, R. W. (1956). Gang-mowing and pasture management. *J. Brit. Grassld Soc. 11*, 1.

(9) COOPER, M. M. (1959). Sheep research at Cockle Park. *N.S.B.A. Yearbk.*, 5.

(10) COOPER, M. M. (1961). 'Creep' grazing of lambs. *Worcs. agric. Chron. 28* (2), 25.

(11) CROFTON, H. D. (1952). The ecology of immature phases of trichostrongyle nematodes. IV. *Parasitology 42* (1 & 2), 77–83.

(12) CROFTON, H. D. (1954). The ecology of immature phases of trichostrongyle nematodes. V. *Parasitology 44* (3 & 4), 313–324.

(13) CROFTON, H. D. (1957). Nematode parasite populations in sheep on lowland farms. III. *Parasitology 47* (3 & 4), 305–318.

(14) DICKSON, G. R. (1959). The more efficient grazing of ewes and lambs. *J. Brit. Grassld Soc. 14*, 172.

(15) EWER, T. K. (1962). The veterinary surgeon in relation to animal husbandry. *Vet. Rec. 74* (43), 1141–1148.

(16) GIBSON, T. E. (1952). The development of an acquired resistance by sheep to infestation with the nematode, *Trichostrongylus axei*. *J. Helminth. 26*, 43–53.

(17) GIBSON, T. E. (1956). The hazards of parasitic gastro-enteritis in sheep running under conditions of intensive stocking. *Emp. J. exp. Agric. 24*, 278.

(18) GIBSON, T. E. (1958). The development and survival of the preparasitic stages of *Nematodirus* spp. on pasture herbage. *J. comp. Path. 68* (3), 338–344.

(19) GORDON, H. MCL. (1957). Helminthic Diseases. *Advances in vet. Sci. 3*, 296.

(20) GORDON, H. MCL. (1958). The epidemiology of helminthosis in sheep in winter rainfall regions of Australia. 2. Western Australia. *Aust. vet. J. 34* (1), 5–19.

(21) HOLMES, W. (1962). Grazing management for dairy cattle. *J. Brit. Grassld Soc. 17*, 30.

(22) HUNTER, R. F. (1960). Aims and methods in grazing-behaviour studies on hill pastures. *Proc. VIII int. Grassld Congr.* (Reading), 454.

(23) JARRETT, W. F. H., JENNINGS, F. W., MCINTYRE, W. I. M., MULLIGAN, W. & SHARP, N. C. C. (1959). Studies on immunity to *Haemonchus contortus* infection—vaccination of sheep using a single dose of x-irradiated larvae. *Am. J. vet. Res. 20* (76), 527–531.

(24) JENSEN, P., MAPES, C. R., WHITLOCK, J. H. (1955). Pasture management and control of the lancet fluke (*Dicrocoelium dendriticum* Rudolphi, 1819). *Cornell vet. 45* (4), 526–538.

(25) LAMBOURNE, L. J. (1956). A comparison between rotational grazing and set-stocking for fat lamb production. *N.Z. J. Sci. Tech.* Section A., *37*, 555–568.

(26) MCMEEKAN, C. P. (1960). Grazing Management. *Proc. VIII int. Grassld Congr.* (Reading), 21.

(27) PARFITT, J. W. (1963). Host specificity of *Dictyocaulus filaria*. *Vet Rec. 75* (5), 124.

(28) SILVERMAN, P. H. & CAMPBELL, J. A. (1958). Studies on parasitic worms of sheep in Scotland. I. *Parasitology 49* (1 & 2), 23–38.

(29) Soulsby, E. J. L. (1961). Immune mechanisms in helminth infections. *Vet. Rec.* *73* (43), 1053.
(30) Southcott, W. H. (1955). Observations on the removal of *Oesophagostomum columbianum* Curtice from sheep grazing on green oats and pastures. *Aust. J. agric. Res. 6*, 456–465.
(31) Spedding, C. R. W. (1954). Production of worm-free lambs at pasture. *Nature (Lond.) 174*, 611.
(32) Spedding, C. R. W. (1956). The control of worm-infestation in sheep by grazing management. *J. Helminth. 29* (4), 179.
(33) Spedding, C. R. W. (1957). Grazing management and the control of parasites. *Agric. Rev. 11* (9), 24–27.
(34) Spedding, C. R. W. (1959 a). Grazing management for fat lamb production. *Esso Fmr. 11*, 4.
(35) Spedding, C. R. W. (1959 b). Grazing management for intensive lamb production. *N.S.B.A. Yearbk.*, 11.
(36) Spedding, C. R. W. (1961). The parasitological implications of intensive sheep grazing. *Vet. Ann.*, 32–37. Bristol: Wright.
(37) Spedding, C. R. W. (1962). The agricultural ecology of sheep grazing. *Brit. vet. J. 118*, 461–481.
(38) Spedding, C. R. W. & Brown, T. H. (1961). The effects of early weaning on the growth rate of lambs. *VIII int. Congr. Anim. Prod.* (Hamburg), p. 180.
(39) Spedding, C. R. W., Brown, T. H. & Large, R. V. (1960). Some factors affecting the significance of internal parasites in the utilization of grass by sheep. *Proc. VIII int. Grassld Congr.* (Reading), 718.
(40) Spedding, C. R. W., Brown, T. H. & Wilson, I. A. N. (1958). Growth and reproduction in worm-free sheep at pasture. *Nature (Lond.) 181*, 168.
(41) Spedding, C. R. W. & Large, R. V. (1959) Sideways creep-grazing for intensive lamb production. *J. Brit. Grassld Soc. 14* (1), 17.
(42) Spedding, C. R. W., Large, R. V. & Brown, T. H. (1961). The early weaning of lambs. *Vet. Rec. 73*, 1428.
(43) Stewart, D. F. (1950). Studies on resistance of sheep to infestations with *Haemonchus contortus* and *Trichostrongylus* spp. and on the immunological reactions of sheep exposed to infestation. 1, 2 and 3. *Aust. J. agric. Res. 1*, 285; 301; 413.
(44) Suckling, F. E. T. (1955). Set-stocking v. Rotational grazing of sheep on hill country. *N.Z. Sheep-farming Ann.*, 165–175.
(45) Taylor, E. L. (1939). Technique for the estimation of pasture infestation by strongyloid larvae. *Parasitology 31* (4), 473–478.
(46) Taylor, E. L. (1961). Control of worms in ruminants by pasture management. *Outlook on Agric. 3*, 141.
(47) Tetley, J. H. (1949). Rhythms in nematode parasitism of sheep. *Bull. N.Z. Dep. sci. ind. Res.* No. 96.
(48) Todd, A. C., Kelley, G. W. & Hansen, M. F. (1949). Winter survival of sheep parasites on a pasture in Kentucky. *Ky agric. exp. Sta. Bull.*, 533.
(49) Wannop, A. R. (1958). Sheep farming in New Zealand. *Proc. Brit. Soc. Anim. Prod.*, 101–103.
(50) Wheeler, J. L. (1960). Field experiments on systems of management for mesophytic pastures. *Div. Rep. 20, Div. Plant Industry*, C.S.I.R.O., Canberra.
(51) Wheeler, J. L. (1962). Experimentation in grazing management. *Herb. Abstr. 32*, 1.
(52) Andersen, F. L., Wang Guang-Tsan & Levine, N. D. (1966). Effect of temperature on survival of the free-living stages of *Trichostrongylus colubriformis. J. Parasit. 52* (4), 713–721.
(53) Brunsdon, R. V. (1963). Studies on the seasonal availability of the infective stages of *Nematodirus filicollis* and *N. spathiger* to sheep in New Zealand. *N.Z. J. agric. Res. 6*, 253–264.

(54) CONDY, J. B. & HANHAM, D. P. (1966). Nematode infestation of sheep in a high-rainfall area of Rhodesia. *Rhod. Zamb. Mal. J. agric. Res.* 4, 39–44.
(55) CONWAY, A. (1968). Intensive fat lamb production. *Span* 11 (1), 47–49.
(56) COOPER, M. M. & THOMAS, R. J. (1965). *Profitable Sheep Farming.* Ipswich Farming Press Books Ltd.
(57) CROFTON, H. D. (1965). Ecology and biological plasticity of sheep nematodes. I. *Cornell Vet.* 55 (2), 242–279.
(58) DONALD, A. D. (1967). A technique for the recovery of strongyloid infective larvae from small sample units of pasture. *J. Helminth* 51 (1), 1–10.
(59) DONALD, A. D. (1968). Ecology of the free-living stages of nematode parasites of sheep. *Aust. vet. J.* 44, 139–144.
(60) GIBSON, T. E. & EVERETT, G. (1968). A comparison of set-stocking and rotational grazing for the control of trichostrongylosis in sheep. *Brit. vet. J.* 124, 287–298.
(61) GIBSON, T. E. & EVERETT, G. (1968). Experiments on the control of trichostrongylosis in lambs. *J. Comp. Path.* 78, 427–434.
(62) GREENWOOD, E. A. N. & ARNOLD, G. W. (1968). The quantity and frequency of removal of herbage from an emerging annual grass sward by sheep in a set-stocked system of grazing. *J. Brit. Grassld Soc.* 23 (2), 144–148.
(63) HODGSON, J. (1966). The frequency of defoliation of individual tillers in a set-stocked sward. *J. Brit. Grassld Soc.* 21 (4), 258–263.
(64) JORDAN, R. M. & MARTEN, G. C. (1968). A note on the management of grazing, non-lactating ewes. *Anim. Prod.* 10 (1), 121–123.
(65) MORLEY, F. H. W. (1968). Pasture growth curves and grazing management. *Aust. J. exp. Agric. Anim. Husb.* 8, 40–45.
(66) MORRIS, R. M. (1967). *Pasture Growth in Relation to Pattern of Defoliation by Sheep.* Ph.D. Thesis, Reading University.
(67) ROSE, J. H. (1965). The rested pasture as a source of lungworm and gastro-intestinal worm infection for lambs. *Vet. Rec.* 77 (26), 749–752.
(68) RUSSEL, A. J. F. (1967). A note on goitre in lambs grazing rape (*Brassica napus*). *Anim. Prod.* 9 (1), 131–133.
(69) SMITH, H. J. & ARCHIBALD, R. MCG. (1965). Cross transmission of bovine parasites to sheep. *Can. Vet. J.* 6 (4), 91–97.
(70) WALKER, S. D. (1968). Some factors influencing the growth of lambs on irrigated pasture. *Proc. N.Z. Soc. Anim. Prod.* 28, 41–48.

X

FLOCK MANAGEMENT

In earlier chapters (II, III, IV and V) the more important attributes of the animal and plant populations involved in the conversion of herbage to sheep products have been described. The interactions which characterize the production process (Chapter VI) and the problems of practical application have also been discussed (Chapters VII, VIII and IX), primarily in relation to grazing.

Sheep do not necessarily graze during the whole year, however, and there are also important aspects of management which have no direct connection with feeding of any kind.

The disposition of the flock in space, in the course of a year, must take into account the marked seasonality of herbage production and the variation in requirement by the flock. Methods of grazing management apply to the area being grazed at any one time; it is a separate matter to determine which area should be grazed and at what time. The first consideration must be a quantitative appreciation of the conservation need.

CONSERVATION

If there is no intention to produce a surplus of conserved herbage for sale or use in other enterprises, then the amount conserved must equal the probable need plus a safety margin. Common forms of conservation, such as silage or hay, may involve considerable losses of dry matter and nutrients.

Hay and dried grass

Preservation of herbage by drying depends for its efficiency on the speed of the process and on careful handling. Fast drying gives little time for continued loss by respiration and careful handling can reduce the physical losses due to shattering of leaf fragments. Of course, the conserved product cannot easily be better than the material from which it is made and there is normally much scope for improvement in the quality of this. As mentioned in Chapter II, the maximum quantity of D.M. cannot always be harvested at the time when quality is at its highest. A compromise is often aimed at but the object should be to grow and harvest the maximum amount of herbage of whatever quality is required. The latter must be based on a knowledge of what the animal needs and how this varies with its size, stage of pregnancy or lactation and so forth.

Dried products might be expected to have some characteristics not shared by silage, quite apart from questions of handling. The absence of large quantities of water, for example, reduces the 'bulk' of the food and may affect intake. In cold weather it may also be an advantage that the animal does not have to heat up a lot of very cold water. Whether this is an advantage in fact depends to some extent on whether the animal on a dry feed has to take in, by drinking, similar quantities of cold water anyway.

Silage

Ensilage is rather more than a process of preservation, even when no additives are used. The quality of the silage made and the losses sustained, which may exceed 20 per cent of the dry matter ensiled*, are affected by the degree of control exercised over the fermentation process (19). Good silage can be made at temperatures between 24° and 38° C, but the aim is usually 25°–30° C and the production of a final product with a dry matter content of about 20 per cent and a pH value of about 4·0.

The losses sustained greatly affect the amount of herbage that must be harvested to satisfy the conservation need: they may also change the composition of the product and its completeness as a food. For a recent discussion of all aspects of conservation, see Wilkins (45).

Conservation Need

Conservation need is difficult to calculate because, although obviously dependent on the number of animals to be fed, every additional sheep on a farm affects both the amount available for conservation and the amount required. Whatever the distribution of herbage

* In addition to the loss of dry matter during ensilage, considerable wastage, at the sides and top of a stack, may bring the total loss of D.M. up to 30 or 40 per cent.

growth, at an extremely low stocking rate there may be no need for conservation. Late autumn growth left standing through the early winter ('foggage' or conservation *in situ*) may provide enough grazing until early growth commences in the following spring. However small the growth per acre, it may be adequate if the stocking rate is low enough. This is often a wasteful procedure but, theoretically, there is no reason why a range of herbage species and varieties should not be grown specifically to provide a sequence of crops in the required quantities throughout the year. In parts of the world with sub-tropical to temperate climatic conditions and a well-distributed rain-fall, it is possible to rely on natural pastures during the summer and on sown, temperate grasses during the winter. In this way, very little gap may remain, and the question becomes one of the optimum proportions of sown to natural pastures. It is still unlikely, however, that no surpluses will occur and, when they do, conservation is required. At higher stocking rates, in Britain, a period occurs in the winter when there is virtually no herbage available for grazing. Snow cover may ensure this in places, whatever the amount of grass on the ground. The length of this winter gap depends on the kinds of herbage being grown but it is normally longer at higher stocking rates. This kind of relationship between the stocking rate and the length of the winter gap emerges when considering a range of stocking rates from under- to over-stocking.

The proportion to be conserved

It is possible that, where the relationship between stocking rate and amount grown per acre is kept constant, the proportion to be con-served will remain approximately constant also. Thus if twice as much herbage is grown and the stocking rate is doubled, the con-servation need will be the same when expressed as a proportion of the total dry matter or total nutrients produced. This will only involve the same area if the distribution of grass growth remains unaltered, however. In this event, the proportion can be derived from empirical examination of situations in which the stocking rate has appeared to be about right, since any correctly stocked situation will require approximately the same proportion of conservation. An experimental example, of a stocking rate of 3 Half-bred ewes per acre per annum, gave the results shown in Table XL. In the example shown the 'grazed' component was estimated by sampling herbage to ground level before and after each grazing; the 'ensiled' portion was calculated from dry-matter contents based on samples taken as the grass entered the silage stack and from the total fresh weight of grass, which was measured in all cases. The total (in lb per acre) varied from year to year and, although the stocking rate of ewes was kept constant, the number of lambs present each season varied somewhat. Fertilizer

TABLE XL

DRY MATTER HARVESTED

From a dominantly perennial ryegrass ley under intensive sheep grazing.

YEAR	METHOD OF UTILIZATION	DRY MATTER IN LB/ACRE	D.M. CONSERVED (AS PERCENTAGE OF TOTAL)
1959	Grazed	4921	
	Ensiled	2338	32·2
	Total	7259	
1960	Grazed	3796	
	Ensiled	2343	38·2
	Total	6139	
1961	Grazed	4126	
	Ensiled	1206	22·6
	Total	5332	
1962	Grazed	2823	
	Ensiled	2525	47·2
	Total	5348	

(Expt. H. 202, Grassland Research Institute)

usage was the same each year but the botanical composition of the sward changed progressively and it will be noted that dry-matter production tended to decrease annually. It cannot be argued that this represented 'correct' stocking over the whole period, therefore, but the quantity of silage made was never excessive and always adequate. The grazing requirement of the flock was always given priority over conservation. Thus the proportion of the total dry matter harvested which was devoted to conservation may not be unrealistic in relation to a practical situation. The mean percentage conserved (35·0 per cent), over four years, is an indication of the high proportion of the total food consumed which may have to be conserved in order that a winter shortage may be avoided.

Naturally, the area required for conservation must depend on the growth rate of the herbage during the time when the conservation crop is grown. A given number of acre/days will be required at any given herbage growth rate. To some extent, this can be achieved by a large number of acres for a short time, with one cut, or a smaller number of acres for a longer time, with several successive cuts. It is not possible, therefore, to generalize about the proportions of the total area or time that conservation will demand. The proportion required will vary with the reproductive rate of the sheep, however, and may also be influenced by the size of the ewe (relative to the size of the progeny), the level of herbage production and the stocking rate (42, 43).

It has been calculated that the proportions of the annual herbage dry-matter production that would have to be conserved for 160 lb ewes with singles, twins and triplets, would be approximately 20 per cent, 16·5 per cent and 13 per cent, respectively (42).

Thus as the lambing percentage rises, so the proportion conserved diminishes, but the quality of the conserved product becomes more important.

The use of conserved herbage

The conserved material may be fed at any time of shortage, during a mid-summer drought, for instance, and may be used in addition to some grazing. Normally, it is fed in the winter; for most sheep in Britain, during the period mid-December to mid-March. For part of this period it may be the sole source of food; for still longer, in many situations, only negligible quantities of food may be obtained by grazing. In these circumstances, the method of wintering may take on added significance.

WINTERING METHODS

Wintering of hill sheep has meant either housing, movement of flocks and shepherds to lowland pastures, a process known as 'transhumance', sending sheep to grazings rented for the purpose ('tack') or simply survival on the open hill with the minimum of assistance.

Extra feeding

Recent work by the Hill Farming Research Organization has shown that supplementary feeding of hill ewes may result in improved lamb crops, chiefly by a reduction in early losses, particularly on the poorer hills of Scotland. It was found that such improvements, obtained from ewes lambing on the open hill, could be doubled where lambing occurred in a hill enclosure and Robinson (17) has suggested that lambing under close control, either on the hill or in-bye, is an ideal complement to supplementary feeding. It has been further established that supplementation can be successfully achieved on the hill without adverse effects on the foraging habits of the sheep. This is a good illustration of the commonsense view that a poorly fed sheep is likely to be an inefficient one. There are times, of course, when overfat sheep benefit from a restricted diet, but late pregnancy on a mountain-side is never likely to be one of them.

Housing

Housing, too, has been increasingly employed (1) in recent years Under lowland conditions, housing may be unnecessarily elaborate and, if bedding is available, yarding may be a simple alternative. Shelterbelts of trees may also be used (9). The housing of mountain sheep in Wales has been pioneered by Bennett Evans (5), using slatted floors raised about 5 ft above ground level. In-wintering ewe hoggs (this normally refers to sheep between the ages of 6 months and 1½ years) required only 5–6 sq. ft per hogg and no loss of hardiness has been observed on their return to the hill. The advantages have included greater live-weight gain, a lower mortality rate and more wool.

In other countries, housing has been used for many years. In Iceland, sheep may be housed for six months of each year and slatted floors are common. In France and Canada, straw bedding is allowed to accumulate. The space allowed per ewe varies with the size of the animal and, of course, whether it is pregnant and whether it is going to lamb indoors. Between 10 and 12 sq. ft would be more appropriate to lowland ewes than the 5–6 sq. ft mentioned above for Welsh Mountain ewe hoggs. Williams (20) has suggested 8 sq. ft in early pregnancy, increasing to 13–14 sq. ft later on, with an allowance of 1·6 ft of trough length per head.

In recent years, however, attitudes to the housing of sheep in Britain have changed a good deal.

Considerable experience has been gained by pioneer farmers operating on a substantial scale (such as J. B. Cadzow in Scotland and H. R. Fell in Lincolnshire) and from the experimental work of such men as Professor M. M. Cooper at Newcastle University and D. Hurst at the East Riding College of Agriculture in Yorkshire. Several commercial developments have also affected the cost of housing and great efforts have been made to devise very cheap structures, including the use of plastic houses (as pioneered by *The Farmer's Weekly*).

The sheep housing panel of the Farm Buildings Centre (40) has produced a first report in which are summarized recent findings and recommendations relating to most aspects of sheep housing. Full details are given with regard to the design, construction and fittings appropriate to different types of housing.

Diseases of housing

Important aspects of disease control are also dealt with: here general experience is reassuring. With proper precautions and a well-designed house, disease does not appear to present a barrier to such use of housing as may prove economic for other reasons. No major disease problems are yet apparent, although *E. coli* infection in young lambs has been serious in several instances and digestive troubles have been most common in housed hoggs.

Odd cases of pneumonia, foot-rot, orf, pregnancy toxaemia, abortion, urolithiasis, coccidiosis, navel ill, parasitic gastro-enteritis, and infestations with fluke and lice, have also been recorded, but the incidence has probably been no higher than in non-housed situations.

Experience with calves has probably led to somewhat exaggerated fears of disease, particularly respiratory problems, which seem rather less troublesome with sheep (35).

In spite of the fact that disease does not appear to be a serious barrier to housing, it is important to appreciate how big a disaster can rapidly follow if disease does become a problem. It is even more vital in the housed situation to emphasize the importance of preventive measures: curing disease in housed sheep can be both difficult and expensive.

Reasons for housing

Some of the reasons for housing sheep in winter, under British conditions, are concerned with protection of the sheep. The environment may increase the food requirements of sheep (3, 11) and cold may do so markedly when the weather is also windy (4). Nevertheless, there is little evidence of food-saving arising out of housing, largely because sheep out-wintered usually derive a surprising amount of their food from grazing. This results in housed sheep requiring *more* conserved food than those outside. Furthermore, there is usually no other way of harvesting the material collected by the out-wintered sheep.

More important aspects of sheep performance are related to sheer survival, in very harsh climatic conditions, especially of pregnant ewes, and the avoidance of lamb losses. Both of these aspects are greatly affected by the level of performance of the ewes, chiefly their prolificacy.

It is possible that housing would increase the longevity of ewes, or allow older ewes to be kept because they were not subjected to the same kind of stresses, or would reduce the rate of depreciation. There is as yet little precise information on these points.

Housing certainly makes shepherding easier, more comfortable and more efficient. This could be of very great importance and could greatly affect the economics of housing, if losses were thereby reduced or if it allowed one man to look after more ewes.

A major advantage of off-wintering (which may not necessarily imply housing) is often claimed to be greater herbage production from the areas not grazed in the winter (see next section and 'Poaching').

There is no doubt that there are many advantages in housing sheep; the main question is whether they can pay for the additional costs.

For a sheep-only farm, a useful way of calculating the probability of sheep housing being justified for intensive finishing of store lambs, has been proposed by Harkins (30). He has suggested the following equation:

$$M = 0{\cdot}47W_2P_2 - W_1P_1 - FR(W_2 - W_1) - D - V - H - L$$

where M = margin/lamb (in shillings)
W_1 = liveweight as store
W_2 = liveweight finished
P_1 = price per lb l/wt of store ($S.$)
P_2 = return per lb dead weight ($S.$)
F = food conversion ratio
R = cost of ration ($S.$/lb)
D = mortality ($S.$)
V = veterinary charges ($S.$/lamb)
L = labour charges where applicable ($S.$/lamb)
H = annual housing cost per lamb

This may be shortened to

$$M = SP - PP - FR(W_2 - W_1)$$

where SP = sale price and PP = purchase price

It is much more complicated to arrive at a comparable equation for the winter housing of ewes, especially for farms on which sheep are not the only enterprise. In the latter circumstances, it is most important to assess the whole-farm situation, and the possibility of alternative uses for the housing at other times of the year may determine whether it is justified economically or not.

Out-wintering

In relation to specially provided crops (such as turnips or mangels) which can be stored, the alternatives are clear. The sheep can harvest them or the crop can be handled and fed to them. In relation to herbage, the situation becomes very similar as soon as the contribution of grazed grass becomes negligible. In both situations, two aspects must be considered, the effect on the sheep and the effect on the soil and its crop. These are related because both are governed by the extent to which the feet of the sheep churn up the soil surface. Muddy conditions can result in hard lumps developing between the hooves and on the wool, to an extraordinary extent. Both can give rise to much extra shepherding work and, in addition, wool yield may be reduced and foot troubles perpetuated. Ewes' udders may also be damaged in these conditions. It is not surprising that sheep dislike mud; there is nothing whatever to recommend it for sheep.

Poaching

Poaching is a term used to describe the adverse effect of treading on the land. Plate IX (a) shows the damage done in 2 days by 24 ewes in a paddock of $\frac{2}{3}$ acre. The strip which was protected by a feeding trough stands out clearly: the rest of the area was in this same condition only 2 days before. Plate IX (b) shows the extent of this damaged region (in early April, 1963). Plate X (a) shows the much more severe poaching that has occurred around silage racks (seen in background) over a period of several weeks.

This kind of damage may occur every winter because it can take place so rapidly. Only a few wet days are needed on some soils for a heavy stocking rate of sheep to have these effects. On pasture, poaching implies not only some destruction of foliage and, perhaps, whole plants, but also a reduction in subsequent growth. There is no doubt that a situation can arise where heavy concentrations of sheep cause a severe reduction in both plant density and yield. The former may allow the development of weeds, both broadleaved and grass. Reduction of total yield may be accompanied by delay in production in the spring. This may mean that wintered areas give virtually no early spring growth and, when put up for silage, still give a much reduced harvest. Reductions of up to 50 per cent are not unknown.

Such damage as occurs may be very small at low stocking rates but of considerable importance at high stocking rates. The point at which it becomes significant varies with the heaviness of both soil and sheep. It may be unimportant, of course, on stubbles and other areas due to be ploughed up. Where it does matter, there would seem to be every case for avoiding it by removing the sheep during the time when no appreciable grazing would take place. Thus, even under the mildest of lowland conditions, yarding or housing may increase productivity. This is probable under intensive stocking because of the great importance of early spring grazing. If sheep are allowed to wander over a wide area, none of it may produce much early growth. So it seems preferable to damage a little of the area a lot rather than damage the whole area a little. The area wintered on thus has to carry an exceptionally heavy concentration of sheep and it is the weight of this and the period for which it is maintained that is of real importance on any class of land.

There is no doubt that treading can cause damage to the pasture, a reduction in yield and changes in botanical composition (23, 28). The difficulty is to assess the magnitude of the effects within relevant husbandry systems and the extent to which they are related to increasing stocking rates, for example.

Results from experiments at the N.A.A.S. Experimental Husbandry Farms suggest that winter grazing by sheep can sometimes have a marked effect on subsequent productivity but it is difficult to dissociate the effects of treading from other consequences of the presence of the sheep, such as defoliation.

Conway (26), studying intensive sheep production based on high nitrogen usage in Eire, has estimated the total effect of winter treading

PLATE IX

(a)

Close-up of poached area around a strip that has been protected by a trough. This damage occurred in two days.

(b)

The same field as shown above showing the extent of the damaged area.

PLATE X

(a)

The development of mud around silage racks under intensive sheep stocking.

(b)

Housed Ile de France ewes (in France) being turned out for a few hours' grazing.

to be small, involving a reduction of only about 2 per cent in the total dry matter produced, although a reduction of up to 25 per cent occurred on the one-twelfth of the total area that was affected.

Abrupt changes in diet

The consequences of housing or yarding (see Plate X (b)) include a tendency to abrupt changes in diet but, of course, need not do so. It may be more satisfactory to turn ewes out to pasture as they lamb, continuing to feed them some silage or hay, than to put them out for short periods each day. It is worth remembering that ewes in late pregnancy are highly sensitive individuals and metabolic disorders are often associated with abrupt changes in diet or with movement from one place to another (8). An enormous advantage of housing or yarding is the high degree of control that can be exercised over lambing, feeding and the general care of the flock.

CARE OF THE FLOCK

Certain aspects of flock management are so well established that they are often accepted without question. In preceding chapters it has been the aim to take nothing for granted, but to assess the importance of any attribute of pasture, animal or management by its relevance to the production process. The more important ideas about general care of the flock must therefore be mentioned: they are chiefly concerned with wool, feet and teeth.

Wool

Wool is one of the few animal products that is accessible and can be cared for as it grows. The importance of nutrition, worm-infestation and environmental influences, such as muddy ground, have been mentioned. The most frequent reason for attention to the wool is scouring (diarrhoea). Sheep often scour and, as pointed out in Chapter II, for a variety of reasons. It is not always indicative of something wrong, or at any rate significantly so, but it always soils the wool on the hind-quarters. Urine may do so too in female sheep.

The regular removal of such soiled wool ('dagging') is not done simply for the sake of the animal's appearance. It is done for two reasons, one specific and one general. The latter is that any soiled wool can obscure the view and thus prevent observation of what is going on underneath it. The specific reason is the avoidance of fly strike. The greenbottles and bluebottles that cause the trouble lay their eggs on the fleece and they are attracted to damp or soiled wool (16). When the eggs hatch, in some 24 hours, the maggots crawl down to the skin and start to feed on it and, eventually, under it. They cannot develop on dry wool, however, and require additional moisture in order to survive long enough to burrow into the skin, which then exudes fluids itself. Not only does soiled wool attract the flies and protect their progeny, it also keeps all this activity hidden from the shepherd. Behavioural indications were touched upon in Chapter VIII; these vary with the site of the strike, however, and this can occur anywhere on the body. Naturally, the effect on the animal's growth depends upon the severity of the attack. Any wound may be 'struck' by flies and great care must be taken if operations such as tailing and castration are carried out in weather conditions that favour blowfly activity. Joubert (10) has reported that the age at which Dorper lambs were castrated had no influence on either weight or quality of carcase: he concluded that the operation could be safely carried out at any stage between birth and four months. One of the many other factors bearing on the timing of such operations is the likelihood of fly strike. The destruction of the carcases of wild animals was at one time thought to be a useful control measure, but it is now considered that the primary blowflies chiefly multiply on living hosts (16). Maggots removed from sheep should therefore be carefully destroyed. The distances that these flies can cover (14) have been described in Chapter VIII.

Dipping was made compulsory in Britain to eradicate sheep scab (16). It is now being replaced in many counties by spraying. External parasites are not a major problem in lowland sheep in Britain, with the exception of fly maggots, and both dipping and spraying are now chiefly directed to the prevention of fly strike. Many different preparations are used and many give protection for some weeks.

Shearing

Shearing is usually done once a year but twice-yearly shearing has its advocates. Ewes are shorn in late May or early June in a spring-lambing flock. This is at a time when chills are unlikely but generally before very hot weather occurs: the lambs are big enough not to suffer from the disturbance that shearing causes. Pre-lambing shearing has been used with success in New Zealand and one reason may be the avoidance of the disturbance associated with shearing later, when ewes and lambs may have to be gathered over wide areas and in large mobs.

More frequent shearing does not appear to affect significantly the total amount of wool grown, but, in some parts of the world where wool is the main product, lambs are shorn at about 5 months of age (near weaning time). This is done in South America in flocks where lambing takes place in the winter and the ewes are shorn in early summer.

Considerable losses have occurred in Australia, in ewes immediately after shearing, due to cold wet weather. This is one of the best examples of a situation in which plastic coats can provide protection during a critical period and can reduce losses substantially.

A possible development for the future is the practice of chemical shearing (27), assuming that it proves economic and free from major side-effects.

Feet

A grazing animal needs healthy feet. Footrot has been mentioned in relation to foot-damage (Chapter V) and grazing management (Chapter IX). The effect of even mild footrot on lamb performance can be considerable. Infected feet can prevent a young lamb sucking and reduce grazing time in older sheep. In this way it can reduce the milk output of a ewe. Deformed feet can have similar effects and overgrown hooves can harbour stones, mud and faeces in amounts that are sufficient to cripple an animal. Regular trimming of the continuously-growing outer hoof is therefore essential under most grazing conditions, where natural wear is insufficient to keep the feet even underneath.

Regular footbathing in 10 per cent Formalin can control the spread of footrot and, in conjunction with a few other precautions, eliminate it. This has been demonstrated by Littlejohn (12, 13) and proved in practice. Forsyth (6) has discussed the problem in relation to other diseases and other treatments and described some of the simple, mechanical aids to handling the sheep during treatment.

Some of the difficulties of assessing the effect of footrot on productivity are well illustrated in the study by Littlejohn & Hebert (33). They found that in certain flocks the infected lambs also grew faster and it appeared possible that the faster-growing lambs were more susceptible.

TEETH

Sheep have both upper and lower molars but incisors only in the lower jaw. These incisors bite against a pad in the upper jaw. The pad consists of hard connective tissue with a superficial hornified layer (22). Loss of teeth is a serious matter for a ewe and is commonly used as a basis for culling. In Scotland, it has been found that ewes fed on turnips lost more teeth than those fed on hay or concentrates (17). At $4\frac{1}{2}$ years of age, only 37 per cent of the ewes fed turnips had normal incisor teeth, compared with 78 per cent and 68 per cent for ewes fed on hay and concentrates, respectively. Many ewes can live much longer than their teeth; and many can live and reproduce quite happily without any teeth. To have a few missing, however, is a distinct drawback. The extent of the disadvantage depends a good deal on the nature of the diet, and a dense, soft, lowland herbage is probably the easiest to manage if teeth are imperfect. There are many imperfections in teeth and jaw structure which can lead to inefficient grazing or cud-chewing; others accelerate the loss of teeth and shorten the length of the ewe's productive life.

It is possible by dentistry to adjust early imperfections in angle and arrangement so that teeth are not so readily lost and it has been suggested that regular treatment would be a worth-while investment. Markham & Lyle Stewart (15) have reported considerable success in using a technique designed to conserve teeth in sheep. They have also discussed the common types of occlusions met with in sheep.

In all probability intensive sheep-keeping will come to warrant this kind of individual attention.

Sheep teeth are probably best known as a means of gauging the age of sheep: Table XLI shows the age at which teeth erupt.

The importance of a full mouth of sound teeth varies, as stated, with diet: supplementary feeding has a particular significance because of the times at which it is usually given.

TABLE XLI

RELATIONSHIP BETWEEN AGE OF SHEEP AND DENTITION

Age	Number of permanent incisors	Sheep commonly called
6 months . . .	None (6 'milk' teeth)	Lamb
1–1½ years . .	2	2-tooth
1½–2 years . .	4	4-tooth
2½–3 years . .	6	6-tooth
3–4 years . . .	8	Full mouth

SUPPLEMENTARY FEEDING

This can take innumerable forms and it is not intended to discuss them in detail. It is not usual to include hay and silage in this category, but rather to mean foods which supplement the so-called roughages. The latter is an unfortunate term when applied generally to herbage products. It ignores the concentrated nature of young foliage and the low-fibre content which this may have. It may be a useful way of referring to bulky, fibrous foods of low nutritive value but it has caused so much confusion that it is probably better avoided. Supplementary feeding will, in this chapter, be used to describe anything fed in addition to herbage and herbage products. This is nothing more than a convenient grouping, however.

There are three main reasons for supplementing the herbage diet. First and foremost is the need to add to a quantitatively inadequate ration; second, to provide a more concentrated source of nutrients; and third, the correction of specific deficiencies. A possible fourth reason would be to achieve a substitution of one food for another, although this is rarely the object in practice. It is important to appreciate precisely why supplements are fed in a particular case and what other effects they may also have.

Shortage of food

In the simplest case, of an absolute shortage of food each day, additional food can be of any acceptable, balanced nature. The main reasons for it often being dry and concentrated are to facilitate its use, handling, rationing, storage and so on. An example of supplementation to correct a shortage is sometimes to be found in early lactation, particularly in a late spring. If herbage growth is delayed by low temperature, the amount available per ewe per day may be insufficient for her daily needs. If the shortage is simply one of available herbage, supplementary feeding may take many forms quite satisfactorily. If, on the other hand, no amount of available herbage would have met the daily needs of the sheep, the problem may be one of concentration.

The Need for Concentrated Food

When milk yield is high, the necessary supply of nutrients might require an intake of herbage in excess of the ewe's capacity. Assuming that the animal would, as it were, make every effort to satisfy its metabolic needs, such attributes as low digestibility or low dry-matter content might still limit intake to an inadequate level. It has been suggested that at certain times, as, for example, during late pregnancy, intake may be limited by the presence of foetuses. A ewe heavily in lamb might actually have a restricted space in which to put food. If this is so, then the greater the number of foetuses, the greater the need and the smaller the space. Deposits of internal fat may also influence the effective capacity of the rumen.

There are circumstances, then, when what is required is not simply extra food but nutrients in a concentrated form. This is an important distinction, particularly relevant to gestation and lactation. It also underlines how closely related are nutritive value and intake. As noted earlier, food can hardly be said to have a nutritive value at all if stock will not eat it. Even when they will eat it, however, its value is related to the amount that they can eat.

The correction of deficiencies

The correction of specific deficiencies can be undertaken in several ways. In some cases, notably magnesium in the early spring, a deficiency can be met by the incorporation of a substance (in this case, calcined magnesite at ½ oz a head per day) in the concentrate ration. 'Deficiency' is here used simply to describe a situation where an extra supply of the mineral prevents a disorder (hypomagnesaemia) characterized by low values for it in the animal tissues.

Administration of copper to the pregnant ewe for the prevention of swayback is another example. It is also a very good example, however, of the need for care in the provision of such supplements: this has been emphasized by Allcroft (2). Copper can be toxic to sheep in quantities that they will readily consume if given the opportunity (21). Sheep do not always know what is good for themselves or, at any rate, do not always avoid what is bad. The supply of minerals is best controlled, therefore, and a shortage remedied in as specific a manner as possible. Injection of a known quantity of copper is thus preferable to the provision of copper-containing licks to the flock. Common salt (as rock salt) is, however, unlikely to prove harmful and has often been considered an essential supplement for sheep.

A recent approach to mineral supplementation is to place in the sheep's rumen a pellet which will release the required mineral in approximately the right daily amounts. So far, cobalt and magnesium 'bullets' have been used and there is good reason to suppose that such methods will be the most satisfactory means of administration in certain circumstances. These will include range or hill conditions where daily administration would be difficult, laborious or impossible. It is sometimes necessary to place with the 'bullet' a small steel screw, the movement of which prevents the bullet from being coated and thus rendered ineffective (18).

Heavy cobalt oxide pellets have been used successfully in Yorkshire (45), resulting in improved liveweight gain of lambs, and the results of using a magnesium heavy pellet were described as encouraging by Ritchie *et al.* (37), although this has not always been so (32).

Among other specific deficiencies that may occur, that of vitamin D may be relatively common, and responses to injection of vitamin D_2 have been found in most ages of sheep. Deficiency may be related to lack of exposure of the skin to sunlight and is thus something that may be greatly influenced by such practices as housing (36).

The substitution of one food for another

The question of substitution is important in certain circumstances, even when this is not the object of supplementary feeding. Any substantial consumption of a supplement usually decreases the intake of herbage, unless this was initially inadequate. It was pointed out in Chapter V that both milk and concentrate feeding could affect the intake of grass and, consequently, the intake of infective larvae of parasites. The assessment of the value of supplements where herbage is available is often complicated both by the fact that less herbage may be consumed and by the differential influence of parasitism. Concentrates are often fed to lambs in the belief that a high plane of nutrition renders worms innocuous. There may be much to recommend the feeding of an uninfected supplement where parasites are numerous and the numbers ingested can be reduced by a substitution of one feed for another.

There are, of course, more specific additions to the sheep's diet directed at the prevention or control of disease organisms.

Vaccines and Drugs

It is not the intention here to catalogue the vaccinations and injections that can or should be used, but to indicate where they interact with various aspects of management.

Preventive medicine is much concerned with the successful integration of these methods and it is prevention that must be aimed at if the production process is to be efficient. The possibilities of disease elimination are further discussed in Chapter XIV.

Vaccination

Vaccination involves the introduction into the animal body of live or dead organisms treated in such a way that they produce an immune reaction to a disease without producing the disease itself, or at least only producing it in a very mild form. This reaction results in the production, in the sheep, of protective antibodies. Immunization with serum differs in that the antibodies themselves are injected: this confers a temporary protection. The routine injection of ewes and lambs for pulpy kidney (enterotoxaemia) is a good illustration of both procedures. In late pregnancy, the ewe is injected with a vaccine (derived from *Clostridium welchii*, the bacterium concerned). This results in antibodies which the ewe passes on to the lamb in the colostrum. The use of serum in the early days of life can give a similar passive protection to the lamb.

Great strides have recently been made in combining, in one injection, protection against all clostridial infections.

Anthelmintic drugs

Recent advances have also been made in the field of anthelmintics. It would be misleading to select a few from the large number of available worm-drenches, doses and pellets. Gibson (7) has published an up-to-date account of many of these drugs and their efficacy.

There are, however, some aspects of the use of such drugs that must be mentioned.

Anthelmintics can be restricted to a curative role and only used to reduce an excessive worm-burden, but the distinction between prevention and cure cannot be a sharp one. Curing one animal may be an essential part of preventing serious worm-infestation in another. Preventive use can be regarded, therefore, as involving administration before curative treatment is needed or, indeed, may ever be needed. Some of the factors concerned in the probability of serious worm-infestation have been discussed (Chapters V and IX): clearly some assessment of such probability is required before deciding whether the emphasis should be on prevention or cure. It is often sensible, though not of necessity economic, to take all possible steps to prevent excessive infestation by feeding and grazing management. There are circumstances, however, when a real and known risk has to be taken. At such times drugs can be used preventively, in some routine manner, or held in reserve against the need arising. The latter procedure has much to commend it, provided that the need can be promptly assessed.

Drugs are not all equally effective against all species of worm or all stages of development. Those that kill adult worms only, allow a new burden to build up virtually from the next day on. Most of the roundworms (nematodes) require about 3 weeks from ingestion before they reach maturity. Drugs that kill the immature stages too, therefore, may confer some three weeks' freedom from adult worms—or, rather, whatever degree of freedom their degree of efficiency allows.

Whatever the drug, unless it has been developed for persistent effects, more infective larvae are immediately ingested if the sheep stays in the same place. Anthelmintic treatment may therefore be combined with movement of the flock and there is no virtue in staying on the original field whilst worms are 'voided' as a result of anthelmintic treatment. Such adults as are passed out in the faeces will not infect sheep; this is limited to the third stage or infective larvae. In intensive grazing systems, sheep may be frequently moved anyway, but they cannot be moved to a clean area specially just because they have been dosed. Where this is possible, however, it is worth doing.

There is increasing evidence that a move to clean pastures after treatment is a major factor in reaping the benefit of dosing. Gibson & Everett (29) found that the best performance and lowest worm burdens were found in lambs dosed with thiabendazole at weaning time and then moved to clean pasture. The results were less satisfactory if the anthelmintic treatment was omitted and very poor if the lambs were left on the same pasture, whether they were dosed or not. Honer (31) has recently argued that the proper use of anthelmintics requires the achievement of a low infection-pressure and, where the level of infection is high, anthelmintics may make things worse.

As Thomas & Boag (44) have pointed out, the dosing programme should be related to whether the ewes or the pasture are the main source of infection. These authors found that lambs grew faster when their ewes were dosed to suppress the 'spring rise', a result also obtained by Nunns, Rawes & Shearer (34), and Brunsdon (25) in New Zealand. Continuous low-level feeding of thiabendazole has been used (38) to control helminthiasis in lambs and has resulted in improved performance, and Brunsdon (25) has found substantial increases in weight gain of lambs in New Zealand, associated with a 'strategic' drenching programme.

In further experiments with grazing lambs, Ross (39) found that a daily intake of thiabendazole of approximately 5 mg/kg body weight was effective in inhibiting egg production by the parasites. An intake of 8 mg/kg prevented establishment of over 90 per cent of infective larvae of *H. contortus*, *O. circumcincta*, *T. colubriformis* and *N. battus*, but was not 100 per cent effective against any of them.

As mentioned in Chapter IX, lambs develop an immunity to worms; they acquire resistance as a result of moderate exposure to infection. The normal lactation curve of the ewe tends to govern the lamb's grass intake, and thus in many cases its larval intake, so that both increase gradually. Stocking rate and management can greatly influence the process, and so can anthelmintic dosing. It may be argued that too early or too frequent treatment may prevent the development of resistance.

It seems logical to suppose that, if dosing is to be done, it is best done to a pattern based on a knowledge of the likely changes in the parasite population. An example of this pattern is shown in Fig. 10.1.

It is unlikely, though not impossible, that the mature ewe will suffer from worms sufficiently to require dosing on her own account, though this statement is a generalization for conditions of good nutrition and management. Even so, of course, there are exceptions, some of which were described earlier.

Under British conditions, the time when a ewe is most likely to suffer is during late pregnancy, when needs are great and food supplies may be low. Dosing at the beginning of the winter would then have merit in aiming to send the ewe into this period with as few

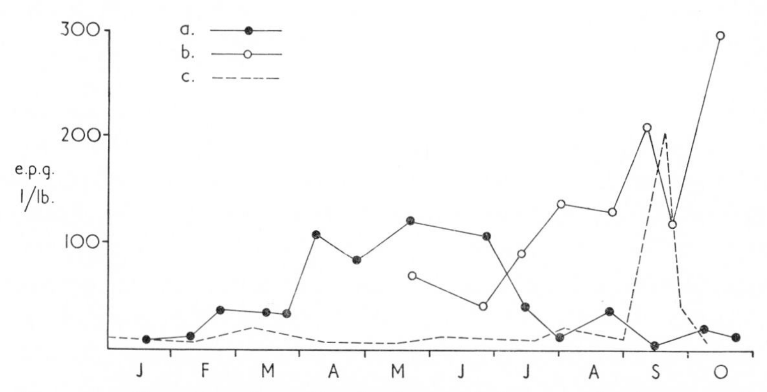

FIG. 10.1—*The pattern of parasitism for sheep stocked at 3 ewes/acre/annum, 1961*

(a) The number of small trichostrongyle eggs per g of ewe faeces
(b) The number of eggs per g of lamb faeces
(c) The number of infective trichostrongyle larvae per lb of herbage

(Source : data from Expt. H. 202, Grassland Research Institute, Hurley)

worms as possible. There is a point of general importance here, too. Dosing at the beginning of a period of prolonged cold or drought may have a persistent influence because reinfection may be slow or absent under such conditions.

Another reason for dosing ewes is, of course, to reduce the number of parasite eggs being deposited and thus the amount of infection available to the grazing lamb. Clearly, this should be aimed at the 'spring rise' (see Chapter IV).

Treatment of lambs appears unnecessary in the first 6 weeks and, except for *Nematodirus* spp., possibly for the first 10 weeks, although the use of drugs that affect immature worms may be justified earlier than those which are restricted to adults. Once begun, treatment may be warranted at intervals of about three weeks, unless grazing management renders this unnecessary.

Methods of administration vary, but usually involve labour. Careful handling may take extra time but it takes fewer risks and damaged palates can cause persistent trouble. Cost is a major factor in treatment but, from a husbandry point of view, voluntary intake from a trough would be most satisfactory, if safe. Vaccination for stomach, intestinal and lungworms could transform the problem of intensive lamb production at pasture, just as would a highly efficient broad-spectrum, easily-administered drug.

Although more efficient anthelmintic drugs are continually being produced, the occurrence of strains of nematodes resistant to them has also been recorded (40).

REFERENCES

(1) Aikman, C. D. (1962). In-wintering ewe hoggs. *Fmg Rev. 21*, 17.

(2) Allcroft, R. (1961). The use and misuse of mineral supplements. *Vet. Rec. 73* (47), 1255–1266.

(3) Armstrong, D. G., Blaxter, K. L., Graham, N. McC., & Wainman, F. W. (1959). The effect of environmental conditions on food utilization by sheep. *Anim. Prod. 1* (1), 1–12.

(4) Blaxter, K. L., Joyce, J. P. & Wainman, F. W. (1963). Effect of air velocity on the heat losses of sheep and cattle. *Nature (Lond.) 198* (4885), 1115.

(5) Evans, G. L. B. (1961). Wintering of hill hoggs. *N.S.B.A. Yearbk.*, 1961, 16–17.

(6) Forsyth, B. A. (1958). Footrot in sheep. *Outlook on Agric. 2* (2), 86–91.

(7) Gibson, T. E. (1962). Veterinary anthelmintic medication. *C.A.B. Helminth. Tech. Comm.*, 33.

(8) Hughes, L. E. & Kershaw, G. F. (1958). Metabolic disorders associated with movement of hill sheep. *Vet. Rec. 70* (4), 77–78.

(9) Hutchinson, J. C. D. & Bennett, J. W. (1962). The effect of cold on sheep. *Wool Technol. Sheep Breed. 9* (1), 11–16.

(10) Joubert, D. M. (1959). Effect of age at which lambs are castrated on carcase weight and quality. *Anim. Prod. 1* (2), 163–165.

(11) Lambourne, L. J. & Reardon, T. F. (1963). Effect of environment on the maintenance requirements of Merino wethers. *Aust. J. agric. Res. 14* (2), 272–293.

(12) Littlejohn, A. I. (1955). The use of formalin in the control of footrot in sheep. *Vet. Rec. 67*, 599.

(13) Littlejohn, A. I. (1961). Field trials of a method for the eradication of footrot. *Vet. Rec. 73*, 773.

(14) Macleod, J. & Donnelly, J. (1963). Dispersal and interspersal of blowfly populations. *J. Anim. Ecol. 32* (1), 1–32.

(15) Markham, J. H. A. & Stewart, W. L. (1962). Dental conservation in sheep. *Vet. Rec. 74* (36), 971–978.

(16) Newsom, I. E. (1952). Sheep Diseases. Baltimore, The Williams & Wilkins Co.

(17) Robinson, J. F. (1958–61). 2nd Rep. Hill Farming Res. Org., 17–22.

(18) Underwood, E. J. (1962). *Trace Elements.* 2nd ed. New York and London, Academic Press Inc.

(19) WATSON, S. J. & NASH, M. J. (1960). *The Conservation of Grass and Forage Crops.* 2nd ed. 758 pp. Edinburgh and London, Oliver & Boyd.
(20) WILLIAMS, S. (1962). Intensive fat lamb. *Agriculture (Lond.)* 69 (9), 426–430.
(21) ROSS, D. B. (1964). Chronic copper poisoning in lambs. *Vet. Rec.* 76 (32), 875–876.
(22) WEINREB, M. M. & SHARAV, Y. (1964). Tooth development in sheep. *Am. J. Vet. Res.* 25 (107), 891–908.
(23) BROWN, K. R. (1968). The influence of herbage height at treading and treading intensity on the yields and botanical composition of a perennial ryegrass-white clover pasture. *N.Z. Jl agric. Res.* 11, 131–137.
(24) BRUNSDON, R. V. (1966). Further studies of the effect of infestation by nematodes of the family *Trichostrongylidae* in sheep: an evaluation of a strategic drenching programme. *N.Z. vet. J.* 14 (7), 77–83.
(25) BRUNSDON, R. V. (1966). Importance of the ewe as a source of trichostrongyle infection for lambs: control of the spring-rise phenomenon by a single post-lambing anthelmintic treatment. *N.Z. vet. J.* 14 (8), 118–125.
(26) CONWAY, A. (1968). Intensive fat lamb production. *Span,* 11 (1), 47–49.
(27) DOLNICK, E. H. (1968). Shearing sheep with chemicals. *Agric. Res. U.S.D.A.* 17 (4), 8–9.
(28) EDMOND, D. B. (1966). The influence of animal treading on pasture growth. *Proc. X int. Grassld Congr.* (Helsinki), 453–457.
(29) GIBSON, T. E. & EVERETT, G. (1968). Experiments on the control of tricho-strongylosis in lambs. *J. Comp. Path.* 78, 427–434.
(30) HARKINS, J. (1968). Assessing new capital investment on hill sheep farms. *Scot. Agric.* 47, 196–200.
(31) HONER, M. R. (1968). An experimental analysis of the strategic treatment of sheep against gastro-intestinal strongylides. *World Rev. Anim. Prod. IV* (16), 35–42.
(32) HVIDSTEN, H. (1967). Studies on hypomagnesaemia in sheep as influenced by fertiliser treatment of pasture. *Zeitschrift fur Tiersphysiol. Tierernahr. Futter-mittelk.* Bd. 22, H. 4, S., 210–219.
(33) LITTLEJOHN, A. I. & HEBERT, C. N. (1968). Foot-rot in unweaned lambs. *Vet. Rec.* 82 (24), 690–695.
(34) NUNNS, V. J., RAWES, D. A. & SHEARER, G. C. (1965). Strategic anthelmintic medication of ewes. *Vet. Rec.* 77 (12), 328–332.
(35) PARKER, W. H. (1968). Housing of ruminants. 1. Its advantages and its health problem. *Vet. Rec.* 83, 208–210.
(36) QUATERMAN, J., DALGARNO, A. C. & ADAM, A. (1964). Some factors affecting the level of vitamin D in the blood of sheep. *Brit. J. Nutr.* 18, 79.
(37) RITCHIE, N. S., HEMINGWAY, R. G., INGLIS, J. S. S. & PEACOCK, R. M. (1962). Experimental production of hypomagnesaemia in ewes and its control by small magnesium supplements. *J. agric. Sci.* 58, 399–404.
(38) ROSS, D. B. (1965). Continuous low-level feeding of thiabendazole to control helminthiasis in lambs. *Vet. Rec.* 77 (24), 672–674.
(39) ROSS, D. B. (1968). Further experiments with low-level feeding of thiabendazole to lambs. *Res. vet. Sci.* 9, 515–520.
(40) Sheep Housing Panel (1967). *Housing Sheep.* Farm Buildings Centre, Stoneleigh Abbey, Rep. No. 7.
(41) SMEAL, M. G., GOUGH, P. A., JACKSON, A. R. & HOTSON, I. K. (1968). The occurrence of strains of *Haemonchus contortus* resistant to thiabendazole. *Aust. vet. J.* 44, 108–109.
(42) SPEDDING, C. R. W. (1967). Practical implications of genetic and environmental influences: Sheep. Pp. 451–465 in *Growth and Development of Mammals.* Ed. G. A. Lodge & G. E. Lamming. (Proc. 14th Easter Sch. Nottingham) 451–465. London, Butterworths.

(43) SPEDDING, C. R. W. (1969). The consequences of intensive stocking rates on high-nitrogen pastures: Sheep. *J. Sci. Fd Agric.* In press.
(44) THOMAS, R. J. & BOAG, B. (1968). Roundworm infestation in lambs. *J. Brit. Grassld Soc.* *23* (2), 159–164.
(45) WATSON, W. A., BODEN, S. M., STOBBS, A. W. & RUTHERFORD, A. (1966). The use of heavy cobalt oxide pellets in the prevention of unthriftiness in lambs in Yorkshire. *Vet. Rec.* *79* (10), 276–280.
(46) WILKINS, R. J. (1967). Fodder conservation. Ed. R. J. Wilkins. *Occ. Symp. No. 3, Brit. Grassld Soc.*

XI

SUSTAINED PRODUCTIVITY

One of the most important aspects of productivity per unit of food is the time for which any particular rate can be maintained. It is obvious that the assessment of rates of production over a short period may be no guide to the actual or possible achievements over a longer period. There are, however, several reasons for this.

First, the conditions governing productivity may vary greatly with time. Thus growth of grass in May is a poor guide to that in July or December because some of the major factors determining growth rate, such as light intensity, temperature and day-length, are quite different at these times. It is not inevitable that they should be so, but, in general, where agricultural processes take place against a background of natural variation, plant growth is seasonal. There is no reason why one should not discuss, or experiment within, quite short periods for many purposes. Sustained productivity, however, is concerned with long periods and has to take into account the variation between years as well as that within them. For plant growth, the year would seem a minimal unit, from this point of view. Equally, the rate of animal production varies with the herbage supply and the age and size of the animal; it also varies with time, therefore. But the minimal period may not be one year. From some points of view it could be an animal's lifetime: from others, it could be linked to the reproductive rate or frequency. In many sheep-farming systems, where ewes lamb once each year, a whole year is a convenient unit for both plant and animal production. If ewes lamb rather more frequently, however, the interval between one mating and another may be more relevant.

Secondly, the accuracy of many determinations may be limited by the shortness of the period studied and the production measured may therefore be a poor sample of the true rate.

Thirdly, many consequences of the particular rate of production measured may not be evident in the short term. If they are not evident they may never be considered, let alone measured.

The long-term effect of such associated changes may be extremely important. The most obvious example of this is age itself. No one would calculate sustained productivity from the performance of one ewe in one year. Quite apart from the other obvious drawbacks, the ewe ages in one year, although the changes associated with aging may not be evident. Given time, of course, the outcome can hardly be ignored. The ewe dies and it is therefore always accepted that she will; the time she takes to do so is less often noted, yet every rate of production with which she is associated during her life must be assessed in terms of a certain length of productive life. This is why longevity must be regarded as a variable in the production model (see Chapter VI) and not simply as a framework within which other variables operate.

There are many changes which can affect long-term productivity, only the more important will be considered here. Broadly speaking, they can be grouped into those that occur in the animal itself and those that form part of its environment.

Animal Factors

These are chiefly relevant to breeding stock. In the very long term, account has to be taken of genetic changes which may have been favoured by methods of husbandry and which could affect subsequent productivity. There are, for example, arguments both for and against the proposition that sheep should be reared in the same kind of environment as that for which they are bred and in which they or their progeny will have to live and produce. Some aspects of the relationship between production and breeding are discussed in Chapter XII.

Sustained productivity by an animal population concerns the question as to whether any particular rate can be maintained and it involves consideration of the effect of any particular rate of growth, performance or production on the rate in some subsequent period. It is necessary to distinguish, however, between the effect of the rate itself and any continued effect of the factors which caused it. The more important interactions between what happens in one part of a lamb's life and what happens subsequently are dealt with in Chapters V and IX: those for the ewe are mainly considered in Chapter IV. The implications of rate of growth as such are discussed in Chapter VI, but the practical difficulty is to describe in sufficient detail the associated changes. Two kinds of change are possible. The animal, producing at any given level, in an environment which, if not constant, is similar from one major period to another, say from year to year, may become increasingly deficient in some nutrient or may gradually accumulate toxic substances. Both of these processes can occur in relation to major or minor elements.

Trace elements

The productivity of a ewe flock might be sustained for a time against a background of declining liveweight. Whether this can actually occur depends somewhat on the method of computing productivity, but, whatever the method, it can only operate on measured changes. This is why the animal problems of sustained productivity are often related to imperceptible changes in such things as trace elements. Probably the most important of the mineral deficiencies are associated with copper, magnesium and cobalt.

Similarly, sheep may accumulate elements, such as copper and lead, in amounts which may ultimately prove toxic. It is often considered that a ewe may also accumulate fat to the detriment of her subsequent productivity.

It will be evident that this distinction between changes in the animal itself and those occurring in its environment is not a very sharp one. If the environment supplied all the sheep's requirements and supplied nothing that could adversely affect animal health, there is little reason to suppose that flock productivity at any initial level could not be sustained.

Stress

The possibility exists, however, that what is sometimes called 'stress' may have a cumulative effect. It has been suggested, for example, that an animal may suffer a stress at high-performance levels even when fed adequately for them.

As mentioned in Chapter IV, breeding from ewe lambs, mated at 6 months of age, does not appear to reduce life-time performance. As may be seen from Table XLII, the performance of Half-bred ewes bred from as lambs was in no way inferior to that of ewes which had their first lambs at 2 years of age. It is of considerable interest that, in both the groups of ewes studied, the mean litter size and average daily gain of lambs, up to 76 days, remained nearly constant from year to year. Those ewes which survived to 7 years of age were still performing at the same rate as the mean for younger age groups. The exception illustrated is the fleece weight, which declined with advancing age. The importance of the length of productive life to the over-all rate and efficiency of production has already

TABLE XLII

THE EFFECT OF AGE ON THE PERFORMANCE OF EWES

	Age (in years)						
	1	2	3	4	5	6	7
Litter size*							
LB† . . .	1·28	1·80	1·87	1·97	1·96	1·98	1·95
SB . . .	—	1·75	1·86	1·85	1·99	1·96	1·77
Fleece wt (lb)							
LB . . .	7·52	7·74	7·69	7·34	7·04	6·15	5·66
SB . . .	8·87	8·48	8·01	7·27	6·15	5·41	4·71
Mean daily L/wt gain (lb) of lambs‡ from birth to 76 days.							
LB . . .	0·64	0·74	0·70	0·71	0·72	0·71	0·72
SB . . .	—	0·74	0·74	0·74	0·74	0·71	0·73

* Litter size: number of live lambs born per ewe put to the ram.
† LB: ewes which first lambed at 1 year old.
SB: ewes which first lambed at 2 years old.
‡ Mean of all lambs (singles and twins).

(Information from records of Border Leicester × Cheviot ewes kept at High Mowthorpe Experimental Husbandry Farm and analysed within the School of Agriculture, King's College, University of Durham. Source : Bichard, M. & Yalcin, B. C. (1))

been stressed. Pàlsson (10) has stated that Icelandic sheep live until 11 years old: British sheep do not normally continue to produce for anything like this time.

Loss of teeth

A common reason for a reduction in the length of productive life is loss of teeth: this has already been referred to (see Chapter X). The work of the Hill Farming Research Organization in Scotland has included long-term studies of productivity under hill conditions and the importance of the relationship between diet and defective teeth has been stressed (11). This relationship may be complicated and susceptible to the influence of plane of nutrition as well as the nature of the diet. It may thus, incidentally, be a factor in the adaptation of breeds to particular environments. McManus (9) found 60 per cent more tooth wear in ewes stocked at 9 to the acre than in those at a stocking rate of only 2 per acre. The grazing behaviour, e.g. closeness of grazing, of different breeds (see Chapter XII) could thus influence the problem. These behaviour patterns are often related to the whole environment and to differences in the dispersion of food supplies, but dietary differences and differences in grazing pressure are also involved. Ruane (12), for example, has noted that Roscommon sheep graze as a group, as do many breeds of the arable or lowland regions, whereas Galway sheep spread out over the whole area, a behaviour pattern characteristic of heath or mountain sheep. In the latter, grazing behaviour patterns are commonly associated with large differences in the nature of the diet.

Lactation

Changes in dentition are frequently irreversible: the udder, by contrast, develops afresh each year. There is little evidence, for sheep, whether there is any relationship between the pattern of lactation in one year and that in another. The fact that lactation occurs in the ewe bred from as a lamb presents an opportunity to see whether later lactations are affected by this. Unfortunately there are many other associated changes and it is difficult to distinguish between them. Similarly, ewes which are barren and therefore non-lactating in one year, may differ in many ways from their flock mates when they next all lactate. Even less is known of any relationship between the magnitude or length of one lactation and that of subsequent lactation curves. The plane of nutrition during rearing has been shown to affect subsequent performance in cattle and clearly it could influence the kind of animal produced. This is not to say that ultimate, mature size or proportions would necessarily be altered but that these attributes could certainly be different for part of the animal's life.

Compensatory growth

Compensatory growth is a phenomenon of this kind. It is a term used to describe the acceleration in growth rate which often occurs when a period of poor nutrition is followed by an improved level of feeding. The animal appears to catch up to some extent, growing faster than if its normal progress had remained uninterrupted. This is illustrated by the growth of early-weaned lambs (Fig. 5.7). In this case, growth rates increase with time but little is known of any changes in the efficiency with which food is converted. Davies & Owen (19) found that restricted milk feeding, during artificial rearing of lambs, reduced growth rate but also greatly reduced feed costs. They pointed out that the check at weaning might be greater to a lamb fed on liquid milk substitute *ad lib.* Subsequently, however (20), it was demonstrated that lambs which had received a restricted quantity of milk replacer, also grew more slowly and had a poor food-conversion efficiency during the subsequent feeding period (from 13·6–34·1 kg liveweight). The hypothesis of mammalian growth proposed by Dickinson (4) assumed that temporary advantage may be taken by the animal, during growth, of variations in the environment, when the latter exceeds a minimal level demanded by the genotype for stable development to normal mature size. Permanent stunting may result, however, if less than this minimal level is provided.

Reproduction

In general, the breeding ewe is productive, apart from the growth of wool, for only part of the year, during gestation and lactation. Within this period most ewes are at a high rate of production for only about 4 months. There is thus a considerable time interval in which deficiencies can be remedied, tissues rested and condition restored, if need be. The kind of questions posed in this section might become more acute in a situation where the ewe was more productive or productive over a longer period.

Obviously it cannot be assumed that because, for example, one ewe can produce twins or triplets, all ewes can do so. Even less can it be assumed that, because one ewe can lamb twice a year, all ewes can do so and produce 3 lambs each, on each occasion. In other words, rates of production over different periods cannot simply be summed, although this may well provide a hypothesis worth testing. The reasons why such summation may be invalid would include those already indicated, some of which might be eliminated if nutrition was adequate in all respects. Others might be related to the possibility of stress due to high production as such. As Maynard &

Loosli (8) concluded: 'there may well be no conflict between rapid growth and length of productive life, provided the growth obtained is complete and correlated in all its aspects'.

Many of the factors bearing on sustained productivity, however, are, or result from, changes in the total environment of the sheep.

Environmental Factors

Under grazing conditions, significant environmental changes can occur as a consequence of some productive process being carried out for a prolonged period only if this is done on the same area. This is an important distinction because it suggests that any adverse consequences, other than those directly concerning the animal, can be avoided by a suitable rotation of land. This is not so different, of course, from the ideas behind rotational grazing management (see Chapter IX) over a short period.

The alternation of crops in such a way that no crop is grown in two successive years on the same ground, or such that there is a long interval between one period in a crop and another period with the same crop, has long been used to avoid the adverse consequences of plant monoculture. These consequences include exhaustion of the plant nutrients particularly required by one crop and the build-up of pests and diseases. The situation is very similar for animal production at pasture. Consequences are expressed either as a reduction in the quantity or quality of herbage grown or as an accumulation of parasites and diseases of the animal.

Changes in botanical composition

Changes in grass crop production, however, may involve considerable changes in botanical composition. The continued presence of the grazing animal may alter very markedly the number and proportions of grass, clover and weed species present (see Table XLIII). The direction in which the pasture is changed varies with the type and numbers of stock, with their management and with the level of fertilizer application. Table XLIV gives a simple illustration of the botanical changes which can occur in a ryegrass/white clover sward heavily stocked with sheep for 3 years. Such changes, while not inevitable, may greatly affect the amount of herbage produced, the

TABLE XLIII

BOTANICAL CHANGES IN A PERMANENT DOWNLAND PASTURE

After six years' controlled sheep grazing with applications of N, P and K.

MONOCOTYLEDONS	PER CENT CONTRIBUTION TO COVER* OF THE SWARD		DICOTYLEDONS	PER CENT CONTRIBUTION TO COVER OF THE SWARD	
	1952	1958		1952	1958
Grasses					
Festuca rubra	20·2	32·7	*Trifolium repens*	0·6	7.2
Agrostis stolonifera	20·0	3·5	*Medicago lupulina*	1·4	—
Arrhenatherum elatius	9·5	7.5	*Ranunculus bulbosus*	8·1	0·1
Dactylis glomerata	2·7	4·9	*Leontodon* spp.	4·9	4·0
Poa pratensis	2·3	0·1	*Prunella vulgaris*	4·5	0·1
Poa trivialis	—	14·9	*Plantago lanceolata*	4·0	1·3
Holcus lanatus	2·1	2·7	*Poterium sanguisorba*	1·6	—
Lolium perenne	0·9	13·6	*Crepis capillaris*	1·0	0·9
Other grasses	4·2	0·8	Other herbs	7·9	5·8
Total grasses	61·9	80·7	Total herbs	34·0	19·4
			Total herbage	100·0	100·0
Sedge					
Carex flacca	3·6	—			

(Source : D. D. Kydd from Experiment H. 23, Grassland Research Institute)

Also associated with these treatments was an increase in herbage yield (the amount of dry matter being doubled) and an increase in the number of sheep grazing days, from 883 in 1952 to 2821 per acre in 1958.

* Cover = ground cover or total area of ground covered by herbage.

TABLE XLIV

BOTANICAL CHANGES IN A PERENNIAL RYEGRASS/WHITE CLOVER LEY

After establishment and three years' intensive sheep grazing.

SPECIES	PER CENT CONTRIBUTION TO COVER* OF THE SWARD	
	1959	1962
Lolium perenne (S. 23)	80	47
Trifolium repens (S. 100)	20	1
Poa (*annua* and *trivialis*)	—	52
Total	100	100

(Source : D. D. Kydd, from an experiment (H. 202) at the Grassland Research Institute, Hurley)

* Cover = ground cover or total area of ground covered by herbage.

seasonal pattern of distribution and the nutritive value of the product. Some of these changes are very noticeable, others much less so. A reduction in the proportion of clover, for example, makes a big visual impression; it is less obvious and, indeed, extremely difficult to determine exactly in what way this may have altered the nitrogen status of the soil.

Fertilizer application

The possibility of exhaustion of trace elements has often been considered in relation to continuous farming of one kind, where it has been assumed that *gross*, and therefore obvious, nutrient removal would be compensated by replacement as fertilizer. In fact, there are two risks here: a shortage of some trace element may develop, or an accumulation of another element may occur. The former possibility indicates that relatively impure fertilizers may have some value unless all requirements are known in detail. The latter indicates the risk of continually adding ill-defined constituents to the soil. It is well known that apparent copper deficiency in the sheep may be associated with high levels of molybdenum and inorganic sulphate in the herbage. Mention was made in Chapter II of the role of potash in depressing the magnesium content of grass. Since the rate of growth of herbage, the stage at which it is eaten and the weather may all influence the consequences to the animal of such changes, these may only gradually become apparent. The incidence of metabolic disorders, for example, even if increasing with time in these circumstances, would not be expected to do so in a uniform manner.

Continuous stocking and 'sheep-sick' land

Similar considerations apply to parasitic diseases, many of which have been regarded as prime components of what is usually called 'sheep-sick' land. The latter term is used to describe a condition, characteristically arrived at after several successive years of sheep grazing, where sheep no longer thrive. This traditional concept of the dangers of continuous sheep-keeping cannot, at the present time, be completely dismissed, explained or endorsed. Yet the question is one of considerable importance. If sheep-keeping on the same ground year after year is impossible or inefficient, it is a matter of some moment to know the reasons for this. Too little experimentation has been carried out on the subject to make categorical statements possible. It is worth considering, however, what is known of the relevant factors.

First, it must be recognized that much confusion can be attributed to a failure to distinguish between the various forms of so-called continuous, i.e. year after year, stocking. Very rarely is it implied that every acre has sheep on it every day of the year. Sometimes it means that every acre is grazed by sheep at some time during the year: sometimes other stock also graze the same acres.

Generally, the problem of continuous stocking is centred on a sheep-only enterprise, in which sheep are the only grazers but in which cutting by machine may occur. One of the major distinctions that should probably be made is between systems of husbandry that keep the sheep on the same land during both winter and summer and those that are only concerned with successive summer grazings. This is, in fact, a major factor in the effect of stocking rate on the consequences of continuous stocking.

Where land has been successfully occupied by sheep for the whole year and for many successive years, it has usually been at a low stocking rate. A good example of this is provided by hill and mountain grazings. Productivity per unit of land is typically low in these circumstances. Where stocking rates have been high for many years, either the sheep have been removed from the land during the winter, thus reducing the stocking rate per annum, or the proportion of older sheep has been high. There have been good examples of both on the Romney Marsh and elsewhere. Where stocking rates of young sheep have been continuously high, supplementary feeding has often been at a high level also. Nevertheless, there exist exceptions where success has been achieved in circumstances little different, apparently, from others in which the health and performance of sheep has declined severely.

Some of the implications of supplementary feeding have been discussed (see Chapter X) and it will be evident that a sufficiently high level of feeding could virtually eliminate the problems of continuous sheep grazing since little would occur or be needed.

The evidence is considerable, however, that continuous, even heavy, stocking with mature sheep carries negligible risk compared to that associated with young animals. There are exceptions to this proposition, but most of them relate to overcrowding or to underfeeding, and this latter, generally, in relation to periods when the young are being supported Additional feeding of the lamb or of the ewe during lactation could so reduce the lamb's direct dependence on its pasture environment that, if mature sheep can be carried continuously, lambs could also. This approach to 'feedlot' conditions, where the land is simply an exercise ground, need not be restricted to non-herbage feeds. Zero-grazing is, in effect, a system for feeding grazing animals on ground which does not produce their food. This destroys many of the opportunities for certain disease-producing organisms to accumulate.

There are diseases, particularly those associated with parturition, that may increase with time and affect the mature ewe, though often only in her reproductive capacity. These may be easier to avoid or eradicate where the acreage is large than where the stock are restricted to small areas. Where sheep are penned, yarded or housed, there are, of course, efficient means of disinfection available to deal with this kind of problem.

As far as the lamb is concerned, internal parasites have frequently been responsible for the problems of continuous stocking with sheep. Nematodiriasis illustrates this in a remarkably clear-cut fashion.

NEMATODIRUS INFESTATION

Nematodirus spp. do not commonly cause disease in older sheep (5). In lambs, they do so quite frequently and, when they do, the disease occurs earlier in the lamb's life than do the troubles due to other worm species. The adult worms lay eggs which, on being voided in the faeces, take a considerable time to hatch and produce infective larvae. Eggs may be ingested by the lamb and occasionally a few worms may become established in this way. Some infective larvae may emerge in shorter than average periods and actually reinfect a lamb in the same season. In general, however, the eggs dropped on to the pasture in one season will not result in infection of lambs until the following year (18). Gibson (6) has found that the eggs of *N. filicollis* may hatch within 8 weeks, resulting in a constant level of larval infestation on herbage from late August onwards. This might be of great importance where lambs were born at that time of the year. For spring-born lambs, however, the level of pasture infection in the autumn is of less consequence because they have normally become resistant by this time (5). Fig. 11.1 illustrates the pattern of infestation for a population consisting of 90 per cent *Nematodirus filicollis* and

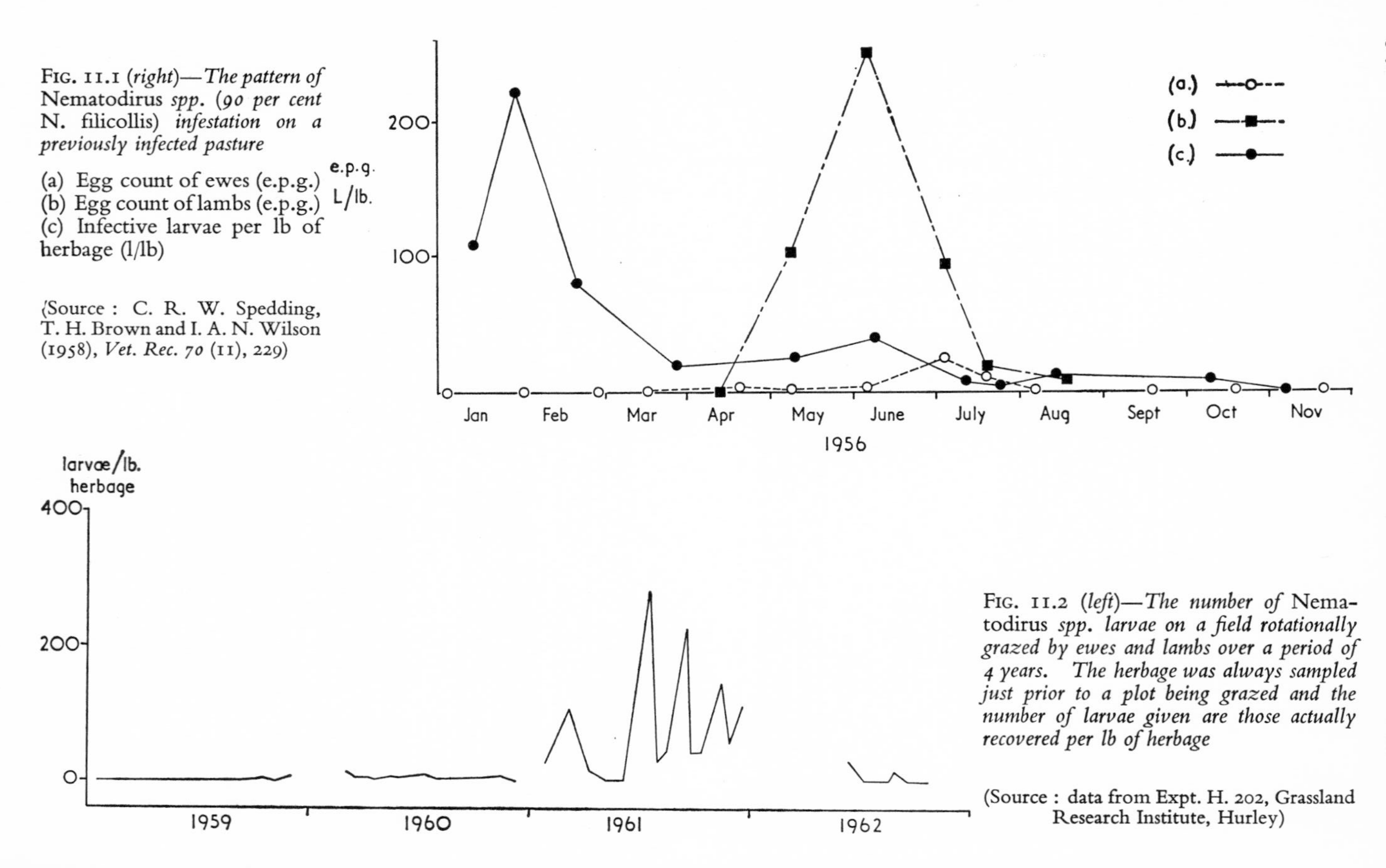

FIG. 11.1 (*right*)—*The pattern of* Nematodirus *spp. (90 per cent* N. filicollis) *infestation on a previously infected pasture*

(a) Egg count of ewes (e.p.g.)
(b) Egg count of lambs (e.p.g.)
(c) Infective larvae per lb of herbage (l/lb)

(Source : C. R. W. Spedding, T. H. Brown and I. A. N. Wilson (1958), *Vet. Rec.* 70 (11), 229)

FIG. 11.2 (*left*)—*The number of* Nematodirus *spp. larvae on a field rotationally grazed by ewes and lambs over a period of 4 years. The herbage was always sampled just prior to a plot being grazed and the number of larvae given are those actually recovered per lb of herbage*

(Source : data from Expt. H. 202, Grassland Research Institute, Hurley)

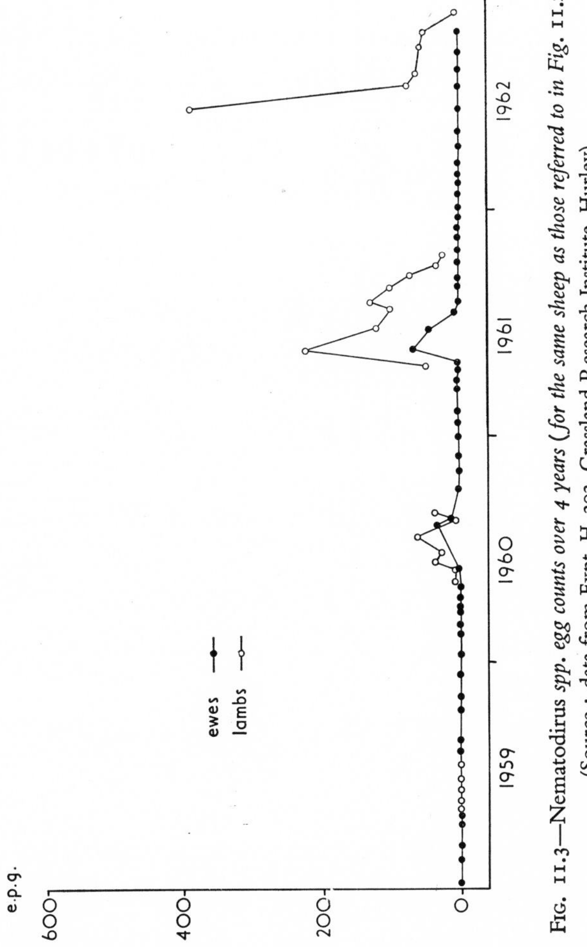

Fig. 11.3—Nematodirus *spp. egg counts over 4 years (for the same sheep as those referred to in Fig.* 11.2)
(Source : data from Expt. H. 202, Grassland Research Institute, Hurley)

10 per cent *Nematodirus battus* in the south of England. The time at which the peak number of infective larvae is reached varies with the year and, possibly, with the species. It may thus occur just when lambs are beginning to graze or much earlier. If it occurs much earlier there is opportunity for larvae to die or to be removed by grazing with resistant sheep, e.g. the ewe flock, or with cattle, before the lambs are exposed. In the north of England *Nematodirus battus* has proved lethal to lambs in the second successive year on a pasture. Thus, in one season, sufficient infection was laid down to result in lamb deaths the following year (2).

With a predominantly *N. filicollis* population in the south, however, even four successive years did not lead to severe infestation. During the first three years the numbers of infective larvae on the pasture increased (Fig. 11.2) and the number of eggs passed in the lambs' faeces also increased (Fig. 11.3) but at a relatively slow rate. This situation occurred whether the stocking rate was maintained for the whole of each year (Figs. 11.2 and 11.3), or whether it was only maintained during successive grazing seasons (Figs. 11.4 and 11.5). In the fifth year of continuous heavy stocking, however, infestation reached a high level and affected lamb growth rate to a marked extent. The pattern for *Nematodirus* spp. is easier to follow because reinfection may not occur to any significant extent within one year.

Control of parasites

With other worm species considerable reinfection takes place within a year and the build-up which occurs from year to year is more difficult to dissociate from other changes which might influence the level of worm-infestation in successive years. For example, an annual decline in milk yield or an annual increase in lambing percentage could lead to more worms each year, even though the number of parasites carried over from one year to the next was constant. This will be clear from the discussion of the significance of the milk supply to the lamb, in Chapter V.

The incidence of infection with tapeworms (*Moniezia* spp.) has also been found to increase where lamb crops successively graze on the same pasture (13). The fact that it is, in all these cases, the lambs that suffer, might suggest that there is little difference between successive seasonal grazings and a maintained annual stocking rate. The presence of the

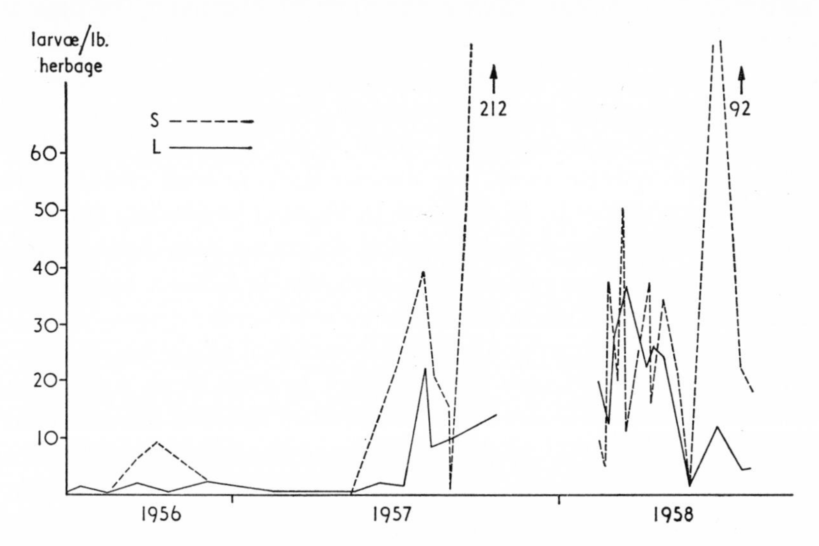

FIG. 11.4—*The number of* Nematodirus *spp. larvae on 'long' (L) and 'short' (S) ryegrass swards grazed for three successive years*

(Source : data from R. V. Large & C. R. W. Spedding (1964), *J. Brit. Grassl. Soc.* *19* (2), 255–262)

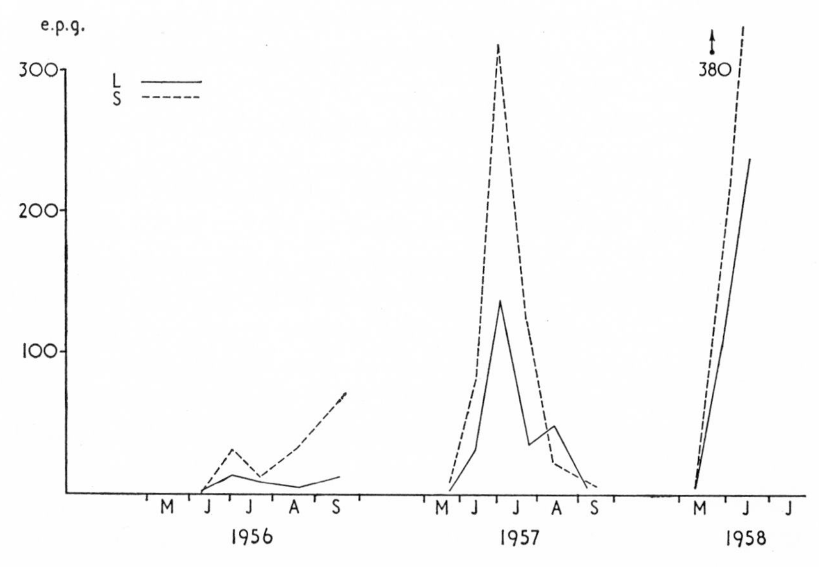

FIG. 11.5—*The number of* Nematodirus *spp. eggs in the faeces of lambs grazing the same 'long' (L) and 'short' (S) swards as are illustrated in Fig. 11.4*

(Source : data from R. V. Large & C. R. W. Spedding (1964), *J. Brit. Grassl. Soc.* *19* (2), 255–262)

ewe during the winter, however, may have effects on the survival of parasites from one year to the next, or on the number of infective larvae on the pasture in the spring. Little is known of these interactions. It is of interest in this connection that the eggs deposited in the winter may fail to develop if passed when the temperature is very low.

So far, curative, anthelmintic treatment has been used where the build-up of a parasite problem has been unavoidable. It would be worth knowing whether supplementary feeding at a critical time could so reduce intake of grass, and therefore larvae, to allow development of resistance without risk of excessive infestation, at high stocking rates.

It would also be worth knowing to what extent milk output could be increased and to what extent such an increase would have a similar effect to that of supplementary feeding. Avoidance of the *Nematodirus* problem has largely rested on complete alternation of pastures used for successive lamb crops (2). There is no doubt that this is effective, even though small numbers of parasites may survive for more than one year (6), and such a practice may well be desirable wherever it is possible. The question remains as to whether this kind of alternation of land use is essential.

It is at this point that the annual stocking rate becomes a vital part of the discussion. Clearly, if an area carries annually a number of sheep which can be supported on half of it during the whole grazing season, then alternation is possible within the system (14). This need not imply a low stocking rate unless such a number turns out to be much less than the maximum possible in any one year. Since a complete rest from grazing of half the area in any one season would mean that as much material was conserved as the amount grazed, there is an obvious disparity (see Chapter X). The requirement for conserved food is unlikely to be as high as this, except where the grazing season is relatively short, as, for example, in Scandinavia. In Britain, far more would be conserved than was required if this was done, at least under lowland conditions of reasonable fertility. An adjustment of pasture production could doubtless create a balanced situation: it would operate at a low level of productivity per acre but a high level per animal.

It is doubtful, however, whether such a half-and-half arrangement is necessary: the information required is how much of the area must be avoided by lambs and for how long. In the case of *Nematodirus* spp. it may be unlikely that intake of larvae during the first month is important unless the milk supply is very low. The period when infected pasture must be avoided would then be from one month of age to slaughter. If the latter occurs at an early age, 3–4 months, then the critical period may be only 2–3 months. The area to be avoided would, of course, be that infected by the lambs during the same period in the previous year, since this is the time when the lambs would have been depositing the maximum number of eggs. At its simplest, therefore, the arrangement has to be a half-and-half affair for the lambs, though not for the ewes. Creep-grazing, provided lambs used the creep areas sufficiently, could have a big influence on this problem.

One method of control, then, is to ensure that one half (A) of the area is not grazed by lambs, at least during the critical period of their lives, in year 1: during this time they will be confined within area B. They need not, of course, graze all over B. In year 2, no lamb grazing will occur in B; all of it will take place within A. A great many variations on this theme are theoretically possible but little has been done in the field to test them as yet. The principle involved is just as relevant to any other disease-producing organism, about which the necessary information is available. This must include where the infection is and when it is at its peak on the pasture; to which sheep and at what ages it is most damaging; and whether or not it is primarily a question of parasite numbers.

In lowland practice, annual stocking rates of up to 3 ewes, of about 150 lb liveweight, per acre, allow the kind of alternation described. The sheep can be concentrated on half the area while lambs are being fattened, since this only requires a density of 6 ewes and their lambs to the acre. Management may have to be adjusted (see Chapter IX) in order that this density does not lead to reduced performance, but it is certainly both possible and practicable. Higher annual stocking rates often occasion some difficulty, though it must be re-emphasized here that the precise number has to be related to the size of the sheep and the amount of herbage grown. The reasons for such difficulties as are encountered, however, are not always related to the stocking rate, but to the plane of nutrition associated with the stocking rate. The ability to arrange a satisfactory alternation for the avoidance of some diseases may greatly depend on the length of time for which lambs are within the system. Now, in general, higher stocking rates tend to result in somewhat reduced lamb growth rates and thus in an extended lamb-grazing season. The latter may easily cover 7 or 8 months, taking into account both the spread of lambing dates and those for slaughter. This is another example of the indirect effect that nutrition may have. An investigation that could prove most rewarding would be into the relationship, particularly in the long term, between disease incidence and stocking rate, with nutrition held constant.

Nematodirus has been much discussed here because it is a clear example of year after year accumulation, where management may have a clear-cut effect. There are a few other examples, such as *Oesophagostomum columbianum* in Canada (16), and doubtless more detailed knowledge of other individual species would place them in the same kind of category.

What is perhaps chiefly lacking at the moment is an adequate definition of the problems of continuous sheep stocking. This requires controlled experimentation of a high order to determine the causes of ill-thrift in sheep kept in this way. It is possible that poor performance is the result of many factors, only one of which may become prominent at any one time. When disease increases from year to year, it appears to be associated with accumulation on the pasture rather than in the animal. Accumulation on the pasture can be avoided by the use of rotations of stock, land and crops but success depends upon an adequate length of rotation. In general, although it is known that some parasites can survive on pasture for longer than twelve months, a complete year seems satisfactory as an interval between, for example, two successive grazings by lambs. Thus, although ploughing and reseeding might, in some circumstances, reduce the survival of parasites, time is probably the dominant factor.

Pasture treatment

Whether pasture treatment can influence the outcome depends on many things. There is no doubt that hot, sunny weather, for example, has a powerful effect in reducing the population of infective larvae on grassland. Probably, practices such as harvesting and gangmowing, which spread faeces about and break up faecal lumps, can accelerate the reduction process in such weather. Nevertheless, worm-infestation is often serious in countries with a hot, sunny climate, if the relative humidity is also high. At times when the weather is more favourable for the larvae, spreading of the faeces may be undesirable since it simply spreads infection. Pasture sprays and fumigants have been little studied, and fertilizer practices have received little attention from the point of view of their effect on the larval population.

In general, the infective larvae of parasites are microscopic and well-hidden. Those that are exposed and readily accessible to a spray are those that will not survive long anyway (3, 7, 17). Those that might be worth special efforts to eradicate are thus the ones that are most difficult to attack. An effective spray, for example, would not only have to be toxic to the parasite and relatively non-toxic to the herbage, it would have to be used in sufficient quantity to saturate the sites where the larvae were. This might require vast volumes and might leave an undesirable residue; it would then have to be completely harmless to sheep as well. Combined with irrigation it is not impossible to visualize a successful spray but it does not look so promising as fumigation.

Fumigation also presents problems but it does allow short-term concentration without a residual difficulty. Gaseous saturation could be visualized which would kill all parasitic stages, diffuse into the air soon afterwards and leave the sward unaffected. Some initial work has been done along these lines with methyl bromide (15). It is usually considered that the large areas normally required for sheep grazing would render any of these procedures uneconomic. Two trends could alter this. First, treatment of these kinds might be more economic than not being able to keep sheep intensively year after year on the same ground. This assumes that one annual treatment would be sufficient to prevent a yearly accumulation—an entirely different problem from that of control within a year. Secondly, some forms of intensive lamb rearing at pasture (see Chapter XIV) may only be possible under relatively worm-free conditions. These systems could be so highly productive that either higher input costs could be sustained or much smaller areas would be involved. Again, treatment would be infrequent. At this stage, speculation about methods of application is only warranted by the fact that these are so readily assumed to present a major difficulty. At its simplest a method might be visualized as involving a gas distributor beneath a large, plastic sheet drawn over the area by tractor.

REFERENCES

(1) BICHARD, M. & YALCIN, B. C. (1963). Personal communication.

(2) BLACK, W. J. M. (1960). Control of *Nematodirus* disease by grassland management. *Proc. VIII int. Grassld Congr.* (Reading), 723.

(3) CROFTON, H. D. (1949). The ecology of immature phases of trichostrongyle nematodes. III. *Parasitology 39* (3 & 4), 274–280.

(4) DICKINSON, A. G. (1960). Some genetic implications of maternal effects—an hypothesis of mammalian growth. *J. agric. Sci. 54*, 378–390.

(5) GIBSON, T. E. (1959). The development of resistance by sheep to infection with the nematodes *Nematodirus filicollis* and *N. battus*. *Brit. vet. J.*, 115, 120–124.

(6) GIBSON, T. E. (1959). The survival of the free-living stages of *Nematodirus* spp. on pasture herbage. *Vet. Rec. 71* (18), 362–366.

(7) KAUZAL, G. P. (1941). Examination of grass and soil to determine the population of infective larval nematodes on pastures. *Aust. vet. J. 17* (5), 181–184.

(8) MAYNARD, L. A. & LOOSLI, J. K. (1962). *Animal Nutrition*. 5th ed. New York, McGraw-Hill.

(9) MCMANUS, W. R. (1961). Stocking rate and animal production. *Wool Tech. Sheep Breed. 8* (2), 69–72.

(10) PÀLSSON, H. (1962). Methods of increasing meat production per sheep in Iceland. *Proc. Ruakura Fmrs' Conf.*, Hamilton, N.Z., 61–75.

(11) ROBINSON, J. F. (1961). Dentition investigations. *H.F.R.O. 2nd Rep.* 1958–61, 24.

(12) RUANE, J. B. (1961). Animal breed ing *Agric. Ireland 18* (9), 182–185.

(13) SPEDDING, C. R. W. (1956). Worm-infestation of sheep in relation to ley farming. *J. Brit. Grassld Soc. 11* (2), 99–103.

(14) SPEDDING, C. R. W. (1961). The parasitological implications of intensive sheep grazing. *Vet. Ann.*, 32–37. Bristol, Wright.

(15) STURROCK, R. F. (1961). *The control of Trichostrongyle larvae (Nematoda) by fumigation in relation to their bionomics.* Ph.D. Thesis. London University.

(16) SWALES, W. E. (1943). Observations on the use of phenothiazine for sheep in Eastern Canada. *Can. J. comp. med. vet. Sci. 7*, 280–284.

(17) TAYLOR, E. L. (1938). Observations on the bionomics of strongyloid larvae in pastures. *Vet. Rec. 50* (40), 1265–1272.

(18) THOMAS, R. J. (1959). A comparative study of the life-histories of *Nematodirus battus* and *N. filicollis* in sheep. *Parasitology 49* (3/4), 387–410.

(19) DAVIES, D. A. R. & OWENS, J. B. (1967). The intensive rearing of lambs. 1. *Anim. Prod. 9* (4), 501–508.

(20) OWEN, J. B., DAVIES, R. A. R., MILLER, E. L. & RIDGMAN, W. J. (1967). The intensive rearing of lambs. 2. *Anim. Prod. 9* (4), 509–520.

XII

THE NEED FOR BREEDS

Superficially, sheep production does not appear to demand the very great number of breeds that are to be found in the world today. The conclusion, often reached, that there are too many, cannot be justified without a great deal more information than is currently available, however. From the point of view of the efficiency of the production process, too little is known about the important attributes of different breeds. From the point of view of the application of this process for long periods of time in different environments, the need for many breeds has rarely been studied experimentally. In any case, the existence of breeds that are not obviously required may be no disadvantage and, against the total world sheep population of 1,027 million (sheep of all classes, 1964/65)*, the numbers involved in some of the less well-known breeds would not justify concern even if such breeds had little productive value.

There is, nevertheless, good reason to wonder why there are so many breeds, whether a large number is needed and, if so, why. In considering the production process, it will have been clear that there are not only several possible products but a variety of possible combinations of them. It should also have been clear that efficiency can be obtained in several different ways with as many different kinds of sheep, and that the optimum point on an input-output curve may vary with, amongst other things, the nature of the input. Thus the sheep required may vary with the feed and, in considering the husbandry of sheep, the way they are kept. The demands of different environments are obvious and often spectacular, but they are difficult to determine experimentally. Finally, in the previous chapter, the long-term influence of husbandry practices has been emphasized. It is here especially that caution is needed. The performance of a breed over a long period in a particular environment may differ markedly from that of small flocks over shorter periods.

* As given in the revised edition of *British Sheep* (45).

First, it is as well to be clear about the existing breeds and what is known of the reasons for them.

THE EXISTING BREEDS

Origins and distribution

Sheep were apparently amongst the earliest animals to be domesticated and their use dates from Neolithic times. Their origins are thought to lie in the Middle East and the present domestic breeds have all proliferated from three primitive kinds of wild sheep. These are still found wild: they are the Urial (*Ovis vignei*) of south-west Asia; the Mouflon (O. *musimom*), found on certain Mediterranean islands; and the Argali (O. *ammon*) of central Asia. The other chief kind of wild sheep, the Bighorn (O. *canadensis*) of North America, does not appear to have resulted in any domestic derivations. This subject has recently been discussed in greater detail by Ryder (26) and by Ryder & Stevenson (47). The present breeds have been listed by Mason (42) for the whole world and described in some detail by Fraser (14) and Thomas (29): the British breeds have recently been described in a new edition of *British Sheep* (produced by the National Sheep Breeder's Association, 1968) and in *British Sheep Breeds, their Wool and its Uses* (published by the British Wool Marketing Board (1967)).

It is not intended here to describe in detail the present breeds or what is known of their history, but a few salient features stand out. Not only has there been a tremendous proliferation of breeds, there has also been a tremendous spread of sheep over the whole world. There are now few major areas where no sheep are kept and the main concentrations of sheep are widely separated (Europe and the Middle East, South Africa, Australia and New Zealand, and South America). Not surprisingly, greater numbers of sheep are found in the major pastoral latitudes (13, 14).

Sheep have therefore filled a variety of ecological niches and have fitted into a great many different environments. The contrast between the Scottish Blackface in winter and the Australian Merino in summer is really an astonishing one for basically similar animals. Furthermore, this distribution has been achieved without cossetting: indeed, it has been a characteristic of the sheep that it has fended for itself to an extent unusual for a domesticated animal. Consider also the range of products obtained and the range in such attributes as prolificacy that are concerned in their production. In Table XLV details are given of a few breeds, selected to illustrate these ranges.

TABLE XLV

VARIATION IN THE IMPORTANT ATTRIBUTES OF DIFFERENT BREEDS OF SHEEP

(compiled from the literature)

Breed	Approximate liveweight of mature ewe (lb)	Lambing percentage* (normal range or average)	Approximate fleece weight (lb)	Source (reference)
Cheviot	130–160	141–158	4·0–5·0	14, 30
Clun Forest	124	150–164	5·4	2
Colbred	180	200	8·0	8
Corriedale	75–85	110–114	7·7–8·8	30, 32
Dorset Horn	160	127–158	6·0	20, 30
East Friesian	150–200	206	6·6–11·0	14, 30
Finnish	110	200–400	2·2–6.6	19, 30
Galway	188	132	9·0	9
Lincoln	200	129–157	14·0	14, 20, 30
Merino	100	103–161	15·0	30
Précoce	—	120–150	—	30
Romanov	—	195–231	—	30
Romney Marsh	180	131–142	8·0–9·0	14, 30
Scottish Half-bred	160	176	7·5	9
Southdown	60	116–170	3·0–4·0	14, 30
Svanka	88	200–400 (twice yearly)	4·4–10·0	17
Welsh Mountain	65–67	95	2·3–2·5	25

* This refers to the number of lambs born per 100 ewes mated, put to the ram or lambing (according to the source).

Breed requirements

One very good reason for the number of breeds then is because a number of quite different sheep were required. Whether deliberate or not, breeding and selection for characters such as milk yield, fleece weight and meat production, have been highly successful. This is demonstrated by the differences which exist both between breeds and between improved sheep and primitive original stock. An example of a surviving primitive sheep in Britain is the Soay (3), which has remained 'unimproved' on the remote and inaccessible islands of the St Kilda group, west of the Outer Hebrides. Since it seems unlikely that the perfect sheep has been attained for any particular purpose, breeding and selection will no doubt continue, no matter what number of breeds exists already. This is evident in Britain today where, although there are about 40 pure breeds and innumerable crosses, new breeds are actively being created. The number of breeds

TABLE XLVI

DISTRIBUTION OF SHEEP BREEDS IN THE UNITED KINGDOM

Breed	Percentage of total ewe flock
Blackface	25
Welsh Mountain	15
Half-bred	15
Cross-bred	10–15
Shropshire, Clun and Kerry Hill (together)	10–15
Cheviot	5
Lincoln	5
Massam	5
Devon, Suffolk, Leicester, Kent, Herdwick, Gritstone, Lonk and and the Down breeds	Each less than 5 per cent and together not more than 10 per cent.

(Source : *Farming Facts*, ed. Graham Cherry, 1962 (5))

is the more extraordinary when 40 per cent of the total sheep population of Britain is accounted for by only two breed groups (see Table XLVI). The creation of new breeds is primarily being undertaken to improve productivity but the environment is by no means ignored in this. Indeed some of these efforts originate in the desire to retain the adaptive advantages of a traditional breed but to increase its productivity.

On the Romney Marsh, the Kent is widely regarded as highly adapted to the land and to the heavy stocking frequently practised on it, but its prolificacy is low. In parts of South America, Romneys are also used and again they are regarded as being adapted to a particular type of land. Thus in Uruguaiana, a county in the Brazilian state of Rio Grande do Sul, Romneys are kept on the lower land where the humidity is high; Merinos, on the other hand, are kept on undulating, drier country. Breeds such as the Ideal (Polwarth) and the Corriedale tend to occupy intermediate land. The reasons for these associations are connected with what is believed to be greater susceptibility to footrot in the white-hoofed breeds and poor wool cover in the newly-born Merino lamb.

There is every likelihood that more breeds would have died out if it had always been simply a question of using the most productive sheep, but it is this deeply held belief in the value of the traditional association of breed and environment (12, 34) that has often led to attempts to modify and improve the local stock.

Breed attributes

Many breeds have, in fact, died out or very nearly so, in Britain, and it is interesting that they include several with markedly different values for some of the attributes of importance in production. Amongst the not very numerous, for example, the Cotswold is reputed to be very hardy, the Wart Holm is able to make use of seaweed, and the Herdwick may continue to breed for more than 10 years.

One of the reasons for reluctance to see breeds disappear is the awareness that their value may not be apparent or that they may become valuable in changed circumstances. It is, in fact, extremely difficult to assess the value of any breed or any animal within it. Very often, comparisons of breeds are obscured by reference to ill-defined and unmeasurable characters, by excessive attention to breed 'points' and by consideration of output unrelated to input. It would be incautious to disregard the logic behind all this, however, simply because there are excesses. First, it is not easy to determine productive value. Secondly, ability to perform in a given environment cannot be determined every time a sheep is judged or assessed, so that some outward signs may have to be depended upon as indicators. If these indices are valid, it does not matter whether they have any intrinsic value and their validity as indicators is not shaken by any demonstration that they are otherwise valueless and merely associated with the desirable attributes. Thirdly, for any controlled breeding to take place, breeds must maintain their identities. This requires standards for shape and size as well as for production traits: if it can include the indicators mentioned above, this is all to the good. Thus breed points may arise. It would be far better, of course, if the breeds were assessed and described in economically useful terms and if the value of apparently useless points was put to the test. But the test would not be simple and might require long-term investigation: it would certainly require an open mind.

THE EVALUATION OF A BREED

This must include two aspects: productive value and adaptation to environment.

Productive value

It will be evident from the discussion in Chapter VI that productive value must be expressed as a series of input-output relationships as well as a group of maximum values for attributes of importance to the production process. A ewe would therefore be described, for example, as capable of yielding so much wool, capable of bearing so many lambs and having a given milk output to food input response curve. Part of the last curve should, of course, include the maximum milk yield. Little is known of the fundamental differences between breeds for milk production, although clearly such differences exist. There is evidence of relatively low milk yields, or poor early lamb growth, in some breeds with a heavy mature weight, just as other and lighter breeds are known to give more milk. Since wool yield is to some extent related to body size, however, some compensatory production probably occurs.

The milk yields that have been measured, however, are not necessarily those that are optimal for any of the breeds studied and it is necessary to determine, by feeding at different levels, where the optimum lies. Similarly, it is not sufficient to determine the growth rate of lambs without also knowing the quantity of milk and grass required to produce it. Although it may be argued that high values for both milk yield and growth rate are desirable, the maximum is not necessarily justified. This is particularly so when considering different feeds and it is possible that different breeds might be at their most productive when they are receiving different feeds. Naturally, this would also have a considerable bearing on adaptation to the environment. It is possible that breeds also differ in respect to milk composition and the length and shape of the lactation curve. Regarded purely from a production point of view, a breed of sheep may be looked on as a particular combination of the important factors which govern productivity. These are ewe size, with related maintenance requirement; fleece weight; the relationship between food input and milk output; prolificacy and litter size; longevity, and related disease resistance, and so on; lamb growth rate, carcase quality and conformation in relation to age.

Some breeds are outstanding in one or another of these attributes but less satisfactory for others (see Table XXXIV). In the Soviet Union, it has been reported (24) that Russian Merinos weighing between 100 and 120 lb produce a greasy fleece weight of 12 lb (= $5\frac{1}{2}$ lb scoured), whilst the heavier (130–155 lb) Kayak Arkharo-Merino gives only 8 lb of wool. In Iceland, the native sheep may continue to produce until 11 years old (22). Pàlsson has estimated milk output of ewes with twins at 6·5 lb a day and that of ewes with singles at 4·1 lb a day, during the first fortnight after lambing. At a liveweight of about 130 lb, however, different strains may have lambing percentages varying from 100–200 per cent. Similarly, in Australia, Weston (33) found differences between strains of Merinos. One strain was 22 per cent heavier and gave 39 per cent more clean wool, but it consumed 17 per cent more food. This resulted in 19 per cent more clean wool per unit of food consumed: the efficiency of wool production was significantly related to wool yield per unit of body weight. The use of such ratios for selection in animal breeding has been discussed by Turner (31).

It must be remembered that some of the important factors may change with the age of the sheep. In a comparison in which Hampshires, Merinos and Columbia-Southdale ewes exceeded in prolificacy ewes of the Shropshire and Southdown breeds (27), it was found that prolificacy rose with the age of the ewe from 2 to 9 years.

Pattie & Donnelly (23) compared a large number of breeds and crosses specifically for lamb production on the central-western slopes of New South Wales. They were able to conclude that a combination of Border Leicester × Merino ewes and Dorset Horn rams would give the greatest returns to lamb producers. As in many such experiments, all the ewes were grazed, after mating, as one flock. Coop (10), working with Corriedale and Romney × Corriedale sheep in New Zealand, found that those with wool-covered faces had 3 per cent more barren ewes, 10–15 per cent fewer lambs per ewe lambing, 2 per cent higher lamb mortality and weaned lambs 3 lb lighter. This represented a total reduction of 23 per cent. in lamb production per ewe mated; ewes with intermediate face cover gave a reduction figure of 7 per cent. The ewes with wool-covered faces, however, produced 0·2–0·4 lb more wool and, being smaller, 3 per cent more could be carried per acre. Similarly, Clarke (6) found that Border Leicester×Romney sheep gave a higher production per animal than pure Romneys, but they also had a maintenance requirement estimated to be 17 per cent higher. Clearly, in cases of this kind, the production per acre may be no higher, though this also depends on the relative foraging efficiencies of different breeds. This last point has to be taken into account in extrapolating from pen-feeding experiments (16).

A further attribute which could be of immense importance is the ability to breed at different seasons. It was pointed out in Chapter IV that large differences exist between breeds, not only in reproductive rate (28), but in the seasonality of breeding. This also varies within breeds, and applies even to Merinos (1), which are sometimes thought to be different in this respect. It is likely that individuals differ in the time of year at which they will breed, in any given environment, and that strain differences may therefore exist. There are obviously a great many possible combinations of different values for each of the important factors and it could be argued that each breed represented a unique combination. The special advantages that might be attributed to a particular breed can only be finally evaluated in economic terms or in terms of the context within which the breed is able to operate efficiently.

Adaptation to environment

When considering the environment within which sheep production takes place, it is important to take a very wide view. The total environment of an individual animal includes its neighbours, their number and stocking density, the general level of its food supply, its management and the disease background, as well as the more obvious features of climate, altitude, topography and soil type. These more familiar facets of environment may very well exercise a dominant influence on the ability of a breed to perform, but it is misleading to treat these as if they were entirely different from other environmental influences.

It is important to recognize, for example, that survival and performance in a harsh climate may be greatly affected by the plane of nutrition. This is likely to be true for all sheep of all breeds, but the ease with which a high plane of nutrition can be achieved within such a climate may be quite different for different breeds. Even the same quantity of food may be rendered more or less available by the method of presentation or the extent of its dispersion on the ground. Where the food supply is widely dispersed, with forage very sparse on the ground, for example, a large animal must cover a greater area to obtain its greater needs than that covered by a small one. Needs are related to weight and size and muscular expenditure tends to be proportional to body weight (4). Whether or not the extra effort involved in collecting food reduced productive efficiency to a greater extent in a larger sheep, it could, of course, result in lowered intake if the time required for food collection became excessive. If, however, food supplies are *sporadically* placed, a patch here and a patch there, flock behaviour becomes of vital importance (see Chapter VIII). A typical lowland flock might graze together successive patches and every sheep would have to walk the maximum distance. Mountain breeds, by contrast, would spread out and establish territories, thus reducing the amount of walking to the minimum.

Direct climatic effects are very well illustrated by rainfall and wind. Breeds vary greatly in the gross structure of the fleece. The long, coarse fibres of the Blackface are extremely efficient at shedding water. Short, dense wool, on the other hand, is extremely good at retaining it. Not only does this alter the animal's ability to control its temperature efficiently, with water and wind having a combined effect greater than that of either alone, but the sheer weight of water held by a fleece can prove a lethal burden to a heavily pregnant ewe. In contrast to the last example, where water supply was excessive, Macfarlane, Howard & Siebert (43) have found large differences in the water metabolism of Merino and Border Leicester wethers grazing saltbush (*Atriplex* spp.). The two breeds produced similar urine concentrations but Leicester used 46 per cent more water (as litres per 24 hours) than Merinos: when body weight was taken into account, he Leicesters used 71 per cent more (as ml per $kg^{0.82}$ per 24 hours). The animals drank extra water in order to reduce renal salt concentrations and their needs clearly varied with breed.

It is not possible to catalogue here all the known ways in which breed characters influence adaptation to environment. It would be unwise, however, to underestimate the possible importance of characters which may seem trivial at the present time. By way of example, consider 'face cover', the amount of wool grown on an otherwise hair-covered face. Superficially, this could be regarded as quite trivial, incapable of influencing either the amount or efficiency of production or the adaptation of an animal to its environment, although, as already noted, recent evidence has accumulated to suggest very strongly that, in New Zealand Romneys, the amount of face cover is negatively related to prolificacy (10). However, vitamin D is produced in the skin of a sheep only when it is sufficiently irradiated with ultra-violet light. Production of the vitamin is thus dependent, not only on the amount of sunlight, but on the angle of the sun and the amount of atmospheric interference to penetration by the violet end of the spectrum. All this, even within the British Isles, varies greatly with geographical location, altitude and season, in other words, with the environment. Now, vitamin synthesis only occurs where the skin is sufficiently exposed and, during the critical winter months, it is confined for the most part to the head, including the ears, and the legs. Face cover, then, might conceivably affect the natural vitamin D supply at certain times and in certain places (11).

In addition to breed differences in adaptation to environments, it is possible that differences between individuals may be of significance. It is increasingly being noted that individual sheep vary in the characteristic level of minerals in the blood stream. For example, in the Netherlands (18), 54 out of 73 Texel sheep had red blood cells with low potassium and high sodium levels and 19 had the reverse, high potassium and low sodium levels. It is not possible, at present, to say whether such features have any bearing on the ability of sheep to thrive in particular situations and it would not necessarily be a simple matter to find out.

Methods of Breed Assessment

It is an urgent need that accurate information be collected and collated in relation to breed assessment. A question uppermost in the mind of the enquiring sheep farmer concerns the breed of sheep that he should keep. The enquiring observer, whether research scientist, advisory officer, veterinarian or teacher, is equally concerned with why particular breeds are found in particular places. There is insufficient information to answer these questions and, unfortunately, it is by no means certain yet how the answers should be sought. Since the methods available are numerous and often expensive of time, labour and money, some observations on this problem may not be out of place.

Breed comparisons

The most immediately obvious method of assessment is by comparative experiment. It is equally obvious that the number of possible comparisons is very large: for 30 breeds, to compare every one with each of the others would require $29+28+27+26+\cdots+1=435$ experiments. Add to this the innumerable husbandry factors to be taken into account, such as stocking rate, type of pasture, fertilizer treatment, management, etc., for any one comparison and the number of environments in which each one could be conducted, and it is clear that the experimental comparison of breeds can only be done very selectively. Comparative experience of farming with different breeds can be gained on a much wider scale, of course, and the more controlled and better recorded such experience is, the more likely it is to form a useful basis for the selection of the more profitable comparisons to undertake experimentally. Some such comparisons can be visualized quite readily. Where two or more breeds can be said, by experience or by long tradition, to suit an environment, they can be compared within it to see which is the better. Even here, two further difficulties arise. First, it is necessary to decide on the purpose for which the breeds are being compared, whether early fat lamb production, the sale of breeding ewes, or wool production, for example. Secondly, such things as stocking rate and fertilizer usage have to be settled. It cannot be over-emphasized that breed comparisons which have not been carefully designed with these considerations in mind, could be most misleading. Farming, rough and ready trials, and inadequately designed experiments may all contribute information bearing on the problem. Such data may help in the design of better experiments or limit the range over which experiments need to be carried out, but they will not serve to assess the production value of one breed against another. It is doubtful whether anyone would compare Southdowns (mature weight of a ewe= 60 lb) with Cluns (140 lb) at the same stocking rate: but it is difficult to state the correct ratio to employ.

Indoor experiments

The same problem applies in every breed comparison and would occur, in principle, if it were made indoors on concentrated feedstuffs. In the latter case, however, at least the amount offered and the amount eaten would be known. But how much should one feed? Should Dorset Downs and Southdowns be offered the same amount per head or per pound bodyweight? To settle on one or another of these possibilities is to assume some part of the answer. To decide exactly how much feed, per head or per pound, is to restrict the answer to this amount. Yet one breed may do best when fully fed and another at some lower level. In short, only a measurement of the response curve will suffice, whether in describing one breed or comparing two. This has often been stated in other terms. For example, 'the economic index which is important in breeding is the amount and quality of meat that is produced per annum per ewe per unit resources (e.g. in relation to nutrient consumed)' (36). It is necessary to know the production of a sheep in relation to several levels of food input and it may well be necessary to repeat the measurement on two or three contrasting feeds. At the present time, it is doubtful if either indoor or grazing experiments alone can provide adequate information. Certainly there are important factors omitted in indoor experiments, but they do allow a high degree of control and precision in measurement, neither of which can be achieved easily under grazing conditions. A combination of both approaches appears most desirable. Fortunately, it is not always necessary to conduct the indoor type of experiment in a particular part of the world: the relevance of the grazing experiment, on the other hand, may be greatly dependent on the environment.

Grazing experiments

Much remains to be learnt about the proper conduct and design of grazing experiments and not the least of the problems concerns the choice of stocking rate. Furthermore, this cannot necessarily be solved by using several different rates as treatments within an experiment.

On paper this often looks a very satisfactory design, as, of course, for some purposes, it undoubtedly is, but two major considerations must not be ignored. First, the animals usually grow during the experiment. In the case of young lambs this may mean an increase, in the weight of sheep per acre, of several hundred per cent. In most animal production experiments the change in animal requirement is considerable during the experiment and is usually an increase. Secondly, the supply of herbage is dependent on a variable growth rate: thus it also changes during a grazing experiment and normally decreases. The plain facts, at the present time, however, appear to be as follows: a great many different kinds of grazing experiment can be justified if the information they provide is really what is required. Often it is not but may give the appearance of being more relevant than the experimental design allows.

It has recently been argued (44) that, where practical application is ultimately contemplated, grazing experiments should be relevant to actual or conceptual systems. This implies that grazing experiments should be carried out using whole husbandry systems (to which the results will apply) or using sufficiently independent parts of systems that the results will still apply. In particular, where the object is increased profitability, experiments should vary the major factors governing profit, over a relevant range, within whole systems and on an adequate scale.

Breeds may validly be compared at many different levels. They may be compared, for example, in terms of wool yield or milk output in specified, controlled conditions: such information is of value. To decide which breed to keep, however, in a particular environment and under a particular system of management, requires a comparison of each breed under that management. To decide which breed to keep as the most profitable use of land, requires comparisons in circumstances where each breed receives the optimum management for that breed.

The difficulties are obvious and it is essential to compile from many sources as many data as are available and to make the greatest possible use of calculations to test hypotheses before embarking on the elaborate programme of grazing experimentation implied.

Wiener (48) has reported the results of a comparison of five breeds grazed as a single flock; this study illustrates the kind of data on performance and conformation that can usefully be derived from such comparisons.

Importance of lambing percentage

Finally, it must be recognized that slight variations in the factors of importance in the production process can have very big effects. The best example of this is the lambing percentage. Breed A, represented

TABLE XLVII

STOCKING RATE, LAMBING PERCENTAGE AND PRODUCTION

The effect of stocking rate and lambing percentage (number of lambs per 100 ewes) on production per acre and per head.

Lambing percentage	Stocking rate (per acre)		Gain per acre from birth to 3 months (20.3.62–17.6.62) based on		L/wt gain per head per day	
	Ewes	*Lambs*	*L/wt.*† *lb*	*Carcase*‡ *lb*	*lb*	*kg*
100	8	8	530	294	0·80	0·36
	10	10	637	360	0·76	0·34
	12	12	658	369	0·68	0·31
	14*	14	696	366	0·61	0·28
200	4	8	445	226	0·70	0·32
	6	12	676	321	0·69	0·31
	8	16	809	371	0·61	0·28
	10*	20	729	331	0·46	0·21
150	6	9	492	260	0·72	0·33
	8	12	701	367	0·70	0·32
	10	15	726	385	0·60	0·27

* Additional hay was required by these groups for 2 weeks at the end of the experiment.

† L/wt gain = final fasted L/wt. minus birth weight.

‡ Carcase gain = actual carcase weight − estimated carcase weight at birth.

Since, for experimental reasons, all lambs were slaughtered at approximately the same age, these carcase gains do not necessarily reflect quantities of a saleable commodity.

(Source : Spedding, C. R. W. (1963), *World Conf. on Anim. Prod.* (Rome))

by a flock of ewes with a lambing percentage of 150, cannot validly be compared with Breed B, represented by a flock at 165 per cent, unless these levels of prolificacy are regarded as definitive for the breed. Breeds, however, at present, would have to be characterized by a lambing percentage range and, in any case, as with stocking rate, one figure probably means quite different things in different breeds. Simply to underline the general problem, consider Table XLVII. All the results shown are pounds of liveweight gain of Suffolk × Half-bred lambs on pure ryegrass pasture from birth to approximately 3 months of age. The variations are solely due to stocking rate and lambing percentage. The stocking rate differences, however, were much modified by a differential rate of nitrogenous fertilizer input (12 lb nitrogen per ewe per month). It is difficult from such evidence to describe the productivity of this kind of sheep under grazing conditions.

Clearly the lambing percentage must be specified; probably it should be varied in the same way as feed input and a relationship established. Unfortunately, lambing percentage is itself open to several meanings, quite apart from its method of calculation (see Chapters III and IV). A large number of combinations of ewes with one, two and three lambs can result in the same lambing percentage, in fact. The most satisfactory solution is to study and assess separately ewes with one, two, three or more lambs. It is true that some of the possible mixtures could react differently but, in circumstances where this might matter, it has already been argued that ewes with singles, twins and so on, should be kept separately.

The Function of Breeds

From a discussion of the production process (Chapter VI), a relatively short list of major factors emerged. It is theoretically possible to determine the interrelationships of these factors for any breed and thus to construct a model of the productive process for that particular breed. Similarly, a new breed could be postulated or the result of a modification to an existing one predicted. It might be satisfactory, in many instances, to proceed in this way until a few of the most useful combinations of factors could be erected as hypotheses to be tested in chosen environments. There may be, in the future, great changes in the environments relevant to sheep-keeping, just as there may be great changes in the sheep to be kept. Before considering these, several features of the present use of breeds should be noted, particularly the practice of cross-breeding and stratification.

Cross-breeding

Crossing parents of two breeds and crossing between strains or varieties, have as their objects better or more useful progeny than either parent, for a particular purpose. The genetic basis for this rests on (*a*) the greater efficiency with which a limited number of different characters can be selected for in each parent, and (*b*) hybrid vigour (heterosis), resulting in improved performance in the first cross.

It requires that separate breeds (or strains within breeds) should be maintained, and these must be able to retain their identity. Now, either such breeds also produce other valuable products, or they can be kept cheaply or in particular environments, or they must be somewhat wasteful. Only some of their progeny can be used for crossing; as a rule, this is the best of one sex only. The remaining lambs must be disposed of in some other way.

In fact, of course, some females must be retained as breeding stock replacements; these will be similar numbers for each breed, however. Whichever way it is arranged, deliberate cross-breeding of this kind must produce, in addition to the product aimed at, some less valuable by-products. If this leads to great inefficiency, biologically, it is likely to increase the price of breeding stock, compared with a self-contained flock of one pure breed. The genetic argument for cross-breeding is therefore frequently allied to a system designed to produce breeding stock at lower cost. Stratification, as found in Britain, is just such a mixture.

Stratification

This is rarely defined, though often described (14, 20), usually with the help of an illustrative diagram, like that below:

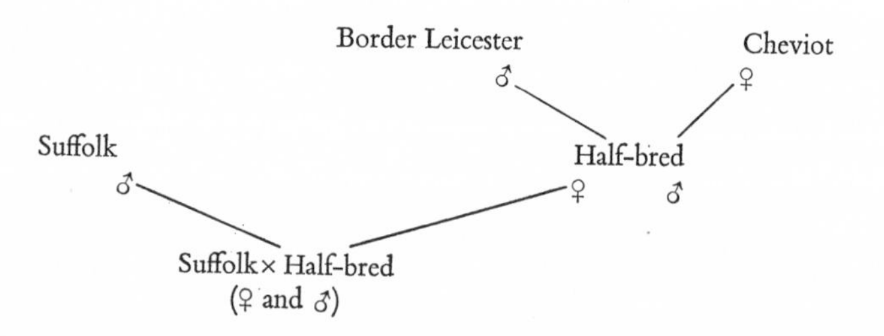

Broadly speaking, the process may be regarded as occurring at lower altitudes as it progresses towards the Suffolk×Half-bred, fat-lamb product. The Border Leicesters and Suffolks need not be numerous since they have only to produce rams sufficient for some of the Cheviots and the Half-breds, respectively. The Cheviots (like the Blackface and the Welsh Mountain) represent in this diagram the basic ewe, which can be cheaply produced in the uplands. Size and growth rate are added in later stages and the actual fat-lamb production occurs on much more expensive acres. This remarkable combination of genetic and spatial stratification can be very efficient but it should not obscure two possibilities; (*a*) it may become inefficient and (*b*) still more efficient systems may exist which will achieve the same objectives. Obviously, if the mountain ewe (the Cheviot, in the diagram) did little more than maintain itself, it would be an inefficient system. This may actually occur in some places in some years where severe winters reduce both the lamb crop and, indeed, the ewe flock. Any system of breeding to which different breeds contribute will also depend for its efficiency on the efficiency with which each breed functions. The extent to which a breed may be dominated by a few rams has been shown by Young & Purser (35). The second possibility (*b*) will be further considered in Chapter XV. There is, of course, a very useful outlet for the draft Mountain ewe in this system of stratification. Her productive life is extended by a move to lower ground and milder climates. At the same time, it is possible in this way to exploit the hills and other environments which might not otherwise produce at an adequate level. The suitability of such areas for the production of sheep breeding stock is a major national asset.

NEW BREEDS AND ENVIRONMENTS

Breeding Objectives

New breeds may be created either to increase the efficiency of the production process or to enable it to be carried out in less familiar or less usual environments. Achievement of the first aim depends on an increased knowledge of the basis of efficiency. In other words, breeding is one way of increasing the efficiency of animal production. Other ways of expanding the animal potential within the production process are considered in Chapter XV. Since the improvements aimed at are basically the same, there are some similarities in the approaches. Breeding, however, is a long-term matter and it cannot be constructively carried out without a clear idea of the precise objectives of the breeding programme. As Colburn (7) has pointed out, one of the difficulties is the wide range of environments in which sheep have to perform. He has suggested that these difficulties may be reduced by thinking in terms of functions within the sheep industry, rather than of breeds. For Britain, he has listed six major functions:

1. Hill sheep
2. Sires of crossbred ewes
3. Self-contained fat-lamb producers
4. Fat-lamb sires
5. Mutton sires
6. Frequent and out-of-season lambing.

By considering the essential characteristics required of sheep in these categories, Colburn concluded that all required some combination of the same five attributes. These he described as prolificacy, milking capacity, growth rate, carcase quality and weight and quality of fleece. It will be clear, from what was said in Chapter VI, that a more precise statement of the optimum values for these attributes is more difficult to arrive at. In the preceding chapters they have been considered in relation to food requirements but it has also been pointed out that, in economic terms, food costs may be relatively unimportant in some circumstances. It is possible that this could be even more relevant in some of the environments in which sheep may be kept in the future.

New Breeds

Selection within a breed is often a slow method of achieving genetic improvement and the stabilization of crosses has been used in various parts of the world to produce new breeds. Ryder (46) quotes the 'Perendale' in New Zealand and the 'Borino' in Australia as examples: the former was developed from the Cheviot and the Romney, the latter from the Border Leicester and Merino. In America, Dr C. E. Terrill has developed a 'Morlam' strain from Dorset Horn × Merino ewes, especially for more frequent lambing. In Britain, Cadzow (39) has crossed the Finnish Landrace with the Dorset Horn to produce the 'Improver' sheep for both lambing frequency and increased lambing percentage. A more complex breeding programme, involving East Friesian milk sheep, Border Leicesters, Cluns and Dorset Horns, was used by Oscar Colburn in producing the 'Colbred' (37). Designed as a replacement for the longwool in British stratification, it has been used to produce a variety of Thornber-Colburn crossbred ewes, sold under brand labels (TC1, TC2 etc.).

The Finnish Landrace is now being widely used in Britain as a crossing ram to increase prolificacy in the crossbred female progeny. Donald, Read & Russell (41) have recently reported the results of 3 years' work at the Animal Breeding Research Organization, comparing crossbred ewes produced by mating Border Leicester, Clun Forest, Dorset Horn, Finnish Landrace and Tasmanian Merino rams to Scottish Blackface ewes. The Finnish and Merino crossbreds were lighter, the former giving the highest mean litter size (1·5, 2·0 and 2·3 lambs born per ewe lambing at ages 1, 2 and 3 years, respectively) and the latter giving about the lowest.

Although less attention has been paid to breeding for the hills, Carter (40) has advocated crossing with Merinos in order to increase both wool yield and quality.

New Environments

New environments may be of two quite different kinds. On the one hand, a greater variety of environments in which sheep are kept may be anticipated. On the other hand, it is to be expected that there will be much greater uniformity within environments, or at least within some of them. This stems from a dichotomy in sheep production, which may well be exaggerated in the future. It is probable that agricultural exploitation of many natural environments may be left to the sheep. At the same time the efficiency of sheep production may be raised to the point where it can compete with any other form of livestock production on the improved agricultural areas.

Exploitation of natural environments

Since the first point touches a very large problem, of great importance in large areas of the world, something must be said to qualify this rather over-simplified view of the role of the sheep. Recent experience in Eastern Africa (15) has prompted the view that the exploitation of a natural environment by one or a few species of animal (or plant) may be less efficient than the use of a large number of species. There are good reasons also for the view that many wild species, native to the area, are better adapted, more profitable and more productive than the existing domesticated ones. This is not only because they may be resistant to disease or tolerant of the climate, but because the several species in the natural population interact to exploit more of the environment. It has been further postulated that in some cases a mixture of wild and domestic species might be the most efficient. This fascinating topic cannot be pursued here: it is introduced as a glimpse of the wider issues behind the exploitation of environments for animal production. Of the existing domestic animals, however, the sheep appears to be most likely to be used to exploit areas which, whilst supporting only the natural vegetation, can still be said to be farmed. This is partly because the use of areas which, with the minimum of inputs of any kind, produce herbage in an exceedingly non-uniform manner, is better achieved with a breeding population whose progeny can be grown to a marketable size in a short period. In this way, breeding stock survive the periods of low herbage productivity and the productive units, the progeny, are grown during the time when herbage growth is at its best. Within Britain, the natural environments of upland, mountain and moor are already chiefly the domain of the sheep. They could be more efficiently exploited than they are, from a biological point of view; but the economic optimum might be quite different again. The choice of grazing animal may also have to be related to integration with other forms of land use, including, for example, forestry and tourism. In upland areas, other possible grazing animals include deer, currently being investigated in Scotland.

Housing

By contrast with all this, there is an increasing trend towards more control of the environment in which sheep are kept. The extreme example of this is housing. Beyond a certain point in intensity of production, whether this is expressed as a very high stocking rate or in very large litters, the grazing situation may lose some of its advantages and a more controllable environment may be required, even at some extra cost. When animals at pasture are near their peak of output, a high degree of control is needed. Thus, within environments, more controlled and more uniform conditions will be achieved where intensive production is practised. But these controlled environments could include contexts within which sheep have not hitherto been kept. Some breeds may be more suited to these contexts than others: altogether new breeds might be better than any of the existing ones. The new environments could be the result of greater pasture production (see Chapter XIII), alterations in the disease component of the existing environments (see Chapter XIV), or deliberate creations in the direction of more efficient management. Housing, for example, could be justified to avoid winter poaching of pasture, to avoid internal parasites, to allow control of reproduction by controlled lighting, or simply to save existing wastage of new-born lambs. Neonatal loss of lambs can reach appalling proportions due to causes other than disease. The inefficiency involved in keeping a ewe for the whole year, or more, and losing its lamb at or shortly after birth, is considerable. In all probability, the control inherent in housing could greatly reduce, if not eliminate, such losses. The likelihood of this kind of inefficiency being eliminated depends greatly on knowing when and where it exists. This is a major argument in favour of recording (21).

Two major environmental changes which may be expected, then, are housing or yarding and heavier stocking at pasture.

CHANGES IN HUSBANDRY METHODS

Finally, changes in husbandry methods may influence the need for breeds. These may not represent changes in the environment in quite the same sense as that previously used. Among the more marked changes that may be anticipated are practices like artificial rearing and early weaning (see Chapter V). Techniques of this type affect both the lambs being reared and the ewes that produce them. For example, lamb size at birth may be less important in artificial rearing, where the milk supply is no longer influenced by the initial sucking demand. Where all lambs are removed at birth, lactation is not merely unnecessary, it may actually be a disadvantage. The probable extension of breeding at different times of the year may present problems of management and environment which may require rather special combinations of attributes.

In view of the variety of possibilities of production/environment interactions which may influence the breeds required for the future, it is scarcely surprising that the great number of existing breeds is often regarded as a gene reservoir of the utmost value (7).

REFERENCES

(1) BARRETT, J. F., REARDON, T. F. & LAMBOURNE, L. J. (1962). Seasonal variation in reproductive performance of Merino ewes in northern New South Wales. *Aust. J. exp. agric. Anim. Husb.* 2, 69–74.

(2) BICHARD, M. & YALCIN, B. C. (1963). Personal communication.

(3) BOYD, J. M., DONEY, J. M., GUNN, R. G. & JEWELL, P. A. (1963). Unpublished observations. Personal communication from P. A. Jewell.
(4) BRODY, S. (1945). *Bioenergetics and Growth.* New York, Reinhold Pub. Co.
(5) CHERRY, G. (1962). *Farming Facts.* No. 33, 7.
(6) CLARKE, E. A. (1962). Cross-breeding in sheep. *Proc. Ruakura Fmrs' Conf.* (Hamilton, N.Z.), 42.
(7) COLBURN, O. (1961). Progress in sheep. *Jl R. agric. Soc. 122*, 54–63.
(8) COLBURN, O. (1963). Personal communication.
(9) CONWAY, A. (1963). Personal communication.
(10) COOP, I. E. (1956). The significance of face cover in Corriedale and Romney-Corriedale cross sheep. *N.Z. J. Sci. Tech., Sec. A. 37*, 542–554.
(11) CURRAN, S. & CROWLEY, J. P. (1961). Supplementary vitamin D_3 for lambs. *Irish J. agric. Res. 1*, 43–48.
(12) DONALD, H. P. (1963). Modern trends in animal health and husbandry: Livestock genetics. *Brit. vet. J. 119* (1), 12–22.
(13) DAVIES, W. (1960). Temperate (and Tropical) Grasslands. *Proc. VIII int. Grassl. Congr.* (Reading), 1.
(14) FRASER, A. (1951). *Sheep Husbandry.* 2nd ed. London, Crosby Lockwood.
(15) HUXLEY, J. (1962). Eastern Africa: the ecological base. *Endeavour 21* (82), 98–107.
(16) JONES, J. G. W. (1954). The efficiency of mutton production in crossbred sheep. *Proc. Brit. Soc. Anim. Prod.*, 29–34.
(17) KALINOVSKAJA, N. A. (1962). Prolific sheep of the mountains—Svanka. *Ovtsevodstvo 8* (11), 39. (Anim. Breed. Abstr. (1963), *31* (2), 1177.)
(18) KRAAY, G. J., GAILLARD, B. D. E. & BROUWER, E. (1961). On high-potassium and low-potassium sheep in the Netherlands. *T. Diergeneesk.*, 86, afl. 14, 937–946.
(19) MAIJELA, K. (1958). Unpublished records. Personal communication (1963) from G. Wiener & A. F. Purser.
(20) MINISTRY OF AGRICULTURE, FISHERIES & FOOD (1956). Sheep breeding and management. *M.A.F.F. Bull. No. 166.*
(21) MORRIS, I. (1961). *Report on Sheep Recording and Progeny Testing.* London, H.M.S.O.
(22) PÀLSSON, H. (1962). Methods of increasing meat production per sheep in Iceland. *Proc. Ruakura Fmrs' Conf.* (Hamilton, N.Z.), 61.
(23) PATTIE, W. A. & DONNELLY, F. B. (1962). A comparison of sheep breeds for lamb production on the central-western slopes of New South Wales. *Aust. J. exp. agric. Anim. Husb. 2*, 251–256.
(24) ROBERTS, E. M., CLAPHAM, B. M., MCMAHON, P. R. & RIKARD-BELL, L. (1962). A report on sheep and wool production in the Soviet Union. *Wool Tech. Sheep Breed. 9* (2), 61–90.
(25) ROBERTS, R. C. (1959). The use of pedigree and Mountain-type rams. *N.S.B.A. Yearbk.*, 36–42.
(26) RYDER, M. L. (1960). Sheep breeds in history. *N.S.B.A. Yearbk.*, 17–25.
(27) SIDWELL, G. M., EVERSON, D. O. & TERRILL, C. E. (1962). Fertility, prolificacy and lamb livability of some pure breeds and their crosses. *J. Anim. Sci. 21* (4), 875–879.
(28) TERRILL, C. E. (1962). The reproduction of sheep. In *Reproduction in Farm Animals.* Ed. Hafez, E. S. E. London, Baillière, Tindall & Cox.
(29) THOMAS, J. F. H. (1940). *Sheep.* 2nd ed. London, Faber & Faber.
(30) THOMSON, W. & AITKEN, F. C. (1959). Diet in relation to reproduction and the viability of the young. II. Sheep: World Survey of Reproduction and Review of Feeding Experiments. *Comwlth. Bur. of Anim. Nutr. Tech. Commun.* No. 20.
(31) TURNER, H. N. (1959). Ratios as criteria for selection in animal or plant breeding, with particular reference to efficiency of food conversion in sheep. *Aust. J. agric. Res. 10*, 565–580.
(32) VELLOSO, G. (1963). Personal communication.

(33) WESTON, R. H. (1959). The efficiency of wool production of grazing Merino sheep. *Aust. J. agric. Res. 10*, 865–885.
(34) WIENER, G. (1961). Population dynamics in fourteen lowland breeds of sheep in Great Britain. *J. agric. Sci. 57*, 21–28.
(35) YOUNG, G. B. & PURSER, A. F. (1962). Breed structure and genetic analysis of Border Leicester sheep. *Anim. Prod. 4* (3), 379–389.
(36) ZUCKERMAN, S. (1958). The sheep industry in Britain, 1958. *Nat. Resources (Tech.) Committee.* London, H.M.S.O.
(37) BARBER, D. & YOUNG, W. W. (1965). The Colbred story. *Agriculture, Lond. 72*, 2–6.
(38) BRITISH WOOL MARKETING BOARD (1968). *British Sheep Breeds, their Wool and its Uses.*
(39) CADZOW, J. B. (1966). New types of sheep and their future. *Scott. Agric. 45*, 132–136.
(40) CARTER, H. B. (1967). The Merino sheep in Great Britain. *Text. Inst. & Industry 5* (4), 97–99.
(41) DONALD, H. O., READ, J. L. & RUSSELL, W. S. (1968). A comparative trial of crossbred ewes by Finnish Landrace and other sizes. *Anim. Prod. 10* (4), 413–421.
(42) MASON, I. L. (1951). A world dictionary of breeds, types and varieties of livestock. *Tech. Comm. No. 8, C.A.B.*
(43) MACFARLANE, W. V., HOWARD, B. & SIEBERT, B. D. (1967). Water metabolism of Merino and Border Leicester sheep grazing saltbush. *Aust. J. agric. Res. 18*, 947–958.
(44) MORLEY, F. H. W. & SPEDDING, C. R. W. (1968). Agricultural systems and grazing experiments. *Herb. Abstr. 38* (4), 279–287.
(45) NATIONAL SHEEP BREEDERS ASSOCIATION (1968). *British Sheep.* Rev. ed. N.S.B.A.
(46) RYDER, M. L. (1968). Sheep for the 1970s. *Span 11* (1), 40–42.
(47) RYDER, M. L. & STEVENSON, S. K. (1968). *Wool Growth.* London & New York, Academic Press.
(48) WIENER, G. (1967). A comparison of the body size, fleece weight and maternal performance of five breeds of sheep kept in one environment. *Anim. Prod. 9* (2), 177–195.

XIII

THE POTENTIAL FOR PASTURE PRODUCTION

In this and the two succeeding chapters it is intended to consider the potential for sheep production. This can best be done by examining the possible effects of changes in those factors which have emerged as dominant in the production process and its practical achievement. Many factors are important, and some that appear unimportant now may become important in the future, but relatively few can be visualized as capable of greatly enlarging the potential for production. Before coming to these few, however, it is worth elaborating a little on this question of factors changing in importance. This is part of a more general proposition concerning limiting factors. Any factor is potentially limiting and, indeed, as soon as one is made non-limiting another becomes the limiting one. Thus, the efficiency with which a lamb digests and uses its food is important and, if everything else were non-limiting, this would be the barrier to further achievement. In fact, as pointed out in Chapter VI, the quantity of food consumed directly by the lamb is small in relation to the amount used to keep the ewe for a whole year. Improvement in the efficiency of food conversion by the lamb would have to be very great indeed before it would significantly affect the efficiency of the ewe/lamb unit. The number of lambs per ewe has potentially a much greater effect. Suppose, however, that this was expanded to the optimum value and that this, simply to illustrate the point, was six lambs per ewe per year. The proportion of the food used by lambs would be increased approximately six-fold, compared with a ewe and a single lamb, and any increase in the efficiency with which it was used might have quite a large effect.

So it is, then, that the major factors in sheep production may remain the same, but those which offer the most scope for increasing the future potential are a selection relevant primarily to the time at which it is made. At the present time there appear to be three such factors, (*a*) the pasture yield, (*b*) animal health and (*c*) the reproductive rate of the sheep. The first of these is the subject of this chapter.

Pasture Yield

Animal output per acre clearly depends on the production of pasture per acre. This is the foundation on which it is built and whatever the efficiency of the animal population, it cannot yield beyond the limits established by its food supply. For any given pasture yield, there exists some upper limit, therefore, in terms of animal output. It is doubtful if many pastures are used to anything like this extent. On those that are not, animal production can be increased by increasing the stocking rate or by employing a more efficient animal population. It must not be thought, however, that pasture yield, as a determinant of animal output, can be measured in simple terms. The amount of dry matter grown is not necessarily relevant, and the total quantity of digestible nutrients may not be either, although the latter is probably the nearest to a satisfactory single figure expression that can be obtained at present. This total would represent the amount of dry matter multiplied by its apparent digestibility, the latter being the simplest useful estimate of nutritive value (4). In most grazing situations the seasonal distribution of the food supply is extremely important. Where this cannot be matched by organization of the animal needs or by conservation it may be more efficient to supplement the supply at critical times. Where the animal population has been maintained at a less efficient structural level than could be arranged, as, for example, at a very low lambing percentage, quite small increases in pasture output could result in very large increases in animal output. A very simple model of the possible effect of supplementation is shown in Fig. 13.1.

In order to discuss major possibilities for change in the pasture potential, it will be necessary to use one simple expression for pasture yield. In doing this, it is recognized that the meaning to be attached to any change must be related (*a*) to the quantity used representing a proportionate change in the amount of nutrients, in a form required by the relevant animals, and (*b*) to the extent to which the needs of an efficient animal population can be fitted to the pattern of supply. It is also assumed that the attainment of potential agricultural production implies a considerable measure of control of both inputs and the harvesting of output.

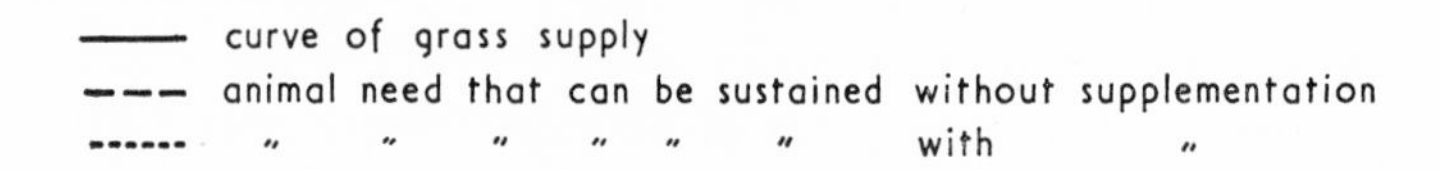

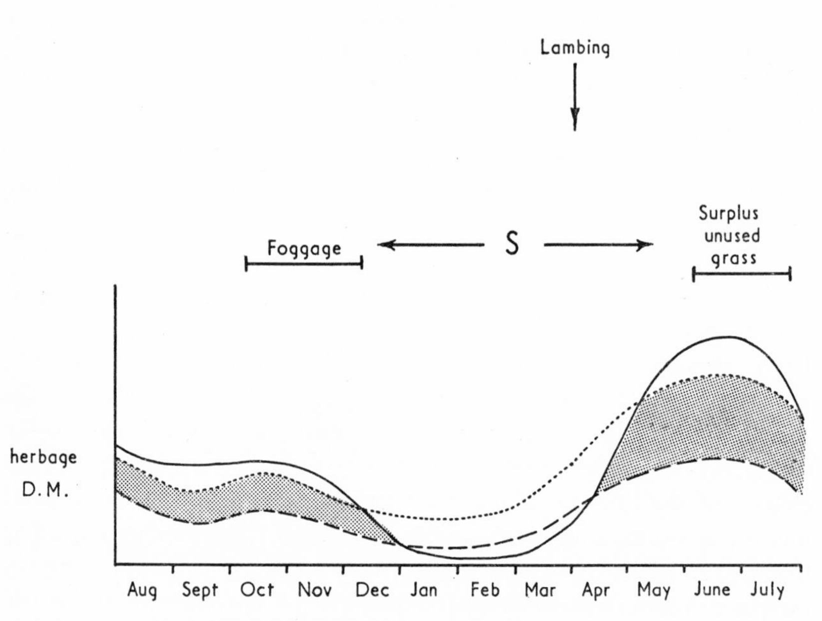

FIG. 13.1—*The effect of supplementary feeding at critical times on the annual carrying capacity of a pasture, without conservation, grazed by an animal population of low productivity (e.g. a mountain ewe flock with a low lambing percentage)*

The curves are hypothetical and illustrate the fact that it is frequently an acute shortage of herbage at one time of the year that may limit the carrying capacity for the whole year. During the period labelled S, supplementary feeding would be required to sustain the upper curve of animal needs: during this time the lower curve would involve loss of body weight in the ewes. The shaded areas represent herbage unused at the lower stocking rate and used at the higher rate.

In the United Kingdom, the average yield of a well-managed grass-clover sward without nitrogenous fertilizer has been estimated as 5,000–6,000 lb of dry matter per acre per annum. Although greater yields than this are sometimes achieved with a pure clover crop or with a mixed sward (7), particularly with irrigation, it has been suggested (5, 10) that dependence on biological fixation of nitrogen sets a limit of about 6,000 lb dry matter per acre to the annual production of a mixed grass and clover pasture. This is clearly not so for pastures in general and in New Zealand, for example, Brougham (24) has estimated that the *average* annual production of *utilized* dry matter from dairy pastures is approximately 3,000 lb D.M. per acre: higher yields, of up to 7,000 lb utilized D.M. per acre per year, are being obtained in quite a wide range of New Zealand environments, with exceptional yields of up to 9,500 lb.

Considering the whole range of data available for New Zealand, Brougham (24) has put the range of production from 9,000 lb D.M. per acre per year to 19–25,000 lb.

The production of lucerne, relatively independent of both nitrogen and water supply, is also capable of reaching up to 20,000 lb D.M. per acre per year (26).

The use of nitrogen

It was pointed out in Chapter II that added nitrogen may result in a poor response, per lb of nitrogen applied, if it merely reduces the proportion of clover and replaces the nitrogen that this previously supplied. In a similar way, the re-circulation of nitrogen via the excreta of the grazing animal may have little effect on herbage yield unless the input of additional nitrogen is high. It appears that major increases in pasture production may have to be sought from a greatly increased application of fertilizer nitrogen to the pure grass sward: this is the kind of sward that results from a high level of nitrogenous fertilizer, anyway. Reid (27) applied 33 rates of fertilizer N, ranging from 0 to 800 lb per acre, to swards of perennial ryegrass (with some timothy). Up to 150 lb of applied N per acre, the response was over 26 lb D.M. per 1 lb of N: thereafter each additional 1 lb of N produced only 13 lb D.M., between 200 and 250 lb N per acre, and less than 1 lb at rates over 500 lb N per acre. It should be unnecessary to point out that this discussion of potential output does not necessarily imply that economic and biological optima will coincide. As Sanders (20) has emphasized, there is often a world of difference between biological possibilities and economic realities in the matter of agricultural production.

Irrigation

Particularly in arid areas, or in dry seasons, the use of water will enormously increase pasture output (19) and in some parts of the world the biggest increases will come from this. The Snowy Mountains irrigation scheme in south-eastern Australia is an example (13). Here, the scheme is expected to add some 500,000 acres to the 1,500,000 acres already irrigated in this region. Fortunately, the ecological implications of the scheme are being considered, including the possible consequences to animal health. It might be expected that irrigation would exert a much greater influence on the incidence of disease-producing organisms than would the use of nitrogen. In some respects their influence might be similar, in that both affect the rate of herbage growth; but, in general, nitrogen in large quantities will be linked with pure grass, whereas water usage is often associated with a high legume content in the sward.

The combined use of nitrogenous fertilizer and water, where required, offers the greatest increase in herbage production. In Britain, it is capable of producing yields of the order of 10,000–12,000 lb dry matter per acre per annum. This doubling of the level of production derived from efficiently managed grass/clover mixtures is clearly a major effect. Nitrogen has also been shown to be a potent force for increasing herbage growth at the ends of the normal growing season (11), and, with some grass species, even in winter. In small, simulated swards of ryegrass, receiving a high level of applied nitrogen and ample water, Cooper (25) found that the total annual production from S.23 was between 18,000 and 19,000 lb D.M. per acre.

Since grass grown with a great deal of nitrogen or water or both does not appear to differ markedly in nutritive value to the grazing animal from that grown with lower inputs (17), probably the crucial question in assessing such an increase in the herbage production base-line, is the effect on the health of the sheep at all stages in its life.

NITROGEN AND SHEEP HEALTH

The application of a nitrogenous fertilizer may represent as many different treatments as there are kinds of nitrogen-containing manure and ways of applying them. The amount of herbage grown and the constituents of that herbage may vary with the total amount used, the frequency of application, the nitrogen status of the soil to which it is being added, the temperature and humidity of the environment and, of course, the input of other plant nutrients. The effect of this herbage on the health of the sheep may further depend on the plane of nutrition of the animal in relation to its metabolic needs. The latter will be quite different according to the size and physiological status of the sheep. This, then, is an important subject but, because of its complexity under grazing conditions, a difficult one to investigate experimentally. This is not to say that it is particularly difficult to carry out experiments which contribute to, or have a bearing on, the subject. Indeed, apart from the interactions with other plant nutrients, it is a relatively easy basis for field experimentation. The difficulty lies in the negative nature of many of the results. Adverse effects on health make it obvious that something is wrong with the grass or the way it is grazed. Even a great many occasions when no ill-health occurs, however, cannot rule out the possibility that a different result could occur under circumstances only slightly different, for example, under slightly different weather conditions.

The best assessment of the effect of the application of nitrogenous fertilizer on sheep health must therefore come from a consideration of both experimental investigations and practical experience. In view of what has been said about the experimental situation, it will be clear that negative results can never exclude other possibilities, but more confidence can be placed in them if they are obtained under more extreme conditions than those likely to be experienced in practice. In recent years, inputs of nitrogen have been studied which certainly come into this category at the present time. One such experiment at Hurley (in 1961 and 1962) concerned the growth and health of early-weaned lambs grazing S. 23 ryegrass receiving 100 lb of nitrogen per acre per month from February to November: a total of 950 lb of nitrogen was in fact used per acre during each grazing season. Three groups of lambs were kept: the first on grass grown in this way and grazed rotationally at intervals of 3–4 weeks; the second on similar grass swards receiving 20 or 60 lb of nitrogen per acre per month; and the third on these treatments with additional irrigation whenever a deficit of 1·0 inch developed. The aim of these treatments was to produce grass in circumstances where neither nitrogen nor water supply limited growth. One result, of course, was to produce grass grown very rapidly, some of it with a high-water content, dark green and as 'lush' in appearance as can be imagined. Such grass has often been regarded as suspect in relation to animal health, particularly in connection with hypomagnesaemia (16). Nitrogen contents were often high, but contents of most other minerals were within the normal range. Sodium was, in some cases, low in the grass receiving 20 lb N per acre a month. The health of the lambs was unaffected by any of these treatments at any time of the year. At certain times, the accumulation of nitrogen, measured as mineral nitrogen present in the soil after incubation (8) reached 200 lb per acre in the surface soil, i.e. the top 3 inches. There were often times, therefore, when a large amount of nitrogen was being added to a soil already containing a

TABLE XLVIII

THE MINERAL CONTENT OF RYEGRASS

(grown with 100 lb N/acre/month) and that of the blood of lambs grazing it

DATE	HERBAGE (D.M.)				BLOOD			
				p.p.m.	mg/100 ml			
	Ca (%)	*Mg* (%)	*Inorganic P* (%)	*Cu* (p.p.m.)	*Ca*	*Mg*	*Inorganic P*	*Cu*
4.5.61	0·79	0·15	0·39	8·5				
17.5.61					11·1	2·40	5·90	0·10
29.6.61	0·73	0·18	0·27	7·5				
14.7.61					12·1	2·36	5·72	0·09
27.7.61	0·78	0·19	0·29	9·0				
10.8.61					11·1	2·50	5.45	0·08

(Source: data from Expt. H. 287 at the Grassland Research Institute)

large quantity, and this under grazing conditions where faeces and urine patches could increase the nitrogen concentration still further in particular areas. The conclusion in these two years was clear, that grass grown with, and containing, unusually large amounts of nitrogen, does not necessarily involve any adverse effect on the health of lambs. Table XLVIII gives a typical series of analyses on the blood of the lambs and the grass that they grazed. The lambs used were injected against pulpy kidney disease and, for experimental purposes, free from most internal parasites; this should be remembered in interpreting the results. The implications of freedom from disease will be considered in Chapter XIV, but the possibility should be noted here that interactions between, for example, parasites, herbage and scouring are possible. It must also be noted that the application of other fertilizers was at a low level. In fact, in the December of the year preceding the experiment each acre received 67 lb of potassium and 42 lb of P_2O_5. No other fertilizer was added in the year of grazing and there was no evidence that any was required.

In view of this kind of result, and the caution with which it must be accepted as a basis for generalization, it is profitable to consider what health hazards are thought to be associated with the use of high rates of application of nitrogenous fertilizers in practice. They include scouring, nitrate poisoning, hypomagnesaemia and hypocalcaemia, in addition to suspicions of deficiencies and poor performance caused by reduced intake of 'unpalatable' grass.

Scouring

This is a symptom of many diseases and disorders, including those due to worm-infestation, but it may not necessarily be a bad thing in itself. Certainly, instances are known where a scouring animal has grown as well as its non-scouring companions. The consistency of the faeces frequently reflects the water content of the feed, and sheep faeces are always more fluid on grass and silage than they are on hay. There is little evidence, however, that lush grass, rapidly grown with nitrogen or irrigation, causes scouring, in the accepted sense of abnormally-watery faeces. At present, it would appear probable that scouring of sheep in practice occurs as a result of infections, probably often interacting with herbage of high-water content. In experiments with 'worm-free' sheep on dominantly clover or dominantly grass pastures, whether receiving heavy applications of nitrogen or not, scouring has been observed to occur sporadically in individual sheep, but has only lasted for about 24 hours in any one animal.

Nitrate poisoning

The excessive intake of nitrates can undoubtedly cause poisoning in sheep, due to the absorption of nitrites produced by rumen organisms. This has been found to occur on green oats and has also occurred due to a high nitrate water supply (12). The evidence for its occurrence on grass is negligible and it has frequently not occurred, for example, on grass with nitrate contents higher than those sometimes regarded as lethal (9). More evidence is required, however, particularly in relation to periods such as late pregnancy.

Hypomagnesaemia

In the field this is often thought to be associated with quickly grown grass and with the use of nitrogenous fertilizer. In fact, grass may be regarded as a rather poor supplier of magnesium at the best of times, but the use of nitrogen does not, by itself, greatly alter the magnesium content of grass (see Table XLIX). It may reduce the mean *herbage* magnesium content if it results in less clover, however. In any event little is known as to the availability of the magnesium in grass and the factors that may effect this. The well-recognized influence of the environment, including temperature, on the incidence of tetany has sometimes been thought to act partly through the herbage, either by altering its rate of growth and thus, perhaps, its magnesium content, or by affecting the amount eaten. Intake, of course, must be of the first importance. Anything which reduces intake is disadvantageous to performance and often to health, unless toxic substances are involved. There is no clear evidence, however, that nitrogen applications have any such effect on the grass intake of sheep.

In making such statements about the use of nitrogen in fertilizers, it is essential to recognize that it is only as a matter of convenience in discussion that one element can be isolated in this way. There is evidence that concurrent use of fertilizers containing nitrogen and those containing potassium tends to increase the incidence of hypomagnesaemic tetany (15, 18). It has frequently been shown that high rates of application of potassium result in a lowered magnesium content of grass (14) and it has been suggested (6) that it is the balance of minerals that is important, particularly the ratios of sodium and potassium to calcium and magnesium. Now the effect of applications of nitrogen on the mineral balance within the plant material must also depend on the application of other plant nutrients and their availability in the soil. The use of nitrogen can be said to result in no adverse consequences in some circumstances, but it is unlikely that there are no circumstances where adverse effects would occur. It is very important, when assessing the effect of any mineral addition to the soil, to confine the conclusion to the total context in which the assessment was made.

The state of the animal must also be borne in mind. In general, the incidence of hypomagnesaemia is highest during the peak of lactation and is considered to be greater in ewes suckling more than one lamb. In this connection, it should be remembered: (*a*) that such ewes commonly lose weight at this time so that only part of their nutrient requirement is met by current food intake; (*b*) that the magnesium reserves that can be withdrawn from the body of a mature ewe are negligible; (*c*) that the magnesium content of herbage is low in early spring because clover is not plentiful and the magnesium in grass is at its lowest; and (*d*) that the quantity of nutrients needed for

TABLE XLIX

THE EFFECT OF NITROGENOUS FERTILIZER APPLICATION ON THE MAGNESIUM CONTENT OF RYEGRASS

Values given are means for all samples taken between May and October (Expt. H. 126, 1957).

	N (LB PER ACRE)	Mg (% D.M.)	$\frac{K}{Ca+Mg}$ (M.EQ.)
Ryegrass (S. 23)	0	0·16	1·4
	120	0·18	1·5
	240	0·18	1·5
	480	0·20	1·1

(Source : T. E. Williams, unpublished data, Grassland Research Institute)

production by the sheep at this time may easily require a dry-matter intake beyond that which is possible on many herbages.

Hypocalcaemia

The results of this are often difficult to distinguish in the field from hypomagnesaemia. Much of what has been said in the previous section also applies here. The grounds are probably even stronger, however, for regarding the level of production of the ewe as a major factor.

It is frequently the case, of course, that high nitrogen usage allows a higher level of production, per individual or per acre, and that metabolic disorders are more likely in highly productive ewes. There are no grounds for believing that hypocalcaemia is likely to prove a barrier to high nitrogen usage, however. Certainly it has proved possible in some experiments to apply more than 160 lb nitrogen per acre per month to a pure ryegrass sward without any hypocalcaemia in ewes grazing it, whether with single or twin lambs. In this experiment, however, concentrates, at 1 lb a head a day, were also fed during the first 6 weeks of lactation. As before, it would be unwise to generalize to other contexts.

Deficiencies

It is often suspected that all kinds of mineral deficiencies would result from the intensive use of nitrogenous fertilizers. In many cases, however, such fertilizers contain appreciable amounts of other minerals as impurities. Stojkovska & Cooke (21) calculated that fertilizers containing a low concentration (about 10 p.p.m.) of a micro-nutrient and applied at a rate of about 500 lb an acre will make little contribution to the micro-nutrient status of the soil and will apply considerably less of the elements than may be removed by an annual crop. Many sources of fertilizer nitrogen may contain rather low levels of several of these elements. In general, there is remarkably little evidence of mineral deficiencies occurring when nitrogen is heavily used; and high nitrogen applications do not normally have any great effect on the other mineral constituents of grass (see Table XLVIII).

It has been postulated (1) that the function of a pasture feed is to maintain all animal body functions, including such things as saliva flow (which may be influenced by the content of fibre or minerals). It is often a simple matter, however, to add to the animal's diet whatever may be missing and it is arguable that a pasture should not be expected, necessarily, to supply all an animal's needs if more can be grown with some slight deficiency. The trace or 'micro-nutrient' elements (22) can sometimes be supplied to sheep as 'bullets', as in the case of cobalt, designed to release the correct amount each day.

Increased incidence of swayback has been linked with improvement of marginal grassland (2) but there is no clear association with nitrogen usage. The general notion that deficiencies would follow the use of heavy applications of nitrogen appears to be based largely on the idea that nitrogen applications are in some way forcing the grass to grow, and, furthermore, in an artificial, unnatural manner. It is probably idle to speculate as to whether it is more natural to have sufficient or insufficient nutrients for full growth; but it cannot easily be maintained that nitrogen forces grass to grow. The supply of nitrogen or any other plant nutrient simply removes one factor which may have been limiting the full expression of the plant's growth potential. A sward with more nitrogen is free to use more of the sun's energy and, if the latter is limiting, no amount of nitrogen will force growth. Indeed, the risks inherent in high rates of nitrogen application may be precisely at the times when other factors, such as temperature, limit growth and the sward cannot respond in this way. It is at these times that nitrogenous fertilizer tends to increase the nitrate content of the plant.

PERFORMANCE AND INTAKE OF FOOD

Reduced performance in sheep has not been shown to be unequivocally attributable to the use of nitrogen alone. This does not preclude such effects within certain contexts. At Hurley, in experiments using high nitrogen applications, performance of sheep has not so far been depressed and there is no evidence that food intake has been reduced. Where performance of sheep has been lowered by grazing on grass receiving a great deal of nitrogen, it is possible that worm-infestation has been responsible, or that a heavy stocking rate, made possible by the heavy use of nitrogen, has resulted in excessive soiling. Areas of pasture that have received much high-nitrogen urine may be unpalatable and patches of herbage once rejected quickly become overmature. In general, however, the effect of sheep urine on palatability is transient, lasting only a few hours; this is especially so if rain occurs shortly after urine has been deposited.

It is difficult to see, in any of the foregoing sections, evidence of major barriers to the use of as much nitrogen as is economically warranted. It is not possible, and in any event is beyond the present scope, to predict what the economic level might be. Efficient grassland usage may demand anything from complete dependence on clover to heavy usage of nitrogen at all seasons. It could equally well require very high rates of application but only over short periods. Whatever the desirable pattern of usage, nitrogen remains a major means of increasing herbage yield per acre and per unit of time and it does not appear to be limited in its usefulness by injurious effects on animal health. It should, perhaps, be unnecessary to add that such promising hypotheses are not intended to induce complacency. They should be subjected to the most rigorous and long-term investigation over a wide range of environments. The initial evidence should be interpreted as demonstrating that such exhaustive tests would be worth while.

Finally, it should be pointed out that the foregoing discussion has been confined to the use of existing species and varieties of herbage plants. Breeding better varieties, including those resistant to disease, and the introduction of new species into cultivation, are all procedures which could greatly extend the limits of dry-matter production (3, 23).

REFERENCES

(1) ANON. (1962). *N.Z. Agrcist 14* (3), 1–3.

(2) BARLOW, R. M., PURVES, D., BUTLER, E. J. & McINTYRE, I. J. (1960). Swayback in south-east Scotland. *J. comp. Path.* 70 (4), 396–410.

(3) BELL, G. D. (1958). Improvements in crop production. *The Biological Productivity of Britain.* 43–51. London, Inst. of Biol.

(4) BLAXTER, K. L. (1960). The utilization of the energy of grassland products. *Proc. VIII int. Grassl. Congr.* (Reading), 479.

(5) BROCKMAN, J. S. & WOLTON, K. M. (1963). The use of nitrogen on grass/white clover swards. *J. Brit. Grassld Soc. 18* (1), 7–13.

(6) BROUWER, E. (1961). Mineral relationships of the ruminant. In *Digestive Physiology and Nutrition of the Ruminant.* Ed. Lewis. London, Butterworth.

(7) CASTLE, M. E. & REID, D. (1963). Nitrogen and herbage production. *J. Brit. Grassld Soc. 18* (1), 1–6.

(8) CLEMENT, C. R. & WILLIAMS, T. E. (1962). An incubation technique for assessing the nitrogen status of soils newly ploughed from leys. *J. Soil. Sci 13* (1), 82–91.

(9) CONROY, E. (1961). Effects of heavy applications of nitrogen on the composition of herbage. *Irish J. agric. Res. 1*, 67–71.

(10) COWLING, D. W. (1962). The effect of white clover and nitrogenous fertilizer on the production of a sward. II. *J. Brit. Grassld Soc. 17*, 282–286.

(11) COWLING, D. W. (1962). Nitrogenous fertilizer and seasonal production. *J. Brit. Grassld Soc. 18* (1), 14–17.

(12) DIVEN, R. H., REED, R. E., TRAUTMAN, R. J., PISTOR, W. J. & WATTS, R. E. (1962). Experimentally induced nitrite poisoning in sheep. *Am. J. vet. Res. 23* (94), 494–496.

(13) GORDON, H. McL. (1962). The Snowy Mountains scheme and animal health. *Outlook on Agric. 3* (5), 225–233.

(14) 'T HART, M. L. (1960). The influence of meteorological conditions and fertilizer treatment on pasture in relation to hypomagnesaemia. *B.V.A. Conf. on Hypomagnesaemia* (London).

(15) HENDERSON, R. (1960). The application of potassic fertilizers to pasture and the incidence of hypomagnesaemia. *Tech. Ser. No. 1*, Potash Ltd.

(16) HIGNETT, S. L. (1961). Hypomagnesaemia and some other metabolic disorders of the grazing animal. *J. Fmrs' Club.* Pt. 2, 21–28.

(17) LARGE, R. V. & SPEDDING, C. R. W. (1963). The health and performance of lambs grazing a perennial ryegrass pasture receiving nitrogen and irrigation. *Exp. Grassld Res. Inst.* (Hurley), *15*, 52.

(18) ØDELIEN, M. (1960). Is the fertilizer treatment a contributory cause of hypomagnesaemia and tetany in cattle? *Landbrukshøgskolans Institutt for jordkultur, Saertrykk. 48.*

(19) Penman, H. L. (1958). Water control for increased crop production. *The Biological Productivity of Britain.* London, Inst. of Biol. 91–99.
(20) Sanders, H. G. (1958). Present agricultural production. *The Biological Productivity of Britain.* London, Inst. of Biol. 35–42.
(21) Stofkovska, A. & Cooke, G. W. (1958). Micro-nutrients in fertilizers. *Chem. Ind.*, 1368.
(22) Underwood, E. J. (1962). Soil-plant-animal inter-relations. *Trace Elements in Human and Animal Nutrition.* 2nd ed. New York & London, Academic Press. Inc.
(23) Watson, D. J. (1958). Factors limiting production. *The Biological Productivity of Britain.* London, Inst. of Biol. 25–34.
(24) Brougham, R. W. (1966). Potential of present type pastures for livestock feeding. *N.Z. Agric. Sci. 1* (8), 19–22.
(25) Cooper, J. P. (1967). Rep. of Welsh Plant Breeding Station, 1966, 14.
(26) Langer, R. H. M. (1966). Lucerne development for the South Island. *N.Z. Agric. Sci. 1* (8), 32–33.
(27) Reid, D. (1966). The response of herbage yields and quality to a wide range of nitrogen application rates. *Proc. X int. Grassld Congr.* (Helsinki), 209–213.

XIV

THE PREVENTION OF DISEASE

Disease has not been separately considered in previous chapters, for two main reasons. First, there are excellent works available on sheep diseases and for detailed descriptions of symptoms, causes and cures it is far better to refer to them. Secondly, disease has been regarded as secondary to both the production process and its application in practice.

The whole purpose of trying to build a model of the production process is to simplify a complicated situation by selecting only the most important factors and their more important interactions. With a firm grasp of the essential process, it is possible to embrace the additional facts particular to one environment. The basic ideas thus act as a structural skeleton which can be appropriately clothed for any particular occasion. Disease has no real part in the essential process, but appears most frequently as part of the environment in which production occurs or as a result of interactions between the production process and the environment (4, 6). Particular diseases and disorders have therefore been referred to when it seemed appropriate. Nutritional disorders have thus been discussed in chapters on the animal and the pasture, while infectious diseases have tended to find a place in chapters on management. When mentioned they have been considered as undesirable risks or consequences of some facet of production or of the husbandry practices involved in its application. By recognizing the association between a particular context and the incidence of a disease, it is possible to analyse the context, experimentally, into its constituent parts and identify that part which is primarily responsible for the occurrence of disease. This is often quite different from identifying the cause of the disease. It is required even when the cause is known: it may be a somewhat empirical process when the cause is not known.

The Effect of Disease on Production

Disease, then, has been looked on as something which interferes with the production process. Only when it does so is it of agricultural importance, but it is essential to consider this over an adequate period of time in a relevant environment. When disease interferes the primary aim should be prevention. To be thoroughly realistic, however, the following facts should be recognized. First, prevention may cost more than the cure. This is often true when insufficient is known of the causes to make prevention either easy or certain. A current example might be hypocalcaemia. Administration of calcium is a relatively simple cure, provided the need is recognized in time and arrangements exist for rapid treatment. It will be noted that, in such cases, disease may be deliberately *risked* because it is unlikely to affect more than a small proportion of the animals. There are obviously a number of situations where guaranteed prevention of disease in all sheep would be more expensive than curing the few that might be affected.

Secondly, this argument can be extended to a series of years, where a disease may only occur in some years and in some places. This was probably the situation with 'husk', or parasitic bronchitis due to lungworm infestation, in cattle before a vaccine existed. The severity of the occasional outbreak was such, however, that as soon as a vaccine was available it was widely used. In the short term, it is presumably used unnecessarily on occasions; the relevance of this fact must always depend upon whether or not such occasions can be predicted.

Thirdly, there are diseases which may be tolerated in all sheep because of the low level of damage and the high cost of prevention. Some parasitic infestations may come into this category and, in some such cases, prevention may not be possible in the relevant situation.

Fourthly, some diseases, even when not actually doing a great deal of harm at the time, may be worth curing because of the risk they constitute for other sheep. An example of this is footrot, which is worth curing in one ewe during the winter because of its potential for trouble in the following season. In this case, the cost of the treatment should not be considered simply in relation to the productivity of the treated ewe.

Finally, the human beings involved have to be considered. Even where a disease is present at a level where it would be uneconomic to eliminate it, those who work with the animals may dislike its presence sufficiently to cause some action to be taken; or eradicaton may have to be undertaken because of the risk to human health; this could be so in relation to a disease like orf.

These are all examples of the many different forces which have a bearing on the control of disease. It is probably reasonable to conclude that economic considerations are bound to take first place in most cases but it would be short-sighted not to foresee great changes in the economic values of existing techniques and not to recognize the inevitability of new techniques, once the need is demonstrated. In the remainder of this chapter, economic criteria will not be further considered, although they may happen to coincide with, or parallel, biological criteria.

Causes of Disease

Diseases may be due to inherited factors or to management practices but, for the purposes of this discussion, may be divided into those of nutritional origin and those resulting from infective agents. The first have been mentioned in the previous chapter in so far as they represent possible consequences of a marked increase in pasture production. In general, the aim must surely be proper nutrition of all animals, quantitatively and qualitatively. The object in producing animal foodstuffs must be to achieve a balanced diet, free from excesses and deficiencies. There are many difficulties in achieving this with certain naturally occurring diets and supplements may be required. All that needs to be emphasized here, is that the aim should be to avoid nutritional diseases and disorders, but a great deal of knowledge is required in order to feed, correctly, sheep in different physiological states.

Nature and magnitude of effects

The diseases involving infective agents present a rather more complicated set of problems. Before discussing them it is as well to consider the order of effect that disease of any kind has on productivity at the present time. Tables L, LI, LII and LIII have been compiled or selected from recent surveys (3, 5, 12) of the incidence of sheep diseases. Table L shows that the percentage of barren ewes may be greater than the percentage of ewes which die during pregnancy. The combined effect on the number of lambs born, to

TABLE L

LOSSES OF EWES, ASSOCIATED WITH PREGNANCY AND PARTURITION

In Yorkshire and Lancashire (1953–54).

	Estimated %	
	Yorkshire	*Lancashire*
Ewes barren	4·10	6·40
Ewes dying during pregnancy	2·57	5·09
Ewes aborted	1·80	3·15
Ewes dying at or after lambing	1·44	1·28
Stillborn parturitions	1·92	1·69
Successful lambings	87·89	82·73

(from a survey by Leech & Sellers (5))

TABLE LI

PERCENTAGE PERINATAL MORTALITY IN PREGNANT EWES

According to breed	Single births	Multiple births	Weighted mean
Longwool	0·91	2·95	1·78
Down breeds and crosses	1·35	4·41	2·67
Masham	0·42	1·41	0·84
Scottish Half-bred	0·51	1·76	1·04
Hill breeds and crosses	0·93	3·06	1·84
Mixed	0·55	1·83	1·09
Weighted mean	0·78	2·56	—
According to type of husbandry			
Hill	0·53	1·77	1·06
Grassland	0·55	1·84	1·10
Mixed	1·42	4·62	2·80
Arable	0·09	2·81	1·70
Weighted mean	0·78	2·56	—

(from a survey by Leech & Sellers (5))

TABLE LII

CAUSES OF DEATH IN EWES IN NORTHERN IRELAND

Cause of death	Percentage of total deaths
Pregnancy toxaemia	20·0
Accidents and injuries	17·3
Specific infective diseases (enterotoxaemia, braxy, etc.)	14·2
Parasitic diseases (notably fascioliasis)	11·8
Severe weather conditions	7·5
Diseases of the genital system and mammary glands (chiefly dystokia)	7·2
Unthriftiness, debility, etc.	5·5
Miscellaneous	3·4
Unknown and undiagnosed	13·1
	100·0

(Source : Extracted from the results of a survey carried out in 1954–55 by Gracey (3))

TABLE LIII

CAUSES OF DEATH IN LAMBS IN NORTHERN IRELAND

Cause of death	Percentage of total deaths
Severe weather, non-production of milk by ewe, etc.	39·8
Specific infective diseases (pulpy kidney, tick pyaemia, etc.)	23·1
Accidents and injuries	18·6
Born in weak condition	10·8
Miscellaneous	1·4
Unknown and undiagnosed	6·3
	100·0

(Source : Extracted from the results of a survey carried out in 1954–55 by Gracey (3))

which may be added the effect of the number of ewes aborting, can be considerable. Successful lambings varied from an estimated 83 per cent (Lancs) to 88 per cent (Yorks). Table LI illustrates the influence of the number of lambs on the perinatal mortality of ewes. From a quite different survey, Table LII analyses the causes of death in ewes and underlines the importance of pregnancy toxaemia. For hill conditions, Gunn (18) has recorded a mean ewe mortality rate of 3 per cent to 4 per cent, mostly during pregnancy and lactation, on three experimental farms. The general level of ewe mortality in the U.K., however, is likely to be higher than this. A survey for England and Wales in 1958–9 (19) estimated percentage deaths at 5·6 per cent to 10 per cent, depending on the region. Records over a 5-year, intensive grazing experiment (21) showed much variation from year to year, but the mean for 5 years was higher (12·4 per cent) for the very heavily stocked group, compared with that for a less heavily stocked group (8·2 per cent). In Table LIII the losses of lambs are separated into various classes. The very high proportion of deaths due to bad weather or weak condition shortly after birth emphasizes in a different way the effect of nutrition on the ewe during late pregnancy and early lactation. The data illustrate one main kind of influence, the untimely death of either ewe or lamb. It will be recalled (from Chapters IV and VI) that both longevity of the ewe and litter size exert a marked effect on productivity. All diseases which result in the premature death of the ewe lower her life-time performance and increase the food costs of what she has produced. Whenever lambs die, before it is intended, the potential production per ewe is decreased and, again, the efficiency with which her remaining output is produced is lowered. It is perhaps accepting a very wide interpretation of disease to attribute to it the very large loss of lambs which often occurs within a short period after birth. Actual death of lambs is probably not so common after they are 3 to 4 weeks old. Purser & Karam (20) found that early lamb mortality was associated with fine birthcoat and Bichard & Cooper (17) reported losses of lambs, varying with age of the dam, from 23·7 per cent for lambs from 1-year-old ewes and 16·6 per cent (2 years old) to 12·9 per cent from older ewes. Although such losses may in fact result from a great variety of diseases, including, for example, mycotic dermatitis (8), diseases beyond this age exert their effects chiefly on growth rate. In the lamb (see Chapters V and VI), growth rate is probably the major factor influencing the efficiency of production. A reduction in rate of growth generally increases the amount of food required per unit of growth. Diseases which reduce the milk output of the ewe affect productivity in the same way, of course, by reducing growth rate in the lamb. In addition, they may have the effect of increasing the food cost of the milk that is produced. The important point is that disease not only reduces output per acre or per animal, it is responsible for a considerable reduction in the efficiency of production.

At any point in time, in any given situation, it may be desirable to limit output per unit of any one natural resource used. It may, for example, be better to produce less than the maximum possible per acre and to reduce the inputs of fertilizer, labour and so on. In other circumstances, it may be better to maximize output per acre at the expense of output per animal. It is unlikely that it will ever be sensible to produce less than the maximum possible from the total inputs actually employed, however. Once these inputs have been decided upon, it must be profitable to obtain the maximum output possible without increasing them. Thus disease limits potential production (13) and also reduces the efficiency of the production process. It does so by limiting potential rates of production and by decreasing the food conversion ratio. In fact, viewed from the point of view of production from a sheep population, disease simply reduces the efficiency with which food is converted into product. It is worth stressing that disease can reduce efficiency at any level of production: the elimination of disease is, in fact, one way of increasing the efficiency of animal production without necessarily increasing total output. In a grazing flock, the elimination of disease could have 3 possible effects: (*a*) more product would result from the same sheep consuming the same amount of food; (*b*) the same product could be produced by the same sheep consuming less food. This would be more efficient than it was but, under grazing conditions, would leave surplus grass. This would have to be utilized in other ways or less would have to be grown; (*c*) the same product could be produced by fewer sheep, each sheep eating as much as before but, in total, consuming less food. Again, under grazing conditions, this would result in surplus grass.

In both (*b*) and (*c*), output could remain constant while the inputs required for growing grass, including acreage, could be reduced. Only in (*c*), however, would efficiency be high so that, in this case, the reduction in inputs would be greatest.

Disease is thus a factor limiting current efficiency: in the future it could be a major limiting factor to production. Assuming that nutritional disorders can be eliminated by applying increased knowledge of sheep nutrition, it is still probable that the incidence of other diseases would be more likely to increase with increasing output per acre, per sheep, per unit of food or time. This is not the same thing as assuming, for example, that higher stocking rates inevitably lead to increased parasitism, which is not necessarily true. It is simply a recognition of the importance of the rate of reproduction and of the fact that the young animal is generally the more susceptible to infectious diseases. Many of the latter do increase with increased stocking rates of young animals at pasture or with crowding indoors or in outdoor pens. So, an animal population with a much higher proportion of young stock combined with increased use of each unit of space and time, probably represents an environment where infectious diseases could be even more important. It is worth considering, therefore, the possibilities of the elimination or prevention of such diseases.

Elimination and Eradication of Disease

A distinction must be made between the elimination of disease and the eradication of the organisms which cause it. In order to avoid confusion in this discussion the following meanings will be attached to the listed words.

Disease: a morbid condition of the body.

Subclinical: applied to a low level of disease or of infection or infestation with an organism capable of producing disease; also applied to the damage caused by such infestations. No marked, externally visible symptoms.

Clinical: a level of damage at which the animal as a whole may be affected and show marked external symptoms.

These are in no sense proposed as definitions: they are the relatively simple senses in which the words are used here.

There would seem to be little reason why disease should not be eradicated wherever possible and practicable. In this context, practicable implies that the disease-free state can be maintained. If it can, then initial cost and difficulty may possibly be ignored. Much the same might be said of the eradication of disease-producing organisms. Since the only reason for retaining them must be the maintenance of immunity, if they could be permanently and entirely removed from the scene there would be no need for either them or the immunity towards them and their effects.

A detailed consideration of this topic would need to treat each disease and organism separately and much resistance to ideas of eradication can be traced to a rather wholesale advocacy of it in general. It is not proposed to provide a detailed specific treatment here but to distinguish some of the more important general principles in relation to sheep production.

Scale of operation

It is, of course, quite unnecessary to assume that elimination or eradication must apply to the whole world, the whole of Great Britain, to a whole farm or to a whole flock, although it may ultimately be sensible to apply as widely as possible any scheme that proves successful on a smaller scale. Obviously, some small-scale schemes would suffer greatly from being surrounded by a source of re-infection, but this chiefly affects the methods and techniques required. The distinction between possible and practicable, however, must not be too finely drawn. Since the elimination of every disease and every parasite is clearly possible given sufficient scope in choosing and controlling the environment, the whole question is one of practicability—the maintenance of the disease-free or parasite-free state within an agricultural environment.

Control of the environment

It should be noted that given an environment where it can be done initially there may be no further difficulty. Housing, yarding or penning of sheep can confer all the control needed (16) for the maintenance of parasite-free sheep or, at least, for freedom from the vast majority of species. The argument that, one day, immunity might be required loses its force, since there is no reason why it should be needed whilst the controlled environment is retained. In circumstances where the most powerful immunity appears to be of the acquired kind, and not of genetic origin, it would seem unnecessary to retain parasites against some future time when all the housed sheep or their progeny might have to go out to pasture again.

The same is true, of course, of any sufficiently controlled pasture environment. One of the reasons, in the past, why controlled environments and sheep have seemed almost a contradiction in terms, has been the extensive fashion in which sheep have been kept, and it must be pointed out that where extensive sheep husbandry is practised, now or in the future, some diseases may not be sufficiently serious to warrant elimination. In exploring the future potential of sheep production, however, elimination of disease may not only become desirable, but even essential, and the intensity of land use may considerably alter the scale of the problem. This would be further altered by any segregation of stock into separate groups of young and old, for example, or disease-free and infected animals.

Separation of ewes and lambs

There are some reasons for supposing that very intensive production, and reproduction, may involve separation of ewes and lambs at an early age, possibly at birth. There is no great difficulty in arranging that such lambs would be free from disease and from a great many of the disease-producing organisms, at the stage when separation occurs. Before pursuing this possible development, however, it may be noted that separate treatment of ewes and lambs does not necessarily require physical separation. The possibilities of rearing lambs virtually free from worm parasites, by continuous folding of their normally infected ewes over non-infected pasture, were discussed in Chapter IX. It was also pointed out that immunity need only be induced in the stock that might require it and that this need was related to the length of time for which they would be exposed and to the control attainable for that time. On all grounds, this suggests immunity in the breeding stock and a low level of infestation with those parasites to which resistance can only be obtained in this way. Any method of obtaining an immunity without the presence of the organism may prove preferable (7). For the prevention of enterotoxaemias, perhaps the most widespread cause of sudden death in all ages of sheep (11, 15), vaccination already offers a solution. In so far as they may represent diseases which result from the effects of a bacterial population, which would remain completely harmless under other nutritional circumstances, it is obviously possible that greater knowledge and control of the latter could replace vaccination. In some cases, there is no good reason for tolerating the presence, in the ewe, of a parasite that can be eliminated. This is probably true of the liver fluke, for example.

To return to the intensive rearing of lambs which have been

separated at an early age from their ewes, under grazing conditions: it is clear that this must chiefly concern animals intended for meat. It is thus a short-term operation with no special consequences in the longer term. It is probably the situation of greatest need for disease control and of least risk from the attempt. A series of such short-term operations could take place in different fields in different years. Only the meat-producing lambs need be treated in this way and maintained under disease-free conditions; breeding stock and their replacements need not be involved. It is probable that this would prove the most successful application of such complete disease control as can be achieved at the present time and would provide valuable experience on the practicability of going further. Arguments, for example, about a worm-free farm or a worm-free country, will be of limited value until more information is available on the problems and risks of such attempts, on the probable benefit, and on such things as the likelihood of natural populations, of rabbits and hares, for example, maintaining stocks of the very organisms involved in an eradication scheme. The lack of such information is, however, not entirely unconnected with preconceived notions as to the possibility and desirability of eradication.

A DISEASE-FREE UNIT

The first step in an objective assessment might be to consider what is involved in a small-scale, limited attempt at the elimination of disease, and what results are associated with it over a period of several years. It is proposed, therefore, to refer briefly to a special unit, set up at the Grassland Research Institute, for the purpose of improving the precision with which herbage could be evaluated by sheep. It is obvious that a disease-free animal offers considerable advantage in the assessment of the food value of herbage and of methods of utilization, since its performance is less affected by extraneous factors (14).

A detailed description of the 45-acre unit has been published (22), describing the rotational use of land practised within it; a brief summary of the results follows.

Summary of Results, 1959 to 1962

The risks of infection from outside the area are difficult to assess because they cannot easily be separated from those associated with low-level survival within the area. In either event, since very young animals have been used throughout, the biggest risk would seem to be a rapid build-up from a small infection from either source. This build-up would seem more likely at higher stocking rates, with irrigation or in warm, humid seasons. Table LIV shows the number of lambs used each year, the number of grazing days and the length of the grazing season, with an indication of the main managements in use, from 1959 to 1962. Table LV summarizes the most relevant data derived from examination of faeces. This was carried out with

TABLE LIV

NUMBERS OF SHEEP GRAZED IN THE 'WORM-FREE' AREA 1959–62

	1959	1960	1961	1962
Total number of lambs	33	42	51	85
Number of grazing days	4,398	3,705	6,920	11,198
Length of the grazing season	MARCH →		→ DECEMBER	
Number of separate lamb crops	3	3	2	3
Management, etc.	Set-stocked on ryegrass/white clover		Rotationally grazed on irrigated ryegrass receiving differential nitrogenous fertilizer treatments	

variable frequency, depending on the nature of the experiment, but generally using 5–10 g of faeces. Table LVI gives data derived from post-mortem examinations. The results of such examinations refer to the species present at the end of the grazing period and where eggs have occurred in the faeces it has usually been towards this time. Post-mortem counts are thus likely to represent the greatest number that have occurred. For most species a high proportion of the lambs have been completely free. Those that have not, have never carried very many and their performance has not been different from that of their fellows.

At this point, the practical issues involved in the assessment of

TABLE LV

TYPICAL FAECAL EGG-COUNTS FROM LAMBS IN THE 'WORM-FREE' UNIT

Mean number of eggs per 5 g of faeces (1961).

Age of the lamb (months)	*Trichuris ovis*	*Nematodirus* spp.	'Other strongyles'	*Strongyloides papillosus*	*Moniezia* spp.
3	0	0	0	0	0
4	<1	<1	0	<5	0
5	<1	1	0	<5	0
6	1	<1	<1	<5	0
7	<1	1	<1	<5	0

TABLE LVI

RESULTS OF POST-MORTEM EXAMINATIONS OF 65 LAMBS

used in 'worm-free' unit in 1962.

Region of the alimentary tract	Species of nematode	Percentage of lambs in which adults of each species	
		Were found	Were not found
Abomasum	*Haemonchus contortus*	5	95
	Ostertagia sp.	11	89
	Trichostrongylus axei	0	100
Small intestine	*Trichostrongylus* spp.	20	80
	Cooperia sp.	8	92
	Nematodirus spp.	68	32
	Strongyloides papillosus	3	97
Large intestine	*Oesophagostomum* sp.	0	100
	Chabertia sp.	0	100
	Bunostomum sp.	0	100
	Trichuris ovis	83	17

eradication must be considered. Total eradication of a species must mean that none remains. When it comes to measuring this, however, the question is how many represents none, for no matter how often faeces are examined only a small proportion of the total is ever involved. Continually finding no eggs thus entitles one to say that none was found in that proportion of the faeces. This can be expressed as a count of 'less than x eggs per g'. The same holds for the examination of grass for infective larvae: these estimates are even less useful for assessing the absence of infection. The digestive tract is different in the sense that it is perfectly practicable to detect all worms of the larger species, without using a sampling technique at all.

In general, however, it is not possible to prove the negative conclusion: a sheep can rarely be said to have no worms at all, even less can a pasture be proved free from all infection. It may be considered a stringent biological assay of the pasture, however, to graze in the ways shown in Table XLIII. The results are rather more useful where positive, however, and the following conclusions appear justified:

1. It has proved possible to maintain the unit free from diseases due to parasitic worm-infestation of the stomach, intestines and lungs.
2. All indices of the level of worm-infestation show it to have been extremely low.
3. Certain species appear more difficult to eradicate completely than others. These include *Nematodirus* spp., *Strongyloides papillosus* and *Trichuris ovis*. These species, however, show the least tendency to increase rapidly.
4. Certain species can apparently be eliminated from such a unit with relative ease. These include *Haemonchus contortus* and *Ostertagia circumcincta*.

It must be emphasized that all this represents the early experience of one attempt at operating a disease-free sheep unit under grazing conditions. No anthelmintic drugs have been used and clearly the use of highly effective ones could help greatly (1, 2, 9, 10). At present, it is necessary to start with the worm-free, new-born lamb and to try to keep it free. Removal of a worm burden by drugs leaves one very dependent on measurements such as egg-counts to assess success.

Rearing lambs in the way described automatically avoided a number of diseases; most of the common lamb ailments have never been encountered in this unit, once the lamb has been put out to grass. In the developmental stages of artificial rearing particularly, a number of complaints occurred and some proved lethal. Once at pasture, however, a very high standard of health has been maintained. This has been such as to impress because of the relative absence of such things as scouring even on swards of pure clover or grass receiving high levels of nitrogen and irrigation. The incidence of disease is summarized in Table LVII.

TABLE LVII

INCIDENCE OF ALL DISEASES WITHIN THE 'WORM-FREE' AREA, 1959–62

(excluding those associated with the period of artificial rearing). The figures are for the total number of lambs affected.

	Footrot	Fly strike	Pneumonia	Enterotoxæmias	*Coenurus* cysts	Others
1959	Nil	Nil	Nil	Nil	Nil	1†
1960	Nil	1	Nil	Nil	Nil	Nil
1961	Nil	Nil	1*	Nil	1*	Nil
1962	Nil	3	Nil	Nil	1*	Nil

* These occurred while the animals were yarded.

† This was a severe, but unexplained, loss of wool and skin on the back.

All lambs were vaccinated against pulpy kidney (enterotoxaemia) at 6 weeks of age; no other medical treatment was given except for protective spraying with dieldrin suspension, annually, against fly strike.

Summary of Results, 1963 to 1968

The use of the area was continued in substantially the same fashion during the next 6 years, except that adult sheep also grazed within the area between 1963 and 1966 (the number of grazing days rising to a peak of 15,091 in 1965).

The number of lamb grazing days varied between 6,454 in 1963 to a peak of 19,648 in 1965.

Mean egg counts varied from 1 to 12 e.p.g., with the exception of 264 in 1966 when one group of 10 lambs became infected to a noticeable extent. The percentage of zero egg counts dropped to 35 per cent in that year but this was followed by 90·6 per cent in 1967 and 90·0 per cent in 1968.

The general incidence of disease continued to be very low for grazing lambs.

When nematode infection occurred, at a very low level, the species found most commonly at *post mortem* examination were *Ostertagia* sp., *Trichostronglyus* spp. and *Nematodirus* spp.

Strongyloides papillosus was frequently present at a very low level.

The Use of Drugs

As more efficient anthelmintic drugs become available, it becomes increasingly possible to control worm-infestation at a very low level by medication. Reference has been made to recent research towards this end (see Chapters V and X) and it is clear that a very high degree of control can now be achieved by the use of drugs.

It is perhaps appropriate to end a chapter on the elimination of disease with a brief discussion of health.

HEALTH

There is no agreed or entirely adequate definition of health, except for certain limited purposes. Yet it is part of the current change in emphasis, from the curing of disease to its prevention, that health should become a positive aim, and it seems reasonable to suppose that an aim of this kind should be definable. It is not difficult to suggest that the word is well-enough understood and that further argument is trivial hair-splitting. In a sense, the word is understood, but it may be doubted whether this is well enough for the purposes of the future.

The following discussion is undertaken in the belief that health is a neglected agricultural subject and that what we mean by it ought to be the subject of continued debate. It is not intended to propose a definition or to state what ought to be intended by the advocacy of healthy farm stock. It is designed simply to contribute to the debate and to stimulate it further.

Clearly the absence of disease is a prerequisite of health. Equally however, no damage whatever to any part of the body would be a criterion of perfection. Sheer physical damage or injury differs little from the effect of some infections. The latter may be characterized by the production of toxins and by the continued presence of the damaging agents, but the argument still has a physcial aspect to it. Thus a lamb suffering from fly strike may still be healthy if the affected area is small but, if the area increases, there must be a point beyond which the lamb becomes unhealthy. Clearly a distinction must be made here between the health of a cell, a tissue or a limb and the health of the whole body. Damaged or infected parts or tissues may obviously be unhealthy parts or tissues: this does not necessarily make the whole animal unhealthy. The criteria, it seems, would be better related to function. The widespread acceptance of castrated animals as healthy obviates the need to argue that all natural functions must remain unimpaired. To reduce this to a minimum of functions necessary to life is unhelpful and it is possible that we shall mean by healthy simply the ability to perform a given agricultural function. This somewhat depressing and, to some, no doubt, distasteful conclusion, may at least pose the question whether, if health does not include any state in which the desired agricultural function can be sustained, it is, in fact, what is wanted in farm animals. It will be noted that there is a tendency to link health to ability, capability or potential. This is partly because the actual expression of everything that a healthy animal must be capable of, cannot be required at all times. A healthy lamb must be capable, for example, of growing at a rate only limited by its genetic constitution. It cannot possibly do so if inadequately fed, yet when only fed for less-than-maximum growth it may surely be quite healthy. If not, then it must be concluded that there are indeed remarkably few healthy lambs. It is difficult, therefore, to avoid the conclusion that, in agricultural terms, health is a state of the whole animal in which a satisfactory physiological balance can be achieved between the animal's needs and the environment which has to supply them. The idea of such a balance indicates ways in which health could be assessed (the concept of a normal temperature is one recognition of this), as distinct from an elaborate checking against the presence of any one of the innumerable possible diseases. Perhaps the most important conclusion is that health may have only a limited connection with performance. A healthy animal may not be growing fast, for its growth rate will depend upon its level of nutrition. By contrast, an unhealthy sheep may be unable to grow rapidly, however well fed. But, equally, it may be able to do so in spite of ill-health. Thus performance may be a poor guide to health status.

Although an unhealthy sheep may be able to grow rapidly, however, it is unlikely to be able to grow at its maximum rate, or, even if this should prove possible, it is doubtful if it could do so with maximum efficiency. Indeed it seems likely that reduced efficiency in the conversion of food to agricultural products may be regarded as one attribute of ill-health. It is possible that unhealthy animals may prove profitable, but it may be considered unlikely that they will prove biologically efficient in an agricultural sense. It might be expected that only lack of knowledge would prevent greater biological efficiency from leading to more profit so that ultimately the healthy animal will be the one required.

Husbandry, Health and the Community

Since agriculture does not exist in a vacuum, it is necessary to consider the interests of the community. As the purchasers of food, they are interested in its cost of production. Both farmer and consumer are concerned to keep costs down, therefore, and the cost of maintaining any given standard of health has to be considered in the same way as any other cost. Thus, although the healthy animal may be the most efficient it may also be more costly to maintain and uneconomic in certain circumstances.

The community, however, has other interests in the matter.

There is increasing concern, for example, that the easiest way of maintaining health may have dangerous consequences. The routine use of antibiotics is being discouraged, not only because it may lead to a situation where the disease-producing organisms are resistant to the drugs we possess, but also because there is a danger to human health for the same reasons.

Systems of husbandry that depend upon procedures that are, or may become, unacceptable to the community are unlikely to survive and are not worth devising. The same argument can be advanced for husbandry methods that offend against standards of comfort or cruelty. There is no doubt that some actual or possible methods of production are unacceptable and it should be enough that the community does not wish to behave in a particular fashion. A consequent cost may have to be borne for a time but, from a research point of view, the question is merely one of devising a method that is both efficient and acceptable.

One major advantage of the establishment of standards for the keeping of farm stock is that it will channel the considerable efforts now required to produce new husbandry systems for sheep, in constructive and more permanently useful ways.

REFERENCES

(1) GIBSON, T. E. (1959). Controlled tests with various anthelmintics against *Nematodirus* spp. in sheep. *Vet. Rec. 71* (21), 431–434.

(2) GORDON, H. McL. (1961). Thiabendazole: a highly effective anthelmintic for sheep. *Nature (Lond.) 191* (4796), 1409–1410.

(3) GRACEY, J. F. (1961). Losses and disease in Northern Ireland. *N.S.B.A. Yearbk.*, 23–26.

(4) GROOT, T. DE (1963). The influence of heavy nitrogen fertilization on the health of livestock. *J. Brit. Grassld Soc. 18* (2), 112–118.

(5) LEECH, F. B. & SELLERS, K. C. (1959) A second survey of losses associated with pregnancy and parturition in Yorkshire sheep. *J. agric. Sci. 52* (1), 117–124.

(6) MOULE, G. R. (1962). 'Clover disease' of sheep in Australia. *Wool Technol. Sheep Breed. 9* (1), 113–115.

(7) OTTO, G. F. (1962). Developments in active immunization of animals for internal parasites. *Proc. V Ann. Florida Conf. for Veterinarians 5* (2), 4–6.

(8) ROBERTS, D. S. (1962). An approach to the control of Mycotic dermatitis. *Wool Technol. Sheep Breed. 9* (1), 101–103.

(9) ROSS, C. V. & SHELTON, G. C. (1961). Practical tests with anthelmintics for grazing lambs. *Univ. Missouri Res. Bull.*, 773.

(10) ROSS, D. B. (1961). The influence of thiabendazole, a new anthelmintic, on weight gain in lambs. *Vet. Rec. 73* (51), 1455.

(11) SELLERS, K. C. (1959). Research in sheep diseases. *N.S.B.A. Yearbk.*, 42–48.

(12) SELLERS, K. C. & LEECH, F. B. (1955). Survey of losses associated with pregnancy and parturition in Yorkshire sheep. *J. agric. Sci. 46*, 90–96.

(13) SHORT, G. V. (1960). Veterinary science and the world's food. *Outlook on Agric. 3* (1), 39–50.

(14) SPEDDING, C. R. W., BROWN, T. H. & LARGE, R. V. (1960). Some factors affecting the significance of internal parasites in the utilization of grass by sheep. *Proc. VIII int. Grassld. Congr.* (Reading), 718.

(15) STAMP, J. T. (1959). Clostridial diseases of sheep. *Outlook on Agric. 2* (4), 185–191.

(16) WATSON, D. F. (1962). A system for intensive production of lambs. *Rep. Va agric. exp. Sta.*, 58.

(17) BICHARD, M. & COOPER, M. McG. (1966). Analysis of production records from a lowland sheep flock. 1. *Anim. Prod. 8* (3), 401–410.

(18) GUNN, R. G. (1967). A note on hill ewe mortality. *Anim. Prod. 9* (2), 263–264.

(19) LEECH, F. B., VESSEY, M. P. & MENZIES, D. W. (1959). A survey of the losses of breeding ewes in England and Wales in 1958–59. *Animal Disease Surveys, Rep. No. 3*, 12–16. H.M.S.O.

(20) PURSER, A. F. & KARAM, H. A. (1967). Lamb survival, growth and fleece production in relation to birthcoat type among Welsh Mountain sheep. *Anim. Prod. 9* (1), 75–86.

(21) SPEDDING, C. R. W., BETTS, J. E., LARGE, R. V., Wilson, I. A. N. & PENNING, P. D. (1967). Productivity and intensive sheep stocking over a five-year period. *J. agric. Sci., Camb. 69*, 47–69.

(22) SPEDDING, C. R. W., LARGE, R. V., BROWN, T. H. & WILSON, I. A. N. (1965). The use of 'worm-free' sheep in grazing experiments. *J. agric. Sci. 64*, 283–290.

XV

THE POTENTIAL FOR SHEEP PRODUCTION

The two previous chapters have been concerned with two of the major determinants of sheep production from pasture. The control of disease will always be important but it presents special problems in the grazing situation, where the environment cannot be controlled to any great extent. Pasture production will remain a vital factor, provided that pasture continues to form a major part of the sheep's diet. It would be unwise to make any such assumptions for all circumstances, however; other possibilities should also be given objective consideration. It is not easy for people who are interested in sheep to discuss unemotionally these future possibilities. This is understandable, and the feelings of the community cannot safely or sensibly be ignored when selecting a possibility for actual development; but it is important to recognize all possibilities at the outset, even the ones that appear unattractive.

In any event, potential production does not imply what *will* be produced, only what *can*. Equally, however, it is unhelpful to divorce a discussion of potentials entirely away from probabilities. Higher pasture production could be obtained by heating the soil, for example, but this does not at present appear likely to represent an economic probability.

Currently, the sheep produces chiefly milk, wool and meat. Changes in demand for each of these products may convert them into relatively unimportant sources of farm income, or they might become high-priced luxuries. If the latter, such features as efficiency of food conversion might become less important. Large increases in the use of sheep's milk for cheese-making might be considered in this category, but if meat became a luxury in the United Kingdom, for example, consumption would be at a lower level. Wool is already a by-product in British sheep farming and could become so in other areas.

Such considerations affect the importance of questions concerning potentials, rather than the potentials themselves, but they also influence the most appropriate way of expressing any potential.

The potential for sheep as a converter of inaccessible uplands to wearable wool may be unaffected by market requirements, but the latter greatly influences its importance.

Amongst the world's sheep there exist breeds capable of producing a wide range of wool qualities and quantities. There is no real difficulty in changing the quality of the wool produced, therefore, by changing the sheep entirely or partially, by crossing. The potential for greater wool production per unit of land is likely to depend most on increased stocking rates, which in turn must be based on greater herbage production.

Similarly, there exists a wide range of milk-producing capacity, although it may be quite difficult to increase yield to any large extent within a particular breed.

It is most likely that the sheep will have to justify itself in the future primarily as a meat producer, however, and the rest of this chapter will therefore be devoted to an assessment of the potential in this direction.

Clearly, the only useful ways of expressing this potential are in terms of output of saleable product per unit of resources employed. The difficulty here is to distinguish what will be the most important resources. Financial expressions would be of great value if sufficient was known to insert ranges of probable values for all the main components. This approach is rapidly becoming possible but the use of food remains the best starting point at the present time (based on the fact that food conversion is the essential part of the process—other resources are ancillary—and food still represents a very large proportion (>50 per cent) of the total costs of production).

THE EFFICIENCY OF MEAT PRODUCTION

As will be clear from Chapter VI, the main factors affecting the efficiency (E) of conversion from food (F) to meat (M) are

Reproductive rate
Ewe size
and Carcase weight.

It may now be asked, what are the potentials for these attributes and how do they influence E.

Reproductive rate

The absolute maximum for reproductive rate is $2\times$ the maximum litter size, since the maximum lambing frequency is twice per year. The maximum litter size may not be independent of ewe size, so it is better if it can be expressed per unit of ewe weight.

This is most readily done by taking the maximum total birth weight of lambs per kg of ewe (W^1) and dividing by the minimum birth weight for survival in that particular breed.

The information available for this is only sufficient to illustrate the main points.

Data on maximum total birth weights found in five breeds at Hurley are shown in Table LVIII. The values shown do not preclude higher weights being possible in the breeds shown and, indeed, exceptional values have deliberately been omitted.

Minimum birth weights for survival are even more difficult to obtain. Table LIX simply gives the lightest lambs for these same breeds that have been successfully reared at Hurley.

It appears from these figures that the maximum reproductive rate

$$\left(\frac{W\ \text{(Table LVIII)}}{B\ \ \text{(Table LIX)}}\right)$$

lies between 6·5 and 11·3 and, although it may vary with breed, it does not appear to be greatly influenced by ewe size. The main weakness in this argument is that we do not know that smaller lambs could not survive if born to the larger ewes. The fact that there have not been any may suggest, of course, that nevertheless we are dealing with probable values for the litter sizes that are obtainable.

Ewe size

The smallest possible ewe may be deduced from existing breeds, such as the Soay (ewe weight= approx. 23 kg), or by a theoretical calculation. The latter can be based on the relationship used by Dickinson *et al.* (9) (see p. 89). If the minimum mean birth weight is inserted in the formula, the size of the smallest ewe required to produce various litter sizes can be calculated.

The minimum ewe size emerges as about 15 kg, depending upon litter size. This figure is, of course, highly speculative, being derived from extrapolation outside the normal range of values for birth weight.

TABLE LVIII

MAXIMUM TOTAL BIRTH WEIGHT (W) OF LAMBS PER UNIT OF EWE WEIGHT (W^1 g/kg)

Breed of ewe	L.wt (kg) of ewe	Litter size	W (kg)	W^1
Scottish Half-bred	79·7	4	18·1	0·227
Kerry Hill	70·3	3	14·8	0·211
Devon Longwool	77·3	3	13·0	0·168
Welsh Mountain	40·3	3	9·2	0·228
Dorset Horn	50·3	3	12·3	0·245

TABLE LIX

MINIMUM BIRTH WEIGHTS (KG) OF LAMBS BORN AND REARED AT HURLEY

Breed		L.wt (kg) of ewe (EW)	Birth weight of lamb	
Ram	*Ewe*		(B) *kg*	*as a % of EW*
Suffolk	Scottish Half-bred	88·0	1·6	1·82
Suffolk	Kerry Hill	64·2	2·2	3·43
Suffolk	Devon Longwool	73·2	2·0	2·73
Suffolk	Welsh Mountain	29·5	0·9	3·05
Dorset Horn	Dorset Horn	96·4	1·1	1·14

Carcase weight

The maximum potential lamb carcase weight is obviously influenced by the mature size of both parents. Market requirements may impose an upper limit that is less than can be achieved: the use of different breeds can generally overcome difficulties due, for example, to degree of fatness, but the market may place an upper limit on sheer size, whatever the composition. Above this limit, potentials are therefore unimportant and it is better to consider this topic another way. The most important question then becomes, how small can a ewe be without lowering the carcase potential below that imposed by the market?

For the purposes of illustration, a market requirement of not more than 23 kg (just over 50 lb) may be assumed. If it is further assumed that the use of heavy rams could raise the carcase potential to a value halfway between that of each parent, then the influence of ewe size can be estimated as follows:

Potential carcase weight of a heavy ram = 30 kg
Potential carcase weight of a small ewe = 15 kg

Potential carcase weight of cross-bred progeny $= \frac{45}{2}$

$= 22{\cdot}5$ kg

The smallest ewe that would not reduce the carcase potential below the market figure, would thus produce a 15 kg carcase when bred pure, implying a ewe weighing about 30 kg (approx. 66 lb) liveweight. This is of the order of size of a Welsh Mountain ewe and it is known that a large ram crossed with such a ewe is capable of producing progeny with carcases of over 20 kg in a reasonable time (4 to 5 months), if well fed.

The previous calculation as to the minimum theoretical ewe size suggests that this would, in fact, lower the carcase potential but perhaps not greatly, bearing in mind the rather high market requirement assumed above.

The effect on efficiency

The effect of these three major factors on the efficiency of food conversion may now be estimated.

It will be assumed that a carcase of 23 kg is both possible and saleable and that it is reproductive rate and ewe size that will dominate the calculation.

If the maximum reproductive rate is accepted as intermediate (e.g.

9·0× 2) between the values derived earlier and that it can be applied to the smallest possible ewe (e.g. 30 kg) capable of producing the maximum reproductive rate, then the potential annual output from such a combination would be 414 kg (= 18·0× 23 kg) of carcase per ewe per year. The annual food requirement can be estimated as follows:

		kg D.O.M.
Food required for ewe maintenance	=	146
Additional food required during pregnancy	=	40
Total food required by progeny	=	3,456
Total per year	=	3,642

The maximum efficiency would therefore be

$$\frac{2 \times 9 \cdot 0 \times 23 \times 100}{3642} = 11 \cdot 4$$

If it is considered that the assumptions made are likely to be wrong, it is a simple matter to see what effect an adjustment to them would have. Similarly if there are grounds for regarding this level of efficiency as unattainable, for one reason or another, the values can be scaled down.

The calculated value can be compared with the (entirely) theoretical maximum that can be derived from the use of progeny only (5). Since the food requirement of the ewe has to be balanced by lamb output, the efficiency of the whole ewe/lamb unit can never exceed that of lambs alone. The value of *E* for lambs only, from birth to slaughter, is of the order of 11–12 kg carcase per 100 kg D.O.M. (using, not the D.O.M. of milk substitute, but the D.O.M. of grass required to produce it). This figure is open to correction for better lambs, more rapid growth and improved diets: otherwise, the fact that the estimate of 11·4 for maximum efficiency is substantially the same as the lamb-only value of 11–12, suggests that it might be possible to achieve a position in which the ewe had become relatively unimportant as a consumer of food. More accurately, it illustrates that this is so in the particular calculation used.

Production per unit of land

The potential output of meat per ha or per acre can be calculated from the potential herbage production per unit of land (see Chapter XIII) and the potential food conversion efficiency.

Thus, a herbage production of 10,000 kg D.O.M./ha (or approximately 9,090 lb/acre), if converted with the maximum efficiency, would result in an output of 1,100 kg carcase/ha. Half that level of herbage output would give 550 kg carcase/ha; 20,000 kg D.O.M./ha would produce 2,200 kg carcase/ha.

These are, of course, theoretical calculations and they ignore the difficulties of translating research into practice, of achieving the higher levels of performance consistently and in less than optimum environments, and the disproportionate additional costs of labour and equipment that might be required.

It is sensible, therefore, to consider also the potentials as established by the best current practice. Two good examples may be drawn from New Zealand and Eire.

In New Zealand, Elliott (3) estimated (from Hutton (4)) that the upper level of production likely to be achieved in farming practice was 270 lb of carcase per acre (297 kg/ha); in addition to the wool production of 8 breeding ewes.

In Eire, Conway (2) has recorded an output of 320 lb carcase per acre (352 kg/ha) and 38 lb wool per acre from 7 ewes and their lambs per acre.

There is no doubt that achievable potentials are higher than these figures for some farms and some farmers. Campbell (1) refers, for example, to production levels of 450 lb of carcase per acre in New Zealand. The difficulty in relation to farm practice is to estimate what proportion of the area is capable of these levels of production and how long it might take before an appreciable proportion of the total output of a country could be achieved at these rates.

REFERENCES

(1) Campbell, A. G. (1968). Increasing the efficiency of pasture use. *Span 11* (1), 50–53.
(2) Conway, A. (1968). Intensive fat lamb production in Ireland. *Span 11* (1), 47–49.
(3) Elliott, I. L. (1966). Implications and perspectives from earlier assessments. *J. N.Z. Inst. Agric. Sci.* (Sept.), 15–17.
(4) Hutton, J. B. (1963). Efficiency of utilisation and conversion of pasture herbage by dairy cattle, beef cattle and sheep. *Proc. N.Z. Inst. Agric. Sci.*, *9*, 97–110.
(5) Spedding, C. R. W. (1968). The agricultural ecology of grassland. Middleton Memorial Lecture, 1968. *Agric. Prog.* In press.

XVI

SYSTEMS OF SHEEP PRODUCTION

The aim of the previous chapters has been to assemble the information required for the understanding of the sheep production process. The information available is, of course, incomplete and new information emerges non-uniformly over the whole range. To make sense of existing facts it is necessary to construct some kind of model: the absorption of new information also requires models to act as skeletons that can be progressively clothed.

In time, such models will be constructed for parts of the production process, for the whole process, for systems of husbandry and for the whole sheep industry of a country. Gradually, as more information is acquired, especially on the interactions between components, it will be possible to express models in mathematical form, suitable for computer processing. Within any economic framework, it will then be possible to select systems for specific purposes, from models capable of representing with sufficient accuracy the complex agricultural processes involved.

Somewhat surprisingly, at a time when the moon can be circled with astonishing accuracy and most people can recall the time when the whole idea was regarded as well-nigh impossible, many still consider such model-building for agriculture as impossibly complicated and not without risk. It cannot yet be done, but it is probably only a very few years away.

The present need, however, is for a practical synthesis of the knowledge available, into systems capable of being rapidly tested, including economic assessment.

The *current* need is for systems that are both practicable and profitable.

The Main Components

Herbage production

For sheep fed on herbage, clearly the yield of digestible herbage per unit of land is a major factor in determining productivity. It does not, of course, follow that the maximum herbage yield will lead to the most profit; this depends also on the cost involved in obtaining it.

There is no point in selecting, for sown pastures, a species or variety with a poor yield.

In general, a high-yielding variety will not be disproportionately expensive to sow or cultivate. The economic optimum for fertilizer input must vary from one environment to another and with many other factors too. No generalization is likely to be valid for this, although it may be expected that land may become the ultimate limiting factor and yields approaching the maximum obtainable may be required.

The choice of herbage species, either to sow or to encourage in permanent pastures, must also depend on economics, particularly those of nitrogen supply. For individual sheep performance, the evidence points to the legumes rather than to the grasses; in terms of herbage yield, there may be no great difference in practice between a mixture of grasses and legumes and the best of either alone. For most purposes, a mixture of perennial ryegrass and white clover would prove a satisfactory basis.

The ewe

There are many reasons, related to environment, availability, farm policy or personal preference, for choosing one breed as against another. The cost and depreciation of the ewe can greatly affect profit and the variation within breeds may be considerable.

Productivity and profitability are both influenced by ewe size and, at present, there is much to be said in favour of the smaller ewes on both counts.

In terms of food conversion efficiency for meat production, the ideal is the smallest ewe that does not actually limit output by her size. As suggested in the previous chapter, if a large ram is used it is unlikely that the size of the ewe will greatly reduce the size of individual carcase produced. This is especially so if the market requirement is for a carcase weight well below what the breed is capable of producing.

A larger ewe *may* be more prolific but it is likely that judicious crossing, or selection within pure breeds for the ewes that have large litters, could result in adequate prolificacy in the smaller ones. Wool quality may be altered but wool yield per head will always tend to be lower in small sheep.

High milk yield is of great importance and really determines how many lambs a ewe can rear or finish within a given system; this is probably the chief reservation that might be entertained about smaller ewes.

Until much more is known about reducing losses around lambing time, 'mothering' qualities must be rated as of the first importance in a ewe.

Longevity and disease resistance are important qualities, with a considerable bearing on the rate of depreciation.

The ram

For direct use in meat production, the ram contributes size, growth rate, carcase quality and conformation. His qualities (of keenness, activity and so on) at mating time are important and affect the number of ewes he can satisfactorily cover.

The cost of the ram is not negligible, especially when profit margins are low: it would be worth bearing the ram in mind in both biological and economic calculations, although the quantity of food consumed by the ram is a relatively small proportion of the total.

Large rams can be used to increase the potential size of the progeny and to increase their potential growth rate. Carcase composition is largely determined by growth rate, and is thus influenced by nutrition more than by anything else. What the ram may determine is the age and size at which fat begins to be laid down at high growth rates.

The lamb

This is not, of course, chosen, except through the choice of parents, which will have fixed the size of carcase that can usefully be produced. The next question is what growth rate to aim at, but it is not yet possible to answer this with much confidence. Lambs are generally capable of much greater growth rates than are usually obtained, but growth rate is not of tremendous importance in determining food conversion efficiency of the flock. It *is* so for efficiency of the lamb alone and rapid growth is therefore most important where the reproductive rate is high (and most of the food is eaten by lambs) or where the lamb is fed on a relatively expensive diet.

A wide range of growth rates are therefore tolerable but the target must be related to the system of production and, in particular, to the cost of the lamb's diet. The more expensive the diet, the faster the growth rate that must be obtained.

Management

The methods of combining ewe, lamb and pasture are numerous and choices must be related to circumstances. However, some important features may be selected.

Disease control is vital, and management, in the widest sense, has a great part to play: the role of *grazing* management may have been overestimated and, at present, it must be regarded as a minor factor in the sense that different methods can all be operated badly, none automatically solves all problems and most can be operated to produce a successful result in some circumstances.

Management must ensure proper levels of nutrition at all times and it has to be recognized that, at times (for weaned lambs, for example), this may mean lenient grazing, i.e. some apparent inefficiency in pasture utilization.

Systems of Meat Production

Certain standards of management, nutrition and disease are essential for successful meat production, whatever the system used. The features that dominate the productivity of the system are the potential productivity and cost of the ewe and the extent to which high output can be achieved from it without great additional costs. The main categories of systems are therefore characterized by major differences, for a given ewe, in the ways in which its output is assured, within a context of intensive use of resources.

The following can be distinguished:

(a) Intensive flock grazing

This may be thought of as orthodox sheep farming pushed to the limit. It may involve very high levels of herbage production and fertilizer input but, in any event, it involves a high stocking rate per unit of herbage (i.e. a high grazing pressure); the actual stocking rate depends upon the environment.

The choice of ewe also varies with the environment but it has to represent high prolificacy for its size and an adequate milk yield; the importance of the wool yield increases at lower levels of prolificacy.

The object of this kind of system is low-cost production based on maximum use of grazing.

Its weakest points are nutrition of the ewe shortly before and after lambing, finishing the lambs and the accumulation of problems with time on the same land. The last point is greatly eased by the use of clean land each year, or alternation in years with cattle or conservation.

Finishing the lambs is the most difficult problem, because of the several different causes that cannot always be predicted or even diagnosed with any certainty.

Parasitic disease is best controlled by a combination of anthelmintic dosing and a change to clean pasture at weaning or for the whole flock in about July (for spring-born lambs in Britain). Nutrition of the lamb on pasture alone requires young, digestible pasture but, especially, plenty of it.

A satisfactory dosing programme would be two doses for the ewes —one at the onset of winter or in late autumn and the second close to lambing time (to carry as few worms as possible in the winter and to reduce the spring output of worm eggs after lambing)—and one main one for lambs in July. For the control of *Nematodirus*, somewhat more is required; for other worms, it is better to dose those lambs that need it and move them to clean pasture at the same time. The aim should be to approach this programme as far as circumstances allow.

Variations in systems of this kind include the drugs used, the method of grazing management, separation of ewes with singles and those with multiples, and the method of wintering.

Major variations are to include conservation in the system or to purchase all food required for the winter.

This system category is enormously dependent upon skilled choice of levels of resources used, skilled management and ability to combine these factors with the environment, including the rest of the farming system.

If successful results could be guaranteed, there might be no further problem.

(b) Intensive grazing with supplementation

This embraces all the previous versions with the addition of supplementation to ensure adequate performance of ewes and lambs. Since supplementation is expensive, its use has to be based on knowing the needs and when they are being met; it therefore implies enough recording to make this possible.

The best use of supplementation for ewes is clearly to ensure the necessary weight gain in late pregnancy and to prevent too much weight loss in early lactation. The deliberate use of the concentrates needed, and no more, at these times, is a major insurance against low quality in winter feeds and variations in the spring weather.

The use of supplementation for lambs at very high stocking rates, both before and after weaning, has been too little investigated. A systematic study is needed, with ewes of different milk yields, to determine the economic level of supplementation at high stocking rates.

Particularly with less expensive supplements, it is quite possible that profitable systems could be devised.

(c) Intensive grazing with use of arable by-products

This is really a special case of (*a*) or (*b*), with the main aim of reducing the cost of the ewe's food. The use of very cheap grassland, rough hill or land otherwise unusable, has exactly the same purpose and may be used in a similar way to stubble grazing, straw feeding etc.

In most flocks, there are substantial periods when nutrition of the ewe is not so critical and cheap foods can be successfully used. The main periods are after weaning until shortly before mating, and then from after mating until about 6 weeks before lambing. Such integration of grazing and the use of by-products can also greatly aid disease control.

(d) Early weaning

The essential feature of early weaning systems is the removal of some or all of the progeny from a ewe at an earlier age than normal, i.e. any time before about 12 weeks.

The object is to avoid some of the problems of finishing lambs, in situations such as (*a*), (*b*) or (*c*); it is only a successful solution if the problems can be attributed to the fact that the ewes and lambs are together. However, it presents an opportunity to offer better conditions of nutrition and disease control to the lambs alone than could be afforded for the whole flock.

In the main, the improved conditions are represented by clean land or high-quality crops. Supplementation can be achieved without weaning, by creep-feeding the lambs.

There are two main questions to answer: what is the best age to wean and how many of the lambs should be removed.

Circumstances can be visualized in which weaning at any time from 4–10 weeks would be best, but for most purposes, where early weaning is required, the best age will be from 8–10 weeks. This allows the major part of the milk supply to be used and takes fewer risks of severe checks in growth.

The answer to the second question does influence the first, however. Ideally, the lambs removed should be no more than is required to ensure that the remaining ewe/lamb unit will operate successfully at the chosen stocking rate and in the environment concerned. In the most extreme conditions, this could mean early weaning of all lambs. In many circumstances, it would refer to all lambs in excess of twins. In between, it really depends mostly on the milk yield of the ewe and on the stocking rate.

There is no doubt that some ewes will manage to finish twins successfully: many more ewes are capable of finishing singles, even at very high stocking rates and under simple managements. In the latter cases, early weaning of lambs in excess of one may be justified at 4–6 weeks, in order to ensure success in the remaining unit. The economic outcome then depends on good growth of the weaned lambs without excessive additional cost.

This approach of timing the weaning to remove the number of lambs that is necessary to restore a successful ewe/lamb combination, is only of value where the ewe can *rear* all her lambs for a minimum of 3 weeks.

Where this cannot be done, it is necessary to remove some lambs at birth and rear them artificially.

(*e*) *Artificial rearing*

This system, or component of systems, is only required where some lambs cannot even be reared to an early weaning age, *within the chosen farming context*. If ewes are needed for light treatment in order to breed again after a very short interval, there may well be a case for the removal of all lambs at birth. In most circumstances it will not be necessary to rear artificially *all* the progeny of ewes in the systems so far envisaged.

The most obvious application currently is to reduce the number of lambs to no more than two immediately after birth, possibly combined with subsequent early weaning of another one at 4–6 weeks of age.

The variations within this category have been fully discussed in Chapter V and will not be elaborated further here.

Future Systems

The group of systems so far described are certainly capable of high levels of productivity per unit of land and per unit of food. Some versions of each are also capable of high output per man/hour. There can be no certainty that any of them will be highly profitable; this depends upon relative costs and prices.

As techniques of increasing prolificacy and frequency of breeding are developed and become more practicable, so some of these systems (for example (*d*) and (*e*)) could attain extremely high levels of productivity. But it is likely that quite new systems will be developed to suit the use of such techniques. These will include complete housing of the entire flock, with full control of light. There is probably no reason why any management of ewes and lambs that is worth using in the field, cannot also be operated under cover, but the greater the degree of control, the more likely it is that it will be sensible to separate ewes and lambs for quite different treatment.

So cheap can be the grazing of dry ewes that it appears likely that they would be kept in this way, if there was no special reason (e.g. season, food-handling, light control) for housing them. So crucial will be successful lambing, on the other hand, that housing will certainly be indicated at this time in adverse weather conditions, and maybe at all seasons. Unless economics dictates otherwise, it is likely that the systems of the future will be based on very high prolificacy and very early separation of ewes and lambs. Either could be kept intensively indoors or outdoors and it is extremely difficult to be sure what the best combination would be.

It is perhaps no bad thing to discuss practical application, both current and future, after having considered the potentials of which sheep and the sheep production process are capable (see Chapter XV).

A full understanding of sheep production must include an appreciation of the range of performance that is possible and of the ways in which systems could operate. It should also be helpful, in choosing a system relevant to the present, to look ahead sufficiently to judge the direction in which systems may move in the future.

After all, the aim of a scientific study of sheep production is not to determine what anyone should do: it is to provide the knowledge of how to do things and to increase the freedom of the individual choice.

Such choices may then be made in the light of this knowledge but also with other considerations in mind, of economics, of personal preference and of both the needs and the standards of the whole community.

GLOSSARY

abomasum: Ruminant's 4th stomach (corresponding to the non-ruminant stomach) in which true digestion rather than fermentation first begins. Capacity about 7 pints in adult sheep.

bioclimatographs: Graphs in which floral or faunal distribution and climatic factors are plotted simultaneously to show the climatic factors which limit the distribution of a given species.

black disease: Infectious necrotic hepatitis. Characterized by sudden death from a toxin produced by an organism of the Clostridium group, which is transmitted to the sheep by liver fluke. Death is followed by rapid decomposition and blackening of the inner surface of the skin.

bloat: Ruminant condition in which a foam of very small gas bubbles or froth covers the ingesta. The resulting increase in pressure may force the diaphragm against the lungs and prevent breathing. Often associated with moves to succulent grazing.

cistern milk: Milk in the lower 'cistern' cavity of the udder which can easily be withdrawn without stimulation.

climax: The final or stable community in a successional series, which is self-perpetuating and in equilibrium with the physical habitat.

coccidiosis: Disease, mainly of lambs, due to infection with a microscopic protozoal parasite, especially under overcrowded conditions. Causes extensive cell destruction and hence weakening of lambs, which may die from exhaustion.

D.O.M.: Digestible Organic Matter. An assessment of nutritive value of feed calculated from feed dry matter content × per cent organic matter in feed × digestibility of the organic matter fraction.

ecosystems: Systems which include both living organisms and non-living substances interacting to produce an exchange of materials between the living and non-living units.

enterotoxaemia: Disease caused by a *Clostridium bacillus* in the intestine which under certain conditions produces a toxin. Characterized by sudden death. Most common in lambs (where it is often known as pulpy kidney) and in well-fed sheep.

ewe hogg: Spring-born ewe lambs from the October of their first year until their first shearing in the following spring (i.e. ewe lambs from about 6–12 months of age).

foggage: Grass left ungrazed in the autumn to provide winter grazing *in situ*.

foot candle: The intensity of illumination on a surface when a light source, emitting 1 candela of luminous energy, is placed 1 foot away from it.

forage: Variously used to denote foliar crops that are grazed.

gang-mowing: Mowing by 3 or 5 linked cylinder-mowers at the same time to give a very close cut.

genotype: The genetic constitution of an organism, in terms of the particular combination of characters each of its cells inherits from the parental cells. Not necessarily the characters manifested by the organism.

hogget: A ewe, ram or castrated male sheep older than 9 months, but not yet shorn.

hogget ill-thrift: New Zealand term to describe a condition involving lack of growth by hoggets in the autumn.

hypocalcaemia: (Milk fever). Disease, mainly of the ewe in late pregnancy or early lactation. Associated with low level of calcium in the bloodstream, apparently due to inability to mobilize calcium from bone reserves in periods of calcium stress. Causes staggering, coma and eventually death.

hypomagnesaemia: Magnesium tetany characterized by sudden death. Associated with low level of magnesium in the blood. Generally occurs around lambing time or at the peak of lactation, but may also occur in non-pregnant ewes.

metabolic size: The size of an animal to which its metabolic rate is proportional. Mean standard metabolic size of mammals is often expressed as (body weight)$^{0.75}$.

microclimate: The climate of small defined areas as opposed to large ones, e.g. close to the ground or around an individual leaf or plant.

niche: The position occupied by an organism within its community and ecosystem, resulting from its structural adaptations, physiological responses and behaviour.

omasum: Third ruminant stomach in which breakdown of ingested food continues. Capacity about 2 pints in the adult sheep.

orf: Contagious pustular dermatitis. Common in older lambs and young sheep in summer, when the causal virus is activated. Characterized by lesions around the mouth and coronet of the hoof, which may be infected by bacteria causing pus formation.

photoperiodicity: Response of an organism to the relative duration of day and night.

phytoliths: Mineral particles as hydrate of silica, found in plant tissue, especially herbage.

pining: Chronic wasting disease due to vitamin B_{12} deficiency. B_{12} is normally manufactured by rumen microbes but this ceases when cobalt-deficient pastures are grazed.

pregnancy toxaemia: Occurs in late pregnancy, usually in ewes carrying multiple litters. Associated with an increased quantity of ketone bodies and decreased quantity of glucose in blood. Characterized by staggering and loss of appetite followed by a comatose condition and death.

pulpy kidney: Lamb disease characterized by sudden death. Caused by a *Clostridium bacillus* which, under certain conditions, produces a toxin in the intestines (*see also* **enterotoxaemia**).

starch equivalent: The amount of starch which has the same fat-producing value as 100 lb of a foodstuff.

steely wool: Wool from sheep on copper-deficient diets. The crimp is much reduced and the wool becomes a characteristic colour with a steely or glassy sheen.

store animals: Animals fed at a nutritional level which allows only very slow growth, usually in winter.

swayback: Most common in lambs at birth or a few weeks old. Caused by progressive destruction of the white matter of the lamb's brain, while it is in the womb of a ewe with low blood copper content. Mild cases may survive and copper therapy is sometimes effective.

T.D.N.: An (American) assessment of nutritive value calculated from the percentages of apparently digestible nutrients as follows:

T.D.N. per 100 lb of feed = % apparently digestible crude protein
+ % apparently digestible nitrogen-free extract
+ % apparently digestible crude fibre
+ 2·25 (% apparently digestible crude fat)

territory: The part of the home range of an animal which it will defend against other members of its own species (so that maximum utilization of an area's amenities is achieved by the species).

trace elements: Elements present in very small quantities (i.e. as 'traces').

vasectomy: Sterilization of male animals by severance of the *vas deferens*. This blocks the passage of sperm to the exterior, although sperm and male sex hormones are still produced.

INDEX